Rheology of the Earth

Rheology of the Earth

Deformation and flow processes in geophysics and geodynamics

GIORGIO RANALLI

*Department of Geology and Ottawa–
Carleton Centre for Geoscience Studies,
Carleton University, Ottawa*

Boston
ALLEN & UNWIN
London   Sydney   Wellington

Allen & Unwin Inc.,
8 Winchester Place, Winchester, Mass 01890, USA

the U.S. company of

Unwin Hyman Ltd
PO Box 18, Park Lane, Hemel Hempstead, Herts HP2 4TE, UK
40 Museum Street, London WC1A 1LU, UK
37/39 Queen Elizabeth Street, London SE1 2QB, UK

Allen & Unwin (Australia) Ltd,
8 Napier Street, North Sydney, NSW 2060, Australia

Allen & Unwin (New Zealand) Ltd in association with the
Port Nicholson Press Ltd,
Private Bag, Wellington, New Zealand

First published in 1987

Library of Congress Cataloging-in-Publication Data

Ranalli, Giorgio.
  Rheology of the earth.
Bibliography: p.
Includes index.
1. Rock deformation. 2. Geophysics. 3. Geodynamics.
I. Title.
QE604.R36   1986   551.8   86-17311
ISBN 0-04-551110-1 (alk. paper)
ISBN 0-04-551111-X (pbk. : alk. paper)

British Library Cataloguing in Publication Data

Ranalli, Giorgio
  Rheology of the earth : deformation and
flow processes in geophysics and
geodynamics.
1. Geodynamics 2. Rheology
I. Title
551.1   QE501.3
ISBN 0-04-551110-1
ISBN 0-04-551111-X Pbk

Set in 10 on 12 point Times by Mathematical Composition Setters Ltd, Salisbury
and printed in Great Britain by Mackays of Chatham

To T.

... sed tu
ingenio verbis concipe plura meis

and our children

Preface

The meaning of a word lies partly in its history. *Rheology*, in its etymological sense, denotes the study of deformation and flow of matter; however, in common usage it is a general term that also covers the flow properties. Thus, for instance, one talks of the rheology of metals, of ice, or − as in the present case − the *rheology of the Earth*.

As a subdiscipline of geology and geophysics the rheology of the Earth therefore deals both with the *methods of study* of the mechanical properties of our planet, and with the *mechanical properties* themselves and their role in geodynamic processes. Deformation and flow of Earth materials are of primary importance in many branches of Earth science, from seismology to plate tectonics and planetary fluid dynamics. Although some of the topics in which rheology is essential are treated in individual textbooks and monographs, this is the first book to deal with the rheology of the Earth as a coherent whole, from the complementary viewpoints of *continuum mechanics* (analysis of deformation and flow in extended bodies) and *solid-state physics* (microphysics of creep). The rheology of lithosphere and mantle at different timescales − from seismic wave propagation to mantle convection − is analysed, starting from geophysical observables. Several examples of geodynamic processes are included.

The aim of the book is to enable the student of geology and geophysics to acquire a clear understanding of the fundamentals of the subject, a precondition to dealing with the research literature and to carrying out independent work. It could therefore be a suitable text for postgraduate and senior undergraduate courses in Earth rheology, geodynamics, and tectonophysics, and useful additional reading for courses in physics of the Earth and structural geology. The research worker or student who reads it independently will, I hope, also benefit.

The decision to write this book owes much to the encouragement of George Skippen. Most of the material in Part I is a revised version of lecture notes written in co-operation with Alan Gale, who kindly agreed to their use. Keith Bell, Martin Bott, and Patrick Selvadurai reviewed different parts of the manuscript. Richard Ernst helped with some of the proofreading of Part III. I wish to express my thanks to them all, and also to Roger Jones of Allen & Unwin, who has offered support without pressure, and to Peter Olson and David Price, who made many suggestions to improve the manuscript. I am, of course, entirely responsible for the end result.

vii

PREFACE

Carleton University has provided support for the typing and the drafting of figures (mostly done by Elsie Lambton and Harold Ellwand, respectively, with unfailing patience and good humour). The Natural Sciences and Engineering Research Council of Canada financed some of the research work eventually incorporated in the book. Some of the writing was carried out while holding a Fellowship at Van Mildert College, University of Durham.

Finally, I would like to mention with gratitude the memory of Pietro Caloi, who many years ago in my native city – Rome – first kindled my interest in the physics of the Earth.

GIORGIO RANALLI

Contents

CONTENTS

List of tables

Introduction

Processes of deformation and flow are ubiquitous within the Earth, with characteristic times ranging from the order of seconds (seismic-wave propagation) to hundreds of millions of years (geodynamics). An understanding of rheology is therefore a prerequisite for the quantitative study of several geological and geophysical phenomena.

This book endeavours to give a reasonably complete and fairly rigorous account of the fundamentals of the rheology of the Earth. Knowledge of descriptive plate tectonics, and familiarity with mathematics and physics at science undergraduate level, are assumed. The arguments are presented in a progressive way, on the assumption that the book will be studied as a whole; but a reader well-versed in some aspects of the subject may very well go directly to the chapter or section of interest.

The book is divided into three parts. Part I (*Rheology of continua*) covers the continuum-mechanics analysis of the deformation of extended bodies, and surveys the main classes of rheological behaviour. The pace is deliberately slow at the beginning, when notions such as index notation, stress, and strain are introduced. This is followed by a treatment of infinitesimal elasticity, Newtonian and non-Newtonian viscosity, linear rheological bodies, yield criteria, plasticity, and brittle failure. Taken as a unit, this part forms a self-contained introduction to continuum-mechanics principles for Earth scientists, with applications to geodynamics stressed whenever appropriate. Tensor notation is used throughout, and arguably the level of mathematical detail in Part I is higher than that required for the treatment of Earth rheology given in the rest of the book. However, it is my experience that students benefit from having the fundamentals of the subject placed on a firm quantitative foundation at the outset. The systematic treatment of the various models of the deformation of continua clarifies the interrelation between different problems, and gives a better context on which to base subsequent study.

Part II (*The continuum approach to Earth rheology*) deals with the rheological properties of the lithosphere and the mantle over the whole frequency spectrum, as inferred from geophysical evidence. It includes discussions of elastic and anelastic Earth models; thermal properties of silicates, the thermal budget of the Earth, temperature distribution in the lithosphere, and thermal convection in the mantle; flexure of the lithosphere and viscosity of the mantle from surface loading data; and lithospheric stresses, faulting, and foreland belts. Although not intended as a substitute for textbooks on physics of the Earth and geodynamics, the material covers all of the main aspects of the mechanics of the interior of the Earth. A degree of selection of topics has been unavoidable; those presented were selected to convey the flavour of the application of continuum mechanics to the analysis of geophysical and tectonic processes, and to provide examples that will enable the student to analyse more problems from the rheological standpoint.

The picture of Earth rheology obtained from the application of continuum-mechanics methods and geophysical observation is not complete without an analysis of the atomic basic of deformation and flow in polycrystalline materials. This is done in Part III (*The microphysical approach to Earth rheology*), which begins with a survey of crystal defects and their mobility, in order to establish a connection between rheological behaviour in the continuum and microphysical domains, followed by a discussion of microphysical models of creep. Topics of particular interest to geologists and geophysicists (hydrolytic weakening, dynamic recrystallization, and pressure and temperature effects) are included, together with a brief account of experimental results on the rheology of silicate polycrystals. An entire chapter is devoted to deformation maps and isomechanical groups, which are important tools in the extrapolation of theoretical rate equations and laboratory results to conditions pertaining to the interior of the Earth. The final chapter deals with the rheology of the lithosphere and the mantle as inferred from microphysical models, to close, as it were, the circle and to provide information complementary to that obtained from continuum mechanics and geophysical evidence.

Some general notes for the guidance of the reader. The choice of symbols for physical quantities follows established convention. Unavoidably, this means that the same symbol may denote more than one quantity; however, confusion should not arise since symbols are always defined in their proper context. SI units are consistently used, although sometimes CGS units that are widely used in geophysics are given, together with their SI equivalents. Multiples are denoted by the usual prefixes (i.e. $\mu = 10^{-6}$, $k = 10^3$, $M = 10^6$, etc). The symbol 'a' denotes year (annum), as seconds are particularly unwieldy units when geological timescales are involved. Approximate equality and proportionality are indicated by $\simeq$ and $\propto$, respectively; 'ln' is natural logarithm, and 'log' is logarithm to the base 10.

The reference list at the end of the book (up-to-date to the end of 1985) includes all the works mentioned in the text. I have tried to keep it manageable, from the viewpoint of the student, by including only what I consider to be most useful. Classical textbooks on elasticity, fluid dynamics, rheology, and geophysics are given, to provide a guide both to the broad bases of the subject and to further study. The many review papers have been selected with particular care, since they provide perhaps the best avenue to survey the state of the art in any particular field. When a reference is a review paper, this is usually stated in the text, and a few comments are given. I have also included research papers of a more specialized nature when they are essential to the discussion, or when the topic is not covered by a suitable review.

The material presented in the book will, I hope, repay careful study, and it would be good for the student, especially if coming from a geological rather than physical background, to go through arguments and derivations in detail. Perhaps some of the wonder and interest inherent in the study of Earth rheology will be transmitted through the following pages.

PART I

Rheology of Continua

1 Continuum mechanics and rheology

This is an introductory chapter that begins by defining rheology, sketching its importance in the Earth sciences, and describing continuous media. There follows a brief survey of necessary concepts and tools: index notation, tensors of various ranks, dimensional and tensorial homogeneity, rheological equations, Lagrangian and Eulerian frameworks. The importance of temperature, pressure, and time for the rheological behaviour of Earth materials is emphasized.

1.1 Rheology and the Earth sciences

The materials of the Earth may be subdivided on the basis of their atomic structure into crystalline solids, which exhibit an ordered spatial arrangement of atoms, and fluids, in which such an order is absent. Excluding the obvious case of water, examples of fluids in the Earth are the outer core and magmas. However, the bulk of the Earth consists of polycrystalline aggregates of various compositions and properties. These materials are, so to speak, the main characters in the unfolding history of the planet, and the protagonists of the processes shaping the Earth as it now is. From the physical viewpoint they are extended bodies, occupying a continuum in space. As continuous bodies they can be – and are – subject to movements during which their different parts change their relative positions. Such relative changes result in *deformation* and *flow* of matter.

Deformation and flow are processes of primary importance in the Earth. A list of topics in which these processes play a central role includes rock mechanics (deformation, flow and fracture of rocks and their engineering and mining applications), structural geology and tectonics (mechanics of folding, faulting, and tectonic processes at all scales), geophysics (occurrence and mechanism of earthquakes, physical properties of the interior of the Earth), and geodynamics (plate tectonics and its driving mechanisms, creep and convection currents in the

3

mantle). These topics are covered separately in subject-oriented text-books, among which one may mention Jaeger & Cook (1979) for rock mechanics, Hobbs *et al.* (1976) for structural geology, Jacobs (1974), Jeffreys (1976) and Garland (1979) for geophysics, and Bott (1982), Scheidegger (1982) and Turcotte & Schubert (1982) for geodynamics.

Rheology (from the Greek, 'the study of flow') is the branch of physics dealing with the deformation and flow of materials. Its establishment and evolution have been essentially a 20th century phenomenon although, of course, the study of the deformability of materials goes back to much earlier times. A classic, concise and lucid introduction to rheology is given by Reiner (1960). It emerged quite early that a number of rheological behaviours exist – that is, a variety of deformational responses to applied stress – and that the type of response depends not only on the material under consideration, but also on external parameters such as *pressure, temperature*, and *time*.

An understanding of rheology is a prerequisite for the study of geologic processes involving deformation and flow. Indeed, mechanical and thermal processes in geology cannot be unravelled without an appreciation of rheology, and consequently the latter forms a hitherto neglected, but fundamental, part of geophysics and geodynamics.

There are two basic approaches to the study of rheology. In the *continuum-mechanics* approach the rheological properties of materials are described phenomenologically, without formal reference to the atomic processes which govern the macroscopic behaviour. In the *solid-state physics* (or *microphysical*) approach attention is focused on properties at the atomic level, and on how these affect the phenomenological behaviour. These two approaches – which are different, but complementary – are sometimes referred to as macrorheology and microrheology, respectively, although we may note that originally the latter term had a slightly different meaning (Reiner 1960). Both are necessary in order to deal with the rheology of Earth materials, although it will be seen that the continuum-mechanics approach can carry us a long way in the study of geological phenomena.

It is important to point out at the outset that the terms 'solid' and 'fluid' in rheology do *not* correspond to the definitions based on atomic structure. In the rheological sense a material is *fluid*, independently of its atomic structure, when it flows under constant stress. Fluidity is not an intrinsic property of the material, but depends on external conditions, e.g. temperature, pressure, and duration of loading. Consequently, the Earth exhibits a great diversity of rheological response.

It is sometimes said that the first rheologist was the Greek philosopher Heraclitus, who was born in Ephesus in Asia Minor and flourished *circa* 500 BC. This may perhaps be pushing the point somewhat, but he was certainly among the first to understand that the natural world (including, one may note, the human world) is not made up of immutable substances, but of *events* in continuous flux. This is expressed in his saying, 'Everything flows'. The basic law in nature is not permanence, but change; and this is beautifully exemplified in the processes affecting the Earth.

1.2 The continuum and index notation

A *continuum*, or continuous medium, is represented as a continuous aggregation of idealized 'material particles', i.e. elemental volumes. These are small enough that their position can be given in terms of points in some co-ordinate system, yet large enough that the local value of any variable does not depend on fluctuations at the atomic scale in the immediate neighbourhood of the point. As an example, consider the case in which the variable under consideration is the density of a body (density, of course, can vary with position, but for simplicity we take it to be constant in this case), defined as

$$\rho = m/V$$

where m is mass and V is volume. When we speak of density at a point in the context of continuum mechanics, we refer to an average over a volume V sufficiently large that atomic properties (i.e. the presence or absence of an atom in that particular volume) can be neglected. Naturally, since material bodies are made up of atomic particles, this definition breaks down at volumes comparable with interatomic distances (Fig. 1.1). A continuous medium is therefore an idealization which assumes that physical quantities are definable at points in space, considered as lower limits of shrinking elemental volumes. This makes it possible to express the physical variables mathematically in terms of reasonably well-behaved functions of position and, if necessary, time. This description is formally independent of atomic properties, although, of course, the latter determine the macroscopic behaviour.

The description of material bodies as continua is a powerful tool that allows investigation of many aspects of their rheology. There are several books on continuum mechanics, among which one may mention Fung (1977), which is a clear and compact introduction to the subject.

In order to give a quantitative description of rheological processes, it is necessary to refer the body under consideration to a system of co-ordinates. Generally, properties are functions of position and time, which are the basic *independent variables*. (In *steady-state* processes, i.e.

5

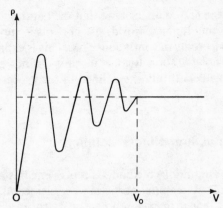

Figure 1.1 Concept of continuum. The macroscopic definition of density breaks down below the critical volume V_0.

those that are independent of time, the properties are fully specified as functions of position only.) The most commonly used co-ordinate system is the orthogonal Cartesian system, consisting of three − or two, if the problem is two-dimensional − right-handed, mutually perpendicular, linear axes x_1, x_2, x_3. In such a system the position of a point in space is univocally determined by $P(x_i^0)$, where $x_i^0 \equiv (x_1^0, x_2^0, x_3^0)$ are the relevant values of the co-ordinates (Fig. 1.2). The subscript i is termed the *index* of the variable x_i^0. As *index notation* is extensively used in continuum mechanics, some of its basic rules are reviewed here.

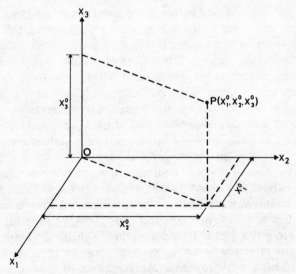

Figure 1.2 Cartesian co-ordinates.

6

Referring again to Figure 1.2, an equivalent way of giving the position of point P is to state that x_i^0 is its position vector. A vector is defined by its components, the number of which is equal to the number of spatial co-ordinates. Therefore, at the most basic level, index notation is simply a way of dealing with quantities with more than one component; in three dimensions, for instance, $x_i^0 \equiv (x_1^0, x_2^0, x_3^0)$ i.e. $i = 1, 2, 3$. This is expressed by saying that the *range* of the index i is three. The range of an index is equal to the number of spatial dimensions of the system.

When using index notation, it is understood that when we write a variable (or an equation) in index form we mean a number of variables (or of equations) equal to the range of the index. For instance, in three dimensions,

$$a_i \quad \text{means} \quad a_1, a_2, a_3$$

$$a_i + b_i = c_i \quad \text{means} \quad a_1 + b_1 = c_1$$

$$a_2 + b_2 = c_2$$

$$a_3 + b_3 = c_3$$

Notice that any letter such as j, k, l or m could have been used for the index, provided it was used in all terms of the equation. If a term has two different indices, then each of them can take the whole range of values. For instance, in two dimensions,

$$\delta_{ij} \quad \text{means} \quad \delta_{11}, \delta_{12}, \delta_{21}, \delta_{22}$$

and, in three dimensions,

$$\sigma_{ij} \quad \text{means} \quad \sigma_{11}, \sigma_{12}, \sigma_{13}, \sigma_{21}, \sigma_{22}, \sigma_{23}, \sigma_{31}, \sigma_{32}, \sigma_{33}$$

There is no theoretical limit to the number of indices that a quantity may have, or to their range. However, in continuum mechanics, when dealing with isotropic bodies, the physical quantities involved usually have zero, one, or two indices; their range is equal to the number of spatial dimensions of the problem, i.e. either two or three.

Notice that, in the above examples, no term contains a repeated index. Examples of repeated indices *in the same term* would be $a_i b_i$, δ_{jj}, σ_{kk}, etc. In index notation a non-repeated index is termed a *free index*. For example, a_i, b_i and c_i have one free index; δ_{ij} and σ_{ij} have two; and $a_i b_i$, δ_{jj} and σ_{kk} have none.

The previous considerations about the meaning of subscripted variables apply only to free indices. A basic rule of index notation is the *summation convention*, which states that the repetition of an index in the same term denotes a summation with respect to the repeated index over its range. Therefore, in three dimensions,

$$a_i b_i = a_1 b_1 + a_2 b_2 + a_3 b_3$$

$$\delta_{jj} = \delta_{11} + \delta_{22} + \delta_{33}$$

$$\sigma_{kk} = \sigma_{11} + \sigma_{22} + \sigma_{33}$$

A repeated index is a dummy index in the sense that it can be changed without any effect on the resultant equation.

It follows from the above that there is a connection between the number of free indices and the number of components of a subscripted variable. A term with no free indices denotes a quantity with one component only; a term with one free index denotes a quantity with a number of components equal to the range of the index; and a term with two free indices denotes a quantity with four (in two dimensions) or nine (in three dimensions) components, that is, the number of components is equal to the range raised to the power two. If R denotes the range and F the number of free indices, then the number of components N is given by the rule

$$N = R^F$$

In an equation written in index notation, each term must obviously have the same number of free indices. If all terms have no free indices, then just one equation is implied; if all terms have one free index, a system of R equations; if all terms have two free indices, a system of R^2 equations.

All of the preceding considerations may be applied to matrices, and therefore index notation may be viewed as a convenient shorthand for matrix calculus. For instance, a quantity with one free index represents a column vector

$$a_i \equiv \begin{pmatrix} a_1 \\ a_2 \\ a_3 \end{pmatrix}$$

$$\delta_{ij} a_j \equiv \begin{pmatrix} \delta_{11} a_1 + \delta_{12} a_2 + \delta_{13} a_3 \\ \delta_{21} a_1 + \delta_{22} a_2 + \delta_{23} a_3 \\ \delta_{31} a_1 + \delta_{32} a_2 + \delta_{33} a_3 \end{pmatrix}$$

8

and a quantity with two free indices represents a square matrix

$$\sigma_{ij} \equiv \begin{pmatrix} \sigma_{11} & \sigma_{12} & \sigma_{13} \\ \sigma_{21} & \sigma_{22} & \sigma_{23} \\ \sigma_{31} & \sigma_{32} & \sigma_{33} \end{pmatrix}$$

The above examples are for $R = 3$, and reduction to the two-dimensional case is immediate.

1.3 Scalars, vectors, and tensors

Index notation is not a purely formal arrangement: the number of free indices is related to a fundamental attribute of physical quantities – their *tensorial rank*.

A tensor of rank zero has no free indices, and is termed a *scalar*. Its value is specified in any co-ordinate system by one component only (its magnitude). The value of a scalar is invariant under a change of co-ordinates. Examples of scalar quantities are density and temperature.

A tensor of rank one has one free index, and is termed a *vector*. Its value is specified by a number of components equal to the number of spatial co-ordinates. In other words, a vector is a quantity having direction as well as magnitude. Besides the position vector, already discussed in Section 1.2, another example is the displacement vector. With reference to Figure 1.3, consider a particle initially at the point P' (x_i') that has moved to the point $P''(x_i'')$. The displacement vector is the

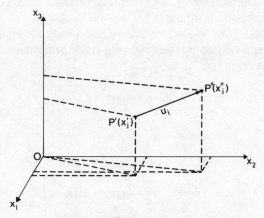

Figure 1.3 Displacement vector.

9

directed line segment u_i, with components (u_1, u_2, u_3) in the three co-ordinate directions given by $u_i = x_i'' - x_i'$.

Unlike a scalar, a vector is not invariant under a change of co-ordinates. Its magnitude is invariant, but its components are not. Mathematically, tensors of various ranks are defined by their laws of transformation under a change of co-ordinates; these laws will be discussed later in this section. Examples of vector quantities include velocity, force, and heat flow.

A tensor of rank two has two free indices, and is termed simply a *tensor* (the term 'tensor' comes from the Latin *tensio*, tension or stress). Tensor components obey specific transformation laws under a change of co-ordinates. As in the case of vectors, there are certain combinations of components that are invariant in any co-ordinate system; these will be discussed as the need arises. The fundamental rheological quantities (stress, strain, and strain rate) are all tensors of rank two.

There are, of course, tensors of rank higher than two, and some of these arise in continuum mechanics (for instance, in describing the elasticity of anisotropic bodies). However, isotropic bodies (i.e. those whose physical properties are independent of direction) usually only require tensors up to rank two. In index notation, as we have already seen, the rank of the quantity can be immediately deduced by the number of free indices. In texts where vector notation is used, a scalar is usually set in italic type, a vector in bold type, and a tensor is denoted by some other arrangement. Alternatively, all components can be written out in full, usually in terms of Cartesian co-ordinates labelled x, y, and z. In this book we shall use index notation wherever necessary, although established convention will be followed when this is different.

Let us briefly discuss *tensor transformation laws*. (We consider exclusively Cartesian tensors, i.e. tensors referred to a system of rectangular Cartesian co-ordinates.)

Assume that x_i and $\bar{x}_i$ are two Cartesian co-ordinate systems with the same origin and related by the following transformation under rotation of co-ordinates:

$$\bar{x}_i = a_{ij}x_j \tag{1.1}$$

that is,

$$\bar{x}_1 = a_{11}x_1 + a_{12}x_2 + a_{13}x_3$$

$$\bar{x}_2 = a_{21}x_1 + a_{22}x_2 + a_{23}x_3$$

$$\bar{x}_3 = a_{31}x_1 + a_{32}x_2 + a_{33}x_3$$

or, in matrix form,

$$\begin{pmatrix} \bar{x}_1 \\ \bar{x}_2 \\ \bar{x}_3 \end{pmatrix} = \begin{pmatrix} a_{11} & a_{12} & a_{13} \\ a_{21} & a_{22} & a_{23} \\ a_{31} & a_{32} & a_{33} \end{pmatrix} \begin{pmatrix} x_1 \\ x_2 \\ x_3 \end{pmatrix}$$

The matrix a_{ij} is called the transformation matrix. Each component of the transformation matrix is related to the rotation of the co-ordinates, being given by the cosine of the angle between the relevant axes, i.e.

$$a_{ij} = \cos(\bar{x}_i, x_j) \tag{1.2}$$

(for instance, a_{11} is the cosine of the angle between $\bar{x}_1$ and x_1, and so on). The inverse co-ordinate transformation is

$$x_i = a_{ji}\bar{x}_j \tag{1.3}$$

The properties of transformation matrices and co-ordinate transformations in general are discussed by Fung (1977), and Jeffreys (1963) also gives a concise treatment of tensor transformations. What is important in the present context is to note that a scalar is a quantity whose only component does not change under co-ordinate transformation, i.e.

$$\rho(\bar{x}_i) = \rho(x_i) \tag{1.4}$$

Equation 1.4 expresses the fact that the density at a point, for instance, does not change if the label of that point changes as a consequence of a co-ordinate transformation.

A vector is a quantity whose three (or two) components transform according to the laws ($\bar{u}_i$ and u_i refer to the components in the $\bar{x}_i$ and x_i co-ordinate systems, respectively)

$$\bar{u}_i = a_{ij}u_j \qquad u_i = a_{ji}\bar{u}_j \tag{1.5}$$

From Equations 1.5 it can easily be proven that the quantity

$$(u_iu_i)^{1/2} = (u_1^2 + u_2^2 + u_3^2)^{1/2} \tag{1.6}$$

is invariant under co-ordinate transformation, that is,

$$(\bar{u}_i\bar{u}_i)^{1/2} = (u_i + u_i)^{1/2}$$

11

Equation 1.6 represents the magnitude of a vector, which does not change under transformation of co-ordinates.

Finally, it can be proven that the nine − or four − components of a tensor of rank two transform according to the laws

$$\bar{\sigma}_{ij} = a_{im}a_{jn}\sigma_{mn} \qquad \sigma_{ij} = a_{mi}a_{nj}\bar{\sigma}_{mn} \tag{1.7}$$

It will be seen later (Section 2.4) that a tensor has three invariants, and that these have important properties that can be related to the state of stress or to the state of strain within a body.

1.4 Principle of dimensional and tensorial homogeneity

We saw in Section 1.2 that in any equation each term must have the same number of free indices, i.e. it must be a tensor of the same rank. This is the principle of tensorial homogeneity. Besides its tensorial rank, another important attribute of a physical quantity is given by its *dimensions*. In mechanics, we choose as primary dimensions the following:

$$\text{mass}\,[m], \text{ length } [l], \text{ time } [t]$$

(The temperature $[T]$ is added when thermal quantities are involved.)

The dimensions of any mechanical quantity can be expressed in terms of the basic dimensions. Here are a few examples (m is mass, V is volume, and u_i is the displacement vector):

$$\text{density} \qquad \rho = m/V \qquad [ml^{-3}]$$

$$\text{velocity} \qquad v_i = \frac{du_i}{dt} \qquad [lt^{-1}]$$

$$\text{acceleration} \qquad a_i = \frac{dv_i}{dt} = \frac{d^2 u_i}{dt^2} \qquad [lt^{-2}]$$

$$\text{force} \qquad f_i = ma_i \qquad [mlt^{-2}]$$

In the SI system (Système International d'Unités), the units for m, l, and t are the kilogram (kg), the metre (m), and the second (s), respectively. In the CGS system, which is sometimes still used in geophysics, the units are the gram (g), the centimetre (cm), and the second (s) − hence centimetre−gram−second (CGS) system. Other mechanical quantities are expressed in derived units.

The *principle of dimensional and tensorial homogeneity* states that all terms in an equation must have the same dimensions and be tensors of the same rank. Consider, for instance, the equation

$$\frac{\partial \sigma_{ij}}{\partial x_j} + \rho X_i = \rho \frac{\partial^2 u_i}{\partial t^2} \tag{1.8}$$

which is the equation of motion for small displacements in continuous media (see Section 6.1; σ_{ij} is stress and X_i is the body force per unit mass). When written in full in three dimensions, it becomes

$$\frac{\partial \sigma_{11}}{\partial x_1} + \frac{\partial \sigma_{12}}{\partial x_2} + \frac{\partial \sigma_{13}}{\partial x_3} + \rho X_1 = \rho \frac{\partial^2 u_1}{\partial t^2}$$

$$\frac{\partial \sigma_{21}}{\partial x_1} + \frac{\partial \sigma_{22}}{\partial x_2} + \frac{\partial \sigma_{23}}{\partial x_3} + \rho X_2 = \rho \frac{\partial^2 u_2}{\partial t^2}$$

$$\frac{\partial \sigma_{31}}{\partial x_1} + \frac{\partial \sigma_{32}}{\partial x_2} + \frac{\partial \sigma_{33}}{\partial x_3} + \rho X_3 = \rho \frac{\partial^2 u_3}{\partial t^2}$$

All terms in Equation 1.8 are tensors of rank one. Given that stress has the dimensions of force per unit area, i.e. $[\sigma] = [mlt^{-2}l^{-2}] = [ml^{-1}t^{-2}]$, and $[X] = [mlt^{-2}m^{-1}] = [lt^{-2}]$, the dimensions of the three terms of Equation 1.8 are

$$\left[\frac{\partial \sigma}{\partial x}\right] = [ml^{-1}t^{-2}l^{-1}] = [ml^{-2}t^{-2}]$$

$$[\rho X] = [ml^{-3}lt^{-2}] = [ml^{-2}t^{-2}]$$

$$\left[\rho \frac{\partial^2 u}{\partial t^2}\right] = [ml^{-3}lt^{-2}] = [ml^{-2}t^{-2}]$$

as required by the principle of dimensional and tensorial homogeneity. (Indices have been omitted in dimensional expressions.)

Dimensional and tensorial homogeneity is a powerful criterion in continuum mechanics, and dimensional analysis plays an important role in some topics in geodynamics. For instance, dimensionless variables, the most well-known being probably the Rayleigh Number (cf. Section 7.8), are of central importance in the problem of thermal convection in the mantle of the Earth.

1.5 Rheological (constitutive) equations

The mechanical state of a body is specified by means of kinematic and dynamic quantities. The former refer to motion (i.e. displacement, velocity and acceleration), the latter to forces of various types. Kinematic and dynamic quantities are related by fundamental laws. For instance, Newton's Second Law of Motion in classical mechanics

$$f_i = ma_i$$

relates a dynamic quantity (force) to a kinematic quantity (acceleration) through a constant of proportionality m, the mass of the body, which (in Newtonian mechanics) is a property of the body that is independent of its mechanical state.

The situation is similar in rheology, where the aim is to establish basic relations between kinematic and dynamic quantities for continuous media, to analyse their meaning and implications, and to make use of their consequences. Such basic relations are termed *rheological equations of state* or *constitutive equations*. Since extended bodies occupy finite domains in space (as opposed to the point-bodies of classical mechanics), knowledge of the kinematic state requires the specification of the position of all of their particles (in the continuum-mechanics sense) and of their variation with respect to time. Similarly, the dynamic state is known when the forces acting on all parts of the body are specified. The first step toward the setting up of rheological equations therefore requires the identification and definition of the basic kinematic and dynamic quantities.

It will be seen in Chapter 2 that the basic dynamic quantity is stress, and the basic kinematic quantity is strain (or strain rate). Both of these are Cartesian tensors of rank two. Stress is related to the *forces* acting on the body, and strain is related to *deformation*. At this stage it is useful to introduce qualitatively the concept of deformation. The occurrence of movement affecting an extended body does not necessarily imply deformation. Translation and rotation (depicted in Figs 1.4a & b) are examples of *rigid-body motion*, i.e. motion that does not involve deformation of the body (the relative position of two points A and B within the body does not change). Expansion or change in volume (Fig. 1.4c), and shear or change in shape (Fig. 1.4d), on the other hand, affect the relative positions of points A and B. Movements of the particles of the body with respect to each other, or (in other words) non-rotational variations of the displacement vector between neighbouring points, result in deformation. However, in practice translation, rotation, and deforma-

14

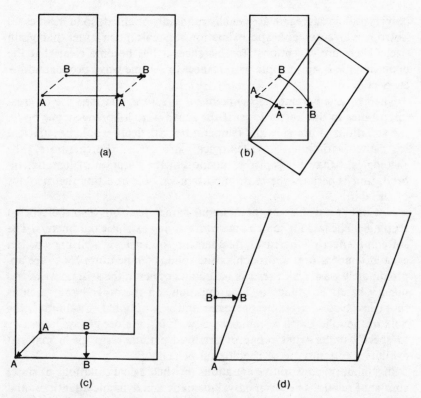

Figure 1.4 Examples of rigid-body motion ((a) translation, (b) rotation) and deformation ((c) change in volume, (d) change in shape).

tion of various kinds often occur together, a problem well known in structural geology and geodynamics.

Just as in classical mechanics there are not only kinematic and dynamic quantities, but also parameters related to the intrinsic properties of a body (e.g. the mass in Newton's Second Law), in rheology there are parameters related to the rheological properties of a body, such as rigidity, compressibility, viscosity, and so on. Although they are sometimes referred to as 'material constants', it is better to denote them as *material parameters* (or *material coefficients*), to emphasize the fact that their values change with changing extrinsic conditions (temperature, pressure, and time), even for the same intrinsic conditions (identical materials). The values of material coefficients have to be determined on the basis of suitable experiments.

Material parameters are scalars in isotropic bodies and they are tensors of higher rank in anisotropic bodies. Although single crystals are anisotropic with respect to physical properties, the grains in a polycrystalline aggregate are usually randomly oriented, so in most cases isotropy is an acceptable approximation at a scale much larger than grain size. (There are exceptions: for instance, it has become clear that the upper mantle is anisotropic with respect to seismic wave propagation – cf. Section 6.6.)

The extrinsic parameters *temperature, pressure*, and *time* are of great importance in the Earth. Even if the Earth were a homogeneous body, the variations of pressure and temperature with depth would be sufficient to cause variations in rheological properties. Furthermore, the rheological behaviour is also a function of the duration of loading. Indeed, time is perhaps the most important variable affecting the rheology of the Earth.

Not only the values of the material parameters, but also the type of rheological behaviour can change with varying extrinsic parameters. The softening effect of increasing temperature on rocks is well known, but even at temperatures approaching the solidus temperature rocks are approximately elastic with respect to loading cycles in the seismic wave frequency band. For loads of long duration, on the other hand, such as those involved in tectonic processes and gravitational instabilities, the bulk of the solid Earth responds by flow. It is therefore always necessary to specify under what range of extrinsic parameters a given material exhibits a certain type of rheological behaviour.

In summary, constitutive equations (or rheological equations of state) consist of relations between basic kinematic and dynamic quantities, and material parameters. They have the general form

$$R(\varepsilon, \dot{\varepsilon}, \sigma, \dot{\sigma}, \ldots, \{M\}) = 0 \qquad (1.9)$$

where ε is strain, σ is stress (indices omitted), an overdot denotes the time derivative (which could be of higher order, or absent), and $\{M\}$ denotes intrinsic material parameters. R is called the *rheological function*.

Several types of constitutive equations can be established on the basis of observation, theory, and experiment. Materials that follow the same constitutive equation are said to belong to the same rheological class, although their material coefficient can, of course, be different. As already emphasized, not only the values of $\{M\}$, but also the form of R may change for a given material under different extrinsic conditions.

The classification of materials into rheological classes such as elastic, plastic, viscous, and so on, reflects the rheological equation governing their behaviour. For example, rocks, steel, and concrete all behave elastically (to a first approximation) at low temperature and pressure, and for short duration of loading. Elastic behaviour implies instantaneity of strain under stress, recoverability of strain when stress is removed,

16

and linear proportionality between stress and strain (cf. Ch. 3). On the other hand, water at room conditions and very fine-grained rocks at high temperature and pressure and long times both behave as viscous bodies (they flow in response to stress, and the strain rate is proportional to the stress). Thus, rocks can behave differently according to their depth and the duration of the relevant geodynamic process.

1.6 Material and spatial descriptions

A concept that appears often in problems dealing with fluid flow (discussed in detail in Ch. 4) is that of *total* (or *material*) *derivative*. As it illustrates some rather fundamental aspects of continuum mechanics, we introduce it here (for a fuller treatment, reference may be made to Fung 1977).

Suppose that a continuous body occupies a given space domain in a Cartesian frame of reference x_i. At time $t = t_0$, let the co-ordinates of any particle within the body be $x_i = c_i$. After a certain time t, during which the body undergoes displacements (translation, rotation, and deformation), the *material particle* initially at c_i will be at a position defined by

$$x_i = f_i(c_1, c_2, c_3, t) \qquad (1.10)$$

If the functions f_i are known for all particles of the body, then the configuration of the body after time t is known. The important thing is that each particle is followed through the successive positions it occupies in space; that is, the initial positions of the particles c_i (and time t) are the independent variables. Similarly, any function describing a property of the body can be given as a function of (c_i, t). Since the basic unit, so to speak, is the material particle, this approach is termed the *material* (or *Lagrangian*) description.

However, the deformation of a body may also be described by considering a fixed point in space, and noting which particles occupy this spatial point at different times. If, as before, c_i is the 'name' or label of different particles, then the particle occupying a given point of co-ordinates x_i at time t is given by

$$c_i = g_i(x_1, x_2, x_3, t) \qquad (1.11)$$

Here the spatial co-ordinates x_i (and time t) are the independent variables; as in the previous case, any other property can be given as a function of (x_i, t). This approach is termed the *spatial* (or *Eulerian*) description.

Intuitively, the two descriptions may be compared with the situation of an observer on the bank of a stream. If the observer follows the path of individual particles in the water (however identified – they could, for instance, be dyed differently), the Lagrangian description (Eqn 1.10) applies; if attention is focused on a fixed point in space and a note is made of the different particles that occupy it in time, then the Eulerian description (Eqn 1.11) applies.

The choice of either mode of description depends on the problem under consideration. If it is necessary to identify the initial position of all particles, then the Lagrangian frame of reference is more suitable; if the instantaneous configuration of the body is under study, then the Eulerian frame of reference is preferable. However, the adoption of either point of view changes the way in which the system is described mathematically, and often these changes cannot be neglected.

In the study of elastic problems the dynamic quantity (stress) must satisfy the equilibrium equations in the deformed state, and therefore requires a spatial description. On the other hand, the kinematic quantity (strain) is related to displacements (i.e. to the changes in the position of particles), so a material description is more suitable. However, as long as strain is infinitesimal (which is the case for most geophysical applications of elasticity), the difference between 'material' and 'spatial' strains is negligible, and the two viewpoints coalesce (this is proved in Section 2.8). The Eulerian framework is therefore used in infinitesimal elasticity.

The situation is different in problems involving velocity fields, where time derivatives of position occur. In the Lagrangian description the velocity and the acceleration of a given particle are simply the relevant time derivatives of its position vector x_i

$$v_i(c_1, c_2, c_3, t) = \dot{x}_i \qquad a_i(c_1, c_2, c_3, t) = \dot{v}_i = \ddot{x}_i \qquad (1.12)$$

where the overdot denotes the time derivative. However, we are usually interested not in the initial configuration of the system, but in the velocity field at time t; so the Eulerian description is commonly used in hydrodynamics. Suppose the spatial velocity field is $v_i(x_1, x_2, x_3, t)$; then a particle at position x_i at time t will have moved to position $x_i + dx_i = x_i + v_i\, dt$ at time $t + dt$. We can therefore write

$$dv_i = \frac{\partial v_i}{\partial t}\, dt + \frac{\partial v_i}{\partial x_j}\, dx_j = \frac{\partial v_i}{\partial t}\, dt + \frac{\partial v_i}{\partial x_j}\, v_j\, dt$$

and, consequently, the acceleration is

$$\dot{v}_i = \frac{\partial v_i}{\partial t} + v_j\, \frac{\partial v_i}{\partial x_j} = \frac{Dv_i}{Dt} \qquad (1.13)$$

18

The operation described by Equation 1.13 is termed *differentiation following the motion*, and Dv_i/Dt is the *material* (or *total*) *derivative* of v_i. The material derivative is the sum of two terms: a *local* derivative $\partial v_i/\partial t$ that represents the time dependence of the velocity field at a fixed point in space, and a *convective* term $v_j \partial v_i/\partial x_j$ that represents the rate of change of velocity due to movement of the particle in a non-homogeneous velocity field.

For any property attached to a particle, the operator

$$\frac{D}{Dt} = \frac{\partial}{\partial t} + v_j \frac{\partial}{\partial x_j} = \frac{\partial}{\partial t} + v_1 \frac{\partial}{\partial x_1} + v_2 \frac{\partial}{\partial x_2} + v_3 \frac{\partial}{\partial x_3} \qquad (1.14)$$

gives the total derivative in the Eulerian description. Identification of total and local derivative is erroneous except in the simple case where the property field is homogeneous in space. The operator D/Dt plays an important role in fluid dynamics, and consequently in problems dealing with mantle flow, since the solid mantle exhibits a fluid behaviour over long timescales.

2 Stress, deformation, and strain

The fundamental dynamic and kinematic quantities of rheology are stress and strain, respectively. Both are tensors of rank two.

Stress is introduced first, and is then used to derive the general equations of equilibrium for continuous media. The various ways of expressing the state of stress at a point – by using Cauchy's formula, isotropic and deviatoric stresses, principal directions and principal stresses, stress quadric, normal and shear stress, and Mohr's circle – are discussed in turn.

The infinitesimal strain tensor is shown to be a measure of small time-independent deformation, and the geometrical meaning of each component of strain is illustrated. Finally, the expressions for finite strain are given in the Eulerian and Lagrangian frameworks, which reduce to identical relations when the deformation is infinitesimal.

2.1 The stress tensor

The definition of stress can be arrived at by considering the interactions between contiguous parts of continuous bodies. Let a continuous medium occupy a volume V in Cartesian space, and let ΔV be an element of volume inside it (Fig. 2.1). This volume element is acted upon by two kinds of forces, i.e. body forces, per unit volume or unit mass, and surface forces acting on the surface ΔS bounding ΔV.

The body force per unit mass X_i in geophysics is the acceleration of gravity. To derive the surface forces, let δS be a small surface element of ΔS, and n_i be the outer normal unit vector to δS. Let δF_i be the surface force that the part of the body outside ΔS exerts on the part inside through δS. This force depends not only on the state of the body, but also on the position and orientation of δS. By Newton's Third Law of Motion, an equal and opposite force is exerted by the material inside ΔS on the material outside. If P is a point on δS, then the surface forces acting on δS are equivalent to a single force acting on P and a couple.

20

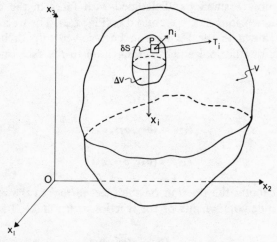

Figure 2.1 Body force X_i acting on element of volume ΔV and surface force (traction) acting on infinitesimal surface element δS.

Assuming that, as δS tends to zero, the couple vanishes and the ratio $\delta F_i / \delta S$ tends to a finite limit, it is possible to define a vector

$$T_{(n)i} = \lim_{\delta S \to 0} \frac{\delta F_i}{\delta S} \qquad (2.1)$$

that represents the force per unit area ($[ml^{-1}t^{-2}]$) acting on the surface with orientation specified by n_i. This vector is termed the *traction*. It should be noted that the subscript n refers to the surface and the only free index is i; the subscript n is retained in parentheses to emphasize that the components of the traction depend on the orientation of the surface element.

A traction vector field can be defined on any surface within a continuous medium, and is usually a continous function of the co-ordinates. Tractions need not be normal to the surface elements on which they act. The traction $T_{(n)i}$ can be resolved into normal and tangential components, or into its three Cartesian components. The convention as to the sign varies – in continuum mechanics, traction is positive when acting in the same general direction as the outer normal to the surface (tension), and is negative in the opposite case (compression). This is the convention we shall follow, since results are then directly comparable with those obtained in a voluminous literature. However, in most works in geology the opposite convention is adopted. This creates no problems, provided the sign convention used in any particular instance is carefully noted.

21

Consider now a small parallelepiped with faces in the co-ordinate planes and dimensions dx_1, dx_2, and dx_3 (Fig. 2.2). The tractions on the three faces shown in the Figure can be resolved into their Cartesian components, one normal and two tangential to the face on which the traction acts:

$$T_{(1)i} = (\sigma_{11}, \sigma_{12}, \sigma_{13})$$

$$T_{(2)i} = (\sigma_{21}, \sigma_{22}, \sigma_{23})$$

$$T_{(3)i} = (\sigma_{31}, \sigma_{32}, \sigma_{33})$$

where $T_{(1)i}$ denotes the traction on the face normal to the axis x_1, and so on. The nine components of the tractions form a 3×3 matrix

$$\sigma_{ij} = \begin{pmatrix} \sigma_{11} & \sigma_{12} & \sigma_{13} \\ \sigma_{21} & \sigma_{22} & \sigma_{23} \\ \sigma_{31} & \sigma_{32} & \sigma_{33} \end{pmatrix} \tag{2.2}$$

where for each component the first subscript denotes the co-ordinate axis to which the surface is normal, and the second subscript denotes the direction in which the component acts. For example, σ_{23} denotes the component acting on the face normal to x_2 in the direction of x_3. The

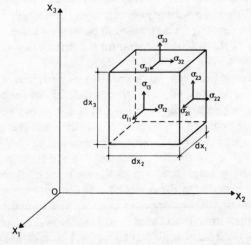

Figure 2.2 Stress components on three faces of infinitesimal parallelepiped of volume dx_1 dx_2 dx_3.

quantity σ_{ij} can be proven to transform under transformation of co-ordinates according to Equation 1.7, and is therefore a tensor of rank two. It is called the *stress tensor*. Its dimensions are force per unit area, $[ml^{-1}t^{-2}]$. Physically, the *normal* components of stress ($i = j$, the on-diagonal components of the matrix) act normally to the surfaces on which they are applied, while the *tangential* or *shear* components ($i \neq j$, the off-diagonal components) act tangentially. The magnitude of the stress components denotes the component of the surface force per unit area (traction) that the part of the body on the positive side (defined by the unit normal) of the surface element exerts on the part lying on the negative side.

The SI unit for stress is the *pascal* (Pa), equivalent to one newton per square metre ($N\,m^{-2}$); the most commonly used multiple in the Earth sciences is the *megapascal* ($1\,MPa = 10^6\,Pa$). Two multiples of the CGS unit ($dyn\,cm^{-2}$) are still in common usage: the *bar* ($1\,bar = 10^6\,dyn\,cm^{-2}$), and the *kilobar* ($1\,kbar = 10^9\,dyn\,cm^{-2}$). Note that $1\,MPa = 10\,bar$.

In keeping with the sign convention for traction, the sign of stress components is determined as follows. If the unit normal to the surface has the same sense as the co-ordinate axis to which the surface is normal, then the stress component is positive if it acts in the positive direction; if the unit normal has a sense opposite to the co-ordinate axis, then the stress component is positive if it acts in the negative direction. The stresses in Figure 2.2 are all positive.

2.2 The equations of equilibrium

The starting point for the study of the state of stress in a continuous body is given by the equations of equilibrium (or of motion, as the case may be). For instance, in the case of tectonic problems, where accelerations may be assumed to be negligible, the equilibrium equations apply. On the other hand, when stresses change rapidly (for instance, in seismic-wave propagation), accelerations come into play and the equations of motion must be used.

For a continuous body to be in equilibrium, the resultant of body and surface forces, and the resultant moment about any axis, must vanish. The procedure for establishing the conditions of equilibrium illustrates a common approach to the analysis of physical problems.

Consider first the equilibrium of forces. The condition that their resultant must vanish implies that the resultant component of force in any co-ordinate direction must be zero. With reference to Figure 2.2, the component of surface force acting on any face of the parallelepiped in

a prescribed co-ordinate direction is given by the relevant stress component multiplied by the area of the face (for clarity, only stress components
acting on three faces are shown). There are three sets of two parallel
faces each; if the stress components meet certain continuity conditions
(they and their derivatives must be continuous), then the components of
surface force acting in, say, the x_1-direction are

$$\sigma_{11}\, dx_2\, dx_3 \text{ and } \left(\sigma_{11} + \frac{\partial \sigma_{11}}{\partial x_1} dx_1\right) dx_2\, dx_3 \quad \text{on the two faces normal to } x_1$$

$$\sigma_{21}\, dx_1\, dx_3 \text{ and } \left(\sigma_{21} + \frac{\partial \sigma_{21}}{\partial x_2} dx_2\right) dx_1\, dx_3 \quad \text{on the two faces normal to } x_2$$

$$\sigma_{31}\, dx_1\, dx_2 \text{ and } \left(\sigma_{31} + \frac{\partial \sigma_{31}}{\partial x_3} dx_3\right) dx_1\, dx_2 \quad \text{on the two faces normal to } x_3$$

(The partial derivatives represent the rate of change of the stress components in the relevant co-ordinate direction.) In addition, if ρ is the density of the material and X_i the body force per unit mass, then the component of the latter in the x_1-direction is $\rho X_1\, dx_1\, dx_2\, dx_3$.

The condition of equilibrium in the x_1-direction is therefore

$$\left(\sigma_{11} + \frac{\partial \sigma_{11}}{\partial x_1} dx_1\right) dx_2\, dx_3 - \sigma_{11}\, dx_2\, dx_3 + \left(\sigma_{21} + \frac{\partial \sigma_{21}}{\partial x_2} dx_2\right) dx_1\, dx_3$$

$$- \sigma_{21}\, dx_1\, dx_3 + \left(\sigma_{31} + \frac{\partial \sigma_{31}}{\partial x_3} dx_3\right) dx_1\, dx_2 - \sigma_{31}\, dx_1\, dx_2$$

$$+ \rho X_1\, dx_1\, dx_2\, dx_3 = 0$$

i.e.

$$\frac{\partial \sigma_{11}}{\partial x_1} + \frac{\partial \sigma_{21}}{\partial x_2} + \frac{\partial \sigma_{31}}{\partial x_3} + \rho X_1 = 0 \qquad (2.3a)$$

and similarly, in the x_2- and x_3-directions,

$$\frac{\partial \sigma_{12}}{\partial x_1} + \frac{\partial \sigma_{22}}{\partial x_2} + \frac{\partial \sigma_{32}}{\partial x_3} + \rho X_2 = 0 \qquad (2.3b)$$

$$\frac{\partial \sigma_{13}}{\partial x_1} + \frac{\partial \sigma_{23}}{\partial x_2} + \frac{\partial \sigma_{33}}{\partial x_3} + \rho X_3 = 0 \qquad (2.3c)$$

In index notation, Equations 2.3a–c become

$$\frac{\partial \sigma_{ji}}{\partial x_j} + \rho X_i = 0 \tag{2.4}$$

A similar procedure is followed for the equilibrium of moments. Obviously, only shearing stresses contribute to moments – they are assumed to act at the mid-point of each face. For instance, equating to zero the sum of moments about an axis parallel to the x_1-direction,

$$\sigma_{23} \, dx_1 \, dx_3 \, \frac{dx_2}{2} + \left(\sigma_{23} + \frac{\partial \sigma_{23}}{\partial x_2} \, dx_2 \right) dx_1 \, dx_3 \, \frac{dx_2}{2} - \sigma_{32} \, dx_1 \, dx_2 \, \frac{dx_3}{2}$$

$$- \left(\sigma_{32} + \frac{\partial \sigma_{32}}{\partial x_3} \, dx_3 \right) dx_1 \, dx_2 \, \frac{dx_3}{2} = 0$$

and, therefore, neglecting infinitesimals of higher order,

$$\sigma_{23} = \sigma_{32} \tag{2.5a}$$

Similarly, with respect to the other two co-ordinate directions,

$$\sigma_{13} = \sigma_{31} \tag{2.5b}$$

and

$$\sigma_{12} = \sigma_{21} \tag{2.5c}$$

i.e. the equilibrium of moments requires that

$$\sigma_{ij} = \sigma_{ji} \tag{2.6}$$

A tensor having the property expressed by Equation 2.6. is a *symmetric* tensor. In a symmetric tensor the corresponding off-diagonal components are equal, i.e. the transpose is identical to the original matrix.

From the symmetry of the stress tensor it follows that Equation 2.4 can be written as

$$\frac{\partial \sigma_{ij}}{\partial x_j} + \rho X_i = 0 \tag{2.7}$$

which is the equation of equilibrium for continuous media. In three dimensions, Equation 2.7 denotes a system of three equations, and in two dimensions it represents a system of two equations. In vector parlance, the first term represents the divergence of the stress tensor.

The equations of equilibrium are sometimes written in terms of body force per unit volume, in which case the second term does not contain the density. However, in the Earth sciences it is more natural to talk in terms of body force per unit mass, which is, of course, gravity. Consequently, if g is the acceleration of gravity, then $X_3 = g$, $X_1 = X_2 = 0$. In several problems in geophysics and geodynamics it is possible to neglect body forces altogether.

Two things should be noted at this stage. First, the symmetry of the stress tensor reduces the number of components that must be specified in order to determine the stress state at a point from nine to six (in three dimensions, or from four to three in two dimensions). Secondly, the equations of equilibrium are valid for all continuous media, regardless of their rheological properties. However, without a knowledge of the rheology they are of very little use, because there are more (unknown) stress components than there are equations. It will be seen in Chapters 3 and 4 that it is precisely a knowledge of the rheology (that is, of the relation between stress, deformation, and flow) that makes it possible to write Equation 2.7 in a more directly usable form.

2.3 Stresses on a surface: Cauchy's formula

If the stress state at a point is known, then it is possible to derive the normal and shear components of stress on any surface with unit normal n_i passing through that point. This is important in several problems in geodynamics, for instance in the theory of faulting (Section 8.6). The relation between traction and stress components is known as Cauchy's formula.

Consider a small tetrahedron OABC inside a continuous body (Fig. 2.3a). Three faces lie in the co-ordinate planes; the fourth (ABC) has outer unit normal n_i and surface area δS. Given the stress tensor σ_{ij} at O (i.e. the nine components of stress acting on the three faces normal to the co-ordinate axes), we wish to derive expressions for the traction $T_{(n)i}$ on ABC. In order to do this we use the conditions of equilibrium, i.e. equate to zero the sum of the forces acting on the tetrahedron.

First, it should be noted that the areas of the three faces perpendicular to x_1, x_2 and x_3 are, respectively,

$$\delta S_1 = n_1 \, \delta S \qquad \delta S_2 = n_2 \, \delta S \qquad \delta S_3 = n_3 \, \delta S$$

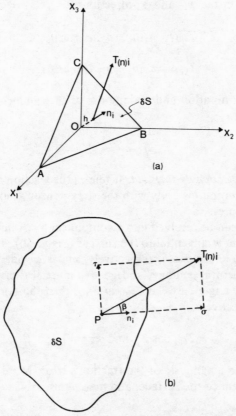

Figure 2.3 (a) Traction on inclined surface ABC (Cauchy's formula); (b) normal and shear stresses on surface δS.

since n_1, n_2, and n_3 are the cosines of the angles between the vector n_i and the co-ordinate axes. Also, if h is the height of the tetrahedron, then its volume is

$$\delta V = \tfrac{1}{3} h \, \delta S$$

Therefore the condition of equilibrium of forces in, say, the x_2-direction is

$$T_{(n)2} \, \delta S - \sigma_{12} n_1 \, \delta S - \sigma_{22} n_2 \, \delta S - \sigma_{32} n_3 \, \delta S + \tfrac{1}{3} \rho X_2 h \, \delta S = 0$$

where X_i is, as usual, the body forces per unit mass. For an infinitesimally small volume ($h \to 0$), this equation becomes

$$T_{(n)2} = \sigma_{12} n_1 + \sigma_{22} n_2 + \sigma_{32} n_3 \qquad (2.8a)$$

27

and, similarly, in the x_1- and x_3-directions

$$T_{(n)1} = \sigma_{11}n_1 + \sigma_{21}n_2 + \sigma_{31}n_3 \tag{2.8b}$$

$$T_{(n)3} = \sigma_{13}n_1 + \sigma_{23}n_2 + \sigma_{33}n_3 \tag{2.8c}$$

that is, in index notation and making use of the symmetry of the stress tensor,

$$T_{(n)i} = \sigma_{ij}n_j \tag{2.9}$$

Equation 2.9 is *Cauchy's formula*. It relates the traction across any surface within a continuous body with the stress tensor given as a function of the co-ordinates.

The traction can be resolved into a component σ normal to the surface and a component τ tangential to the surface (Fig. 2.3b). These are called the *normal* and *shear* stress, respectively, and are widely used in rock mechanics, structural geology, and geodynamics. In terms of the traction, if β is the angle between $T_{(n)i}$ and n_i, then the normal and shear stress are, respectively,

$$\sigma = T_{(n)} \cos \beta \quad \text{and} \quad \tau = T_{(n)} \sin \beta$$

where $T_{(n)}$ is the magnitude of the traction vector. More generally, if t_i is the unit tangent to the surface, and making use of Cauchy's formula,

$$\sigma = T_{(n)i}n_i = \sigma_{ij}n_jn_i \qquad \tau = T_{(n)i}t_i = \sigma_{ij}n_jt_i \tag{2.10}$$

The above equations allow the determination of normal and shear stresses on any surface with specified orientation passing through a point, if the state of stress is known at that point. They will be discussed more fully in Section 2.5. However, before any further discussion it is necessary to introduce other ways of specifying the state of stress within a continuous body.

2.4 Isotropic and deviatoric stress; principal axes and principal stresses

Given a continuous medium (for instance, in a geodynamic context, a rock mass, a folded belt, or a tectonic plate) referred to a set of Cartesian coordinates x_i, complete specification of its state of stress requires

28

knowledge of the stress tensor σ_{ij} at each point within it, i.e. knowledge of its six components as functions of the co-ordinates.

It is often very useful to think of the stress tensor as being composed of two parts; the isotropic and the deviatoric parts. The *isotropic stress* is defined as

$$\sigma_{ij}^0 = \tfrac{1}{3} \sigma_{kk} \, \delta_{ij} \tag{2.11}$$

where σ_{kk} is the sum of the on-diagonal components of the stress tensor and δ_{ij} is the Kronecker delta, or identity matrix ($\delta_{ij} = 1$ for $i = j$, $\delta_{ij} = 0$ for $i \neq j$).

The *deviatoric stress*, on the other hand, is defined by the relation

$$\sigma_{ij}^I = \sigma_{ij} - \sigma_{ij}^0 \tag{2.12}$$

i.e. is what is left of the stress once the isotropic part has been removed. Since $\sigma_{ij}^0 = 0$ for $i \neq j$ (Eqn 2.11), the off-diagonal components of the stress deviator are identical to the corresponding components of the total stress.

From Equations 2.11 and 2.12, it follows that the stress tensor can always be written as the sum of an isotropic and a deviatoric part:

$$\sigma_{ij} = \sigma_0 \, \delta_{ij} + \sigma_{ij}^I \tag{2.13}$$

where we have written $\tfrac{1}{3} \sigma_{kk} = \sigma_0$ for the *mean normal stress*. The deviator therefore represents that part of the state of stress that differs from a hydrostatic state, and is identically zero only when the stress state is hydrostatic (normal stresses equal from all directions). When the stress state is not hydrostatic, some − or all − of the deviatoric components (including on-diagonal components) are different from zero. The stress deviator assumes a particularly large importance in problems involving yielding and flow, and consequently in most geodynamic situations, as it is related to permanent deformation (see Chs 4 & 5).

Although the resolution of the stress tensor into isotropic and deviatoric parts is very important, a perhaps more common representation of the state of stress − certainly in structural geology and tectonics − is often useful, especially when one is interested in the description of the state of stress within an extended body (for instance, within a tectonic plate, or in any given model of a geotectonic feature). This consists of the specification of the stress state in terms of principal stresses and principal axes.

The *principal axes* of stress (or *principal directions*) are the co-ordinate system (or systems) in which the only non-zero stress components at a point are the on-diagonal components (naturally, the principal directions can vary from point to point). The *principal stresses* are the values of these on-diagonal components.

The problem is therefore the following: given the stress tensor σ_{ij} in a Cartesian co-ordinate system x_i, find the co-ordinate system in which σ_{ij} reduces to a diagonal matrix, the *principal stress tensor*:

$$\sigma_{ij}^P = \begin{pmatrix} \sigma_{11}^P & 0 & 0 \\ 0 & \sigma_{22}^P & 0 \\ 0 & 0 & \sigma_{33}^P \end{pmatrix}$$

(In the sequel, the principal stress σ_{11}^P will be denoted simply by σ_1, and so on.) Mathematically, one has to find a suitable transformation of co-ordinates such that $\sigma_{ij} \rightarrow \sigma_{ij}^P$. From the physical viewpoint, the principal axes are that set of co-ordinates such that no shear stresses act on the planes normal to them.

The usefulness of reasoning in terms of principal axes and principal stresses lies in the fact that they give a clear picture of the state of stress at a point (and this picture can be formalized in terms of the stress quadric; see later in this section). However, one still needs six components to determine fully the state of stress (namely, the three principal stresses and the three principal directions).

The derivation of principal stresses and axes is a classic eigenvalue problem. Let an arbitrarily oriented surface element, with unit normal n_i, within a continuous medium be acted upon by a traction $T_{(n)i}$. In general, the direction of the traction and that of the normal do not coincide, unless the latter is a principal axis. The condition for n_i to be a principal direction is therefore, on the basis of Cauchy's formula (Eqn 2.9),

$$T_{(n)i} = \sigma_{ij}n_j = \sigma n_i \tag{2.14}$$

where σ_{ij} is the stress tensor in the system x_i, and σ denotes any one of the three principal stresses. By making use of the Kronecker delta, $n_i = \delta_{ij}n_j$, and Equation 2.14 can be written as

$$(\sigma_{ij} - \sigma\delta_{ij})n_j = 0 \tag{2.15}$$

which becomes, when written in full,

$$(\sigma_{11} - \sigma)n_1 + \sigma_{12}n_2 + \sigma_{13}n_3 = 0$$

$$\sigma_{21}n_1 + (\sigma_{22} - \sigma)n_2 + \sigma_{23}n_3 = 0$$

$$\sigma_{31}n_1 + \sigma_{32}n_2 + (\sigma_{33} - \sigma)n_3 = 0$$

This system of three equations has be to solved for n_1, n_2, and n_3, subject to the additional condition $n_1^2 + n_2^2 + n_3^2 = 1$.

In the theory of matrices (see, for example, Kreyszig 1983) it is proved that a non-trivial solution of Equation 2.15 exists only if the determinant of the coefficients vanishes, i.e. (the double vertical bars denote the determinant)

$$\| \sigma_{ij} - \sigma \, \delta_{ij} \| = 0$$

This results in a third-degree equation in σ:

$$\| \sigma_{ij} - \sigma \, \delta_{ij} \| = \begin{Vmatrix} \sigma_{11} - \sigma & \sigma_{12} & \sigma_{13} \\ \sigma_{21} & \sigma_{22} - \sigma & \sigma_{23} \\ \sigma_{31} & \sigma_{32} & \sigma_{33} - \sigma \end{Vmatrix} = -\sigma^3 + I_1\sigma^2 + I_2\sigma + I_3 = 0$$

(2.16)

where

$$I_1 = \sigma_{11} + \sigma_{22} + \sigma_{33} \tag{2.17a}$$

$$-I_2 = \begin{Vmatrix} \sigma_{11} & \sigma_{12} \\ \sigma_{21} & \sigma_{22} \end{Vmatrix} + \begin{Vmatrix} \sigma_{11} & \sigma_{13} \\ \sigma_{31} & \sigma_{33} \end{Vmatrix} + \begin{Vmatrix} \sigma_{22} & \sigma_{23} \\ \sigma_{32} & \sigma_{33} \end{Vmatrix} \tag{2.17b}$$

$$I_3 = \begin{Vmatrix} \sigma_{11} & \sigma_{12} & \sigma_{13} \\ \sigma_{21} & \sigma_{22} & \sigma_{23} \\ \sigma_{31} & \sigma_{32} & \sigma_{33} \end{Vmatrix} \tag{2.17c}$$

It can also be proved that any real-valued symmetric matrix (such as σ_{ij}) always has three real-valued *eigenvalues,* which in our case are the principal stresses $\sigma_1, \sigma_2,$ and σ_3; and at least one set of orthogonal *eigenvectors,* which in our case are the principal directions.

Solution of Equation 2.16 therefore yields the three principal stresses $\sigma_1, \sigma_2,$ and σ_3 (by convention, $\sigma_1 > \sigma_2 > \sigma_3$). The principal directions are then obtained by solving Equation 2.15 for n_1, n_2, and n_3, successively for the case $\sigma = \sigma_1, \sigma = \sigma_2, \sigma = \sigma_3$. The solution n_i gives the orientation of the relevant principal axis in the co-ordinate system x_i in which σ_{ij} is given. In the principal axes the stress tensor reduces to σ_{ij}^P.

The quantities I_1, I_2, and I_3 appearing in Equation 2.16 and defined by Equations 2.17a, b, and c are the *invariants* of the stress tensor: that is, those combinations of stress components whose values do not change under transformation of co-ordinates (refer to Section 1.3). They are termed the first, second, and third invariant, respectively. The first invariant is simply an expression of the fact that the mean normal stress is a constant independent of the choice of co-ordinate system; the second invariant plays an important role in the theories of flow (see especially Section 4.4) and of plastic yielding (Section 5.5).

A general description of the state of stress at a point, which contains the principal axes and stresses as a particular case, can be given in terms of the *stress quadric* of Cauchy, defined as

$$\sigma_{ij} x_i x_j = \pm k^2 \tag{2.18}$$

where k is an arbitrary real constant. If the co-ordinate axes are taken to coincide with the principal directions, then Equation 2.18 becomes

$$\sigma_1 x_1^2 + \sigma_2 x_2^2 + \sigma_3 x_3^2 = \pm k^2$$

which, for $k = 1$, $\sigma_1 \neq \sigma_2 \neq \sigma_3$ and all of the same sign, is the equation of a triaxial ellipsoid with semi-axes $(\sigma_1)^{-\frac{1}{2}}$, $(\sigma_2)^{-\frac{1}{2}}$, and $(\sigma_3)^{-\frac{1}{2}}$ along the principal directions. If $\sigma_1 \neq \sigma_2 = \sigma_3$, then there is one principal direction normal to a plane in which any two orthogonal unit vectors can be chosen as principal directions, and the ellipsoid becomes a surface of revolution. If $\sigma_1 = \sigma_2 = \sigma_3$, then the stress state is hydrostatic and the stress ellipsoid is a sphere; in this case, of course, there are no defined principal directions.

A modification of the stress quadric often used in the Earth sciences is *Lamé's stress ellipsoid*, which is a quadric surface with axes coinciding with the principal directions and semi-axes $\sigma_1, \sigma_2, \sigma_3$, i.e.

$$\frac{x_1^2}{\sigma_1^2} + \frac{x_2^2}{\sigma_2^2} + \frac{x_3^2}{\sigma_3^2} = 1 \tag{2.19}$$

An example of Lamé's stress ellipsoid is given in Figure 2.4. The ellipsoid represents the locus of the end-points of the traction vector across planes of all orientations through its centre.

The stress ellipsoid gives a clear image of the state of stress at a point, both in magnitude and orientation. In most situations of geological

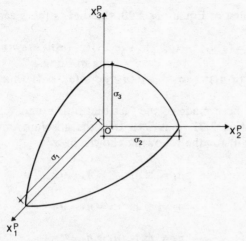

Figure 2.4 Lamé's stress ellipsoid.

interest all three principal stresses are compressive, and it is customary to arrange them in terms of decreasing absolute value. This is the convention used, for instance, to describe the state of stress during faulting (Section 8.6), and in several other tectonic problems.

2.5 Maximum shear stress; Mohr's circle

At the end of Section 2.3 we discussed the resolution of stress into normal and shear components (Eqns 2.10). As this resolution is used to analyse many problems of geodynamical significance (for example, some fracture processes – Section 5.3 – and theories of faulting – Section 8.6), we explore it further here, also in the light of considerations on principal stresses and principal directions developed in the previous section.

Since shear stress is related to fracture, it is important to determine the planes on which shear stress is maximum, since these are the planes of potential fracture. For this purpose, recall Equations 2.10. If the coordinate axes coincide with the principal directions, then the normal and shear stresses become

$$\sigma = \sigma_1 n_1^2 + \sigma_2 n_2^2 + \sigma_3 n_3^2 \qquad \tau = \sigma_1 n_1 t_1 + \sigma_2 n_2 t_2 + \sigma_3 n_3 t_3 \qquad (2.20)$$

where, as before, σ_1, σ_2, and σ_3 are the principal stresses, and n_i and t_i are the unit normal and unit tangent, respectively, to the element of surface. Using the Pythagorean Theorem together with Cauchy's for-

33

mula and the first of Equations 2.20, the shear stress τ can be written as

$$\tau^2 = T_{(n)}^2 - \sigma^2 = (\sigma_1 n_1)^2 + (\sigma_2 n_2)^2 + (\sigma_3 n_3)^2 - (\sigma_1 n_1^2 + \sigma_2 n_2^2 + \sigma_3 n_3^2)^2$$

$$= (\sigma_1 - \sigma_2)^2 (n_1 n_2)^2 + (\sigma_2 - \sigma_3)^2 (n_2 n_3)^2 + (\sigma_1 - \sigma_3)^2 (n_1 n_3)^2 \qquad (2.21)$$

where use has been made of the fact that n_i is a unit vector.

From Equation 2.21, it is seen that the absolute value of τ has a minimum ($\tau = 0$) for the following choices of n_i:

$$n_1 = \pm 1 \quad n_2 = n_3 = 0$$

$$n_2 = \pm 1 \quad n_1 = n_3 = 0$$

$$n_3 = \pm 1 \quad n_1 = n_2 = 0$$

which simply goes to show that no shear stresses act on the three planes with normals in the co-ordinate directions (principal planes), as one would expect from the choice of the principal axes as co-ordinate system.

The planes on which the shear stress is maximum are obtained from the condition

$$\frac{\partial \tau}{\partial n_1} = \frac{\partial \tau}{\partial n_2} = 0$$

(there are only two independent variables since $n_3^2 = 1 - n_1^2 - n_2^2$). Applying this to Equation 2.21, we obtain, after some algebra, that the extreme values of τ are

$$\tau = \tfrac{1}{2}(\sigma_1 - \sigma_2) \quad \text{for } n_1 = n_2 = 1/(2)^{1/2}, \, n_3 = 0$$

$$\tau = \tfrac{1}{2}(\sigma_1 - \sigma_3) \quad \text{for } n_1 = n_3 = 1/(2)^{1/2}, \, n_2 = 0$$

$$\tau = \tfrac{1}{2}(\sigma_2 - \sigma_3) \quad \text{for } n_2 = n_3 = 1/(2)^{1/2}, \, n_1 = 0$$

Since $\sigma_1 > \sigma_2 > \sigma_3$, the magnitude of the *maximum shear stress* is one-half of the maximum stress difference

$$\tau_{max} = \tfrac{1}{2}(\sigma_1 - \sigma_3) \qquad (2.22)$$

and this value is attained on two conjugate planes containing the direction of the intermediate principal stress (σ_2) and bisecting the directions of the maximum (σ_1) and minimum (σ_3) principal stresses. The situation

34

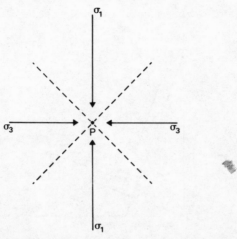

Figure 2.5 Planes of maximum shear stress (broken; normal to the plane of figure).

is depicted in Figure 2.5. This result forms the basis of the theory of faulting.

Geological problems involving stress can often be considered to be approximately two-dimensional. In this case the stress tensor reduces to

$$\sigma_{ij} = \begin{pmatrix} \sigma_{11} & \sigma_{12} \\ \sigma_{21} & \sigma_{22} \end{pmatrix}$$

and any surface is defined by its unit normal and unit tangent

$$n_i = \begin{pmatrix} \cos\theta \\ \sin\theta \end{pmatrix} \qquad t_i = \begin{pmatrix} -\sin\theta \\ \cos\theta \end{pmatrix}$$

where θ is the angle between n_i and the x_1-axis (Fig. 2.6). Then, from Equations 2.10,

$$\sigma = \sigma_{11}\cos^2\theta + \sigma_{22}\sin^2\theta + \sigma_{12}\sin 2\theta$$

$$\tau = -\sigma_{11}\sin\theta\cos\theta - \sigma_{12}\sin^2\theta + \sigma_{21}\cos^2\theta + \sigma_{22}\sin\theta\cos\theta$$

which, using the symmetry of stress tensor and the trigonometric identities

$$\cos^2\theta = \tfrac{1}{2}(1 + \cos 2\theta)$$

35

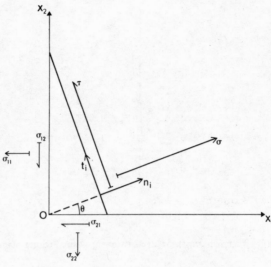

Figure 2.6 Two-dimensional state of stress.

$$\sin^2 \theta = \tfrac{1}{2}(1 - \cos 2\theta)$$

$$\sin \theta \cos \theta = \tfrac{1}{2} \sin 2\theta$$

may be written as

$$\sigma = \tfrac{1}{2}(\sigma_{11} + \sigma_{22}) + \tfrac{1}{2}(\sigma_{11} - \sigma_{22}) \cos 2\theta + \sigma_{12} \sin 2\theta$$

$$\tau = \tfrac{1}{2}(\sigma_{22} - \sigma_{11}) \sin 2\theta + \sigma_{12} \cos 2\theta$$

(2.23)

From the second of Equations 2.23, it follows that $\tau = 0$ for

$$\theta = \tfrac{1}{2}\arctan\left[2\sigma_{12}/(\sigma_{11} - \sigma_{22})\right]$$

(2.24)

which gives the orientation of one of the principal axes (the other is $\theta + \pi/2$).

Expressions for σ and τ referred to the principal stresses can be obtained immediately from Equations 2.23. Furthermore, it is possible to express the principal stresses in terms of the stress tensor σ_{ij} by noting that, for the critical angle given by Equation 2.24, Equations 2.23 become

$$\sigma = \sigma_1 = \tfrac{1}{2}(\sigma_{11} + \sigma_{22}) + \tfrac{1}{2}(\sigma_{11} - \sigma_{22}) \cos 2\theta + \sigma_{12} \sin 2\theta$$

$$\tau = 0 = \tfrac{1}{2}(\sigma_{22} - \sigma_{11}) \sin 2\theta + \sigma_{12} \cos 2\theta$$

(2.25a)

36

and, for $\theta + \pi/2$,

$$\sigma = \sigma_2 = \tfrac{1}{2}(\sigma_{11} + \sigma_{22}) - \tfrac{1}{2}(\sigma_{11} - \sigma_{22})\cos 2\theta - \sigma_{12}\sin 2\theta$$

$$\tau = 0 = -\tfrac{1}{2}(\sigma_{22} - \sigma_{11})\sin 2\theta - \sigma_{12}\cos 2\theta$$

(2.25b)

Equations 2.25a and b, with θ given by Equation 2.24, give

$$\sigma_1 = \tfrac{1}{2}(\sigma_{11} + \sigma_{22}) + [\sigma_{12}^2 + \tfrac{1}{4}(\sigma_{11} - \sigma_{22})^2]^{\tfrac{1}{2}}$$

$$\sigma_2 = \tfrac{1}{2}(\sigma_{11} + \sigma_{22}) - [\sigma_{12}^2 + \tfrac{1}{4}(\sigma_{11} - \sigma_{22})^2]^{\tfrac{1}{2}}$$

(2.26)

Taken together, Equations 2.24 and 2.26 give principal axes and principal stresses (in the two-dimensional case) in terms of the stress tensor in any Cartesian co-ordinate system.

The two-dimensional state of stress on any plane can be represented graphically by means of *Mohr's circle*. Since in practically all applications of Mohr's circle to geological problems compressive stresses are taken as positive, we follow the same convention here. Shear stress is taken as positive when it yields a clockwise moment about the centre of the element (see Fig. 2.7a; this convention applies only to Mohr's circle construction).

In the construction of Mohr's circle, the state of stress at a point is projected on a *stress plane* with co-ordinates σ and τ (Fig. 2.7b). If the normal and shear stresses acting on the plane normal to x_1 are σ_A and τ_A, then they plot on the stress plane as point A of co-ordinates (σ_A, τ_A); similarly, the stresses σ_B and τ_B acting on the plane normal to x_2 plot as point B (σ_B, τ_B). Now a circle with diameter $\overline{AB}$ and centre at C $((\sigma_A + \sigma_B)/2, 0)$ can be drawn; the radii $\overline{CA}$ and $\overline{CB}$, forming an angle π in stress space (Fig. 2.7b), are the equivalent of the segments $\overline{OA}$ and $\overline{OB}$ (Fig. 2.7a), forming an angle $\pi/2$ in physical space. Thus, in general, the angles between normals to planes in physical space are half the angles between the corresponding radii in the stress plane. For instance, the normal and shear stresses on a plane whose normal makes an angle θ with the x_1-axis are represented by the end-point (σ_S, τ_S) of the radius $\overline{CS}$, making an angle 2θ with $\overline{CA}$.

The intersections of Mohr's circle with the σ-axis give the principal stresses $\sigma_1 = \overline{OP}$ and $\sigma_2 = \overline{OQ}$; the principal directions may be determined from the angle between $\overline{CA}$ and $\overline{CP}$. The maximum shear stress, which in physical space occurs on the two planes bisecting the σ_1- and σ_2-directions, occurs at an angle $\pm \pi/2$ from $\overline{CP}$ on the stress plane.

Mohr's circle therefore characterizes graphically the stress state on planes of any orientation passing through a point, once the stress tensor

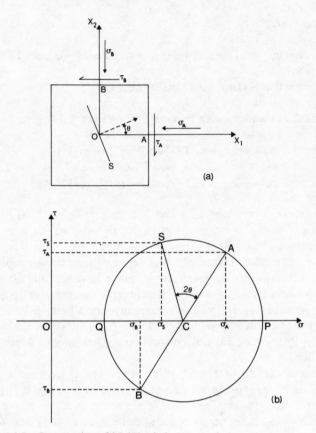

Figure 2.7 Construction of Mohr's circle: (a) physical space; (b) stress space.

is specified at that point. It can be extended to three dimensions if required (see, for example, Jaeger & Cook 1979). It is a useful tool in the analysis of stress distribution and fracture, and in the study of geodynamic processes involving fracture, such as faulting and earthquakes.

2.6 The infinitesimal strain tensor

Now that the dynamic quantity (stress) entering the rheological equation of state (Eqn 1.9) has been defined, it remains to specify the kinematic quantity, strain or strain rate. The former applies in processes involving time-independent deformation, and the latter in processes involving flow. (For the present purpose 'flow' may be deemed to occur if the time derivative of the deformation is non-zero; indeed, strain rate is the time derivative of strain.) In the sequel we introduce the concept of strain, which is sufficient to deal with elasticity (considered in Ch. 3). Strain rate will be defined in Chapter 4 when dealing with time-dependent deformation.

It was shown in Section 1.5 that, in the general case, the displacement of a particle results from translation, rotation, and deformation. Consider a continuous body occupying a domain D referred to a Cartesian co-ordinate system, and let the co-ordinates of two neighbouring points in this initial state be $P(x_i)$ and $Q(x_i + dx_i)$ (Fig. 2.8). After displacement the body occupies the domain D', and accordingly the segment $\overline{PQ}$ has been mapped into $\overline{P'Q'}$, i.e. the particles originally at P and Q have undergone the following changes of position

$$P(x_i) \rightarrow P'(x_i + u_i)$$

$$Q(x_i + dx_i) \rightarrow Q'(x_i + dx_i + u_i + du_i)$$

where u_i and $u_i + du_i$ are the displacements – that is, the differences between the final and initial positions of the relevant particles.

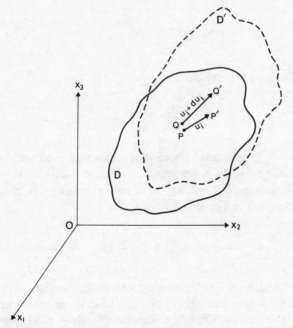

Figure 2.8 Displacements of neighbouring points P and Q. The difference du_i is the combination of strain and rotation.

Let it now be assumed that the displacement u_i is a continuous function of the co-ordinates, and that its first derivatives $\partial u_i/\partial x_j$ – forming a tensor known as the *displacement gradient* – are also continuous and very small. In practice this means that displacements of contiguous particles differ only by an infinitesimal amount, and it allows us to neglect higher-order terms. (Incidentally, this makes the distinction between Lagrangian and Eulerian reference frames disappear, since it is immaterial whether $\partial u_i/\partial x_j$ is calculated in the initial or in the deformed frame; see also Section 2.8.) Under this assumption the difference between the displacements of the material points P and Q can be written as

$$du_i = \frac{\partial u_i}{\partial x_j}\,dx_j \qquad (2.27)$$

It is always possible to resolve the displacement gradient in Equation 2.27 in two parts:

$$\frac{\partial u_i}{\partial x_j} = \tfrac{1}{2}\left(\frac{\partial u_i}{\partial x_j} + \frac{\partial u_j}{\partial x_i}\right) + \tfrac{1}{2}\left(\frac{\partial u_i}{\partial x_j} - \frac{\partial u_j}{\partial x_i}\right) \qquad (2.28)$$

where

$$\tfrac{1}{2}\left(\frac{\partial u_i}{\partial x_j} + \frac{\partial u_j}{\partial x_i}\right) \equiv \varepsilon_{ij} \qquad (2.29)$$

is the *infinitesimal strain tensor* and

$$\tfrac{1}{2}\left(\frac{\partial u_j}{\partial x_i} - \frac{\partial u_i}{\partial x_j}\right) \equiv \omega_{ij} \qquad (2.30)$$

is the *rigid-body rotation*. (Note that strain is a symmetric tensor, i.e. $\varepsilon_{ij} = \varepsilon_{ji}$, while rotation is antisymmetric, $\omega_{ij} = -\omega_{ji}$.)

Using Equations 2.27–2.30, the displacement of any material point can be expressed as the sum of three terms

$$u_i + du_i = u_i + \frac{\partial u_i}{\partial x_j}\,dx_j = u_i + \varepsilon_{ij}\,dx_j - \omega_{ij}\,dx_j \qquad (2.31)$$

the first of which (u_i), being common to neighbouring particles, represents rigid-body translation. We now prove that the other two terms (ε_{ij} and ω_{ij}), as their names indicate, represent strain – a measure of deformation – and rigid-body rotation.

With reference to Figure 2.9, assume that the body D undergoes a

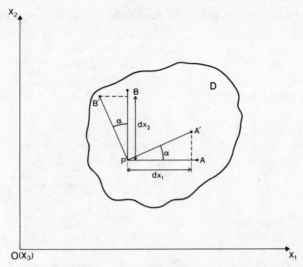

Figure 2.9 Infinitesimal rotation $\alpha \simeq \omega_{12}$ about the x_3-axis (normal to the plane of figure).

small rotation about the x_3-axis: the segments $\overline{PA}$ and $\overline{PB}$, initially parallel to the x_1- and x_2-axes, are therefore rotated to their new positions $\overline{PA'}$ and $\overline{PB'}$. Considering $A\hat{P}A'$, the angle of rotation α is given by

$$\alpha \simeq \tan \alpha = \frac{(\partial u_2/\partial x_1)\, \mathrm{d}x_1}{\mathrm{d}x_1} = \frac{\partial u_2}{\partial x_1}$$

and similarly, considering $B\hat{P}B'$,

$$\alpha \approx \tan \alpha = -\frac{(\partial u_1/\partial x_2)\, \mathrm{d}x_2}{\mathrm{d}x_2} = -\frac{\partial u_1}{\partial x_2}$$

Combining these two equations we have

$$\alpha \simeq \tfrac{1}{2}\left(\frac{\partial u_2}{\partial x_1} - \frac{\partial u_1}{\partial x_2}\right) = \omega_{12}$$

Analogous expressions can be derived for rotations about the x_1- and x_2-axes. (Note that, since the body undergoes only rotation, $\omega_{ij} = 0$ for $i = j$, as indeed it must be for antisymmetric tensors.) Therefore the tensor ω_{ij} represents an infinitesimal rigid-body rotation.

Let us now consider the infinitesimal strain tensor. First of all, it may be noted that strain is *dimensionless,* as are displacement gradient and rotation. When written in unabridged notation, its nine components are

(they would, of course, reduce to four components for two-dimensional strain)

$$\varepsilon_{ij} = \begin{pmatrix} \varepsilon_{11} & \varepsilon_{12} & \varepsilon_{13} \\ \varepsilon_{21} & \varepsilon_{22} & \varepsilon_{23} \\ \varepsilon_{31} & \varepsilon_{32} & \varepsilon_{33} \end{pmatrix}$$

$$= \begin{pmatrix} \dfrac{\partial u_1}{\partial x_1} & \dfrac{1}{2}\left(\dfrac{\partial u_1}{\partial x_2} + \dfrac{\partial u_2}{\partial x_1}\right) & \dfrac{1}{2}\left(\dfrac{\partial u_1}{\partial x_3} + \dfrac{\partial u_3}{\partial x_1}\right) \\ \dfrac{1}{2}\left(\dfrac{\partial u_2}{\partial x_1} + \dfrac{\partial u_1}{\partial x_2}\right) & \dfrac{\partial u_2}{\partial x_2} & \dfrac{1}{2}\left(\dfrac{\partial u_2}{\partial x_3} + \dfrac{\partial u_3}{\partial x_2}\right) \\ \dfrac{1}{2}\left(\dfrac{\partial u_3}{\partial x_1} + \dfrac{\partial u_1}{\partial x_3}\right) & \dfrac{1}{2}\left(\dfrac{\partial u_3}{\partial x_2} + \dfrac{\partial u_2}{\partial x_3}\right) & \dfrac{\partial u_3}{\partial x_3} \end{pmatrix} \quad (2.32)$$

Since strain is a symmetric tensor, only six components (three, in two dimensions) are required to specify it at a point. Of these six components, consider first the on-diagonal ones, $i = j$; for instance $\varepsilon_{22} = \partial u_2/\partial x_2$. With reference to Figure 2.10a, suppose that an elemental volume is stretched (or compressed) in the direction of the x_2-axis (a rigid-body motion may be superimposed on the deformation without affecting the argument). The distance between the two faces perpendicular to the x_2-axis (initially dx_2) becomes, after deformation, $dx_2 + du_2$ and, since u_2 is a function of x_2 only in this case, Equation 2.27 yields

$$du_2 = \frac{\partial u_2}{\partial x_2}\, dx_2$$

i.e. the distance between the two faces perpendicular to the x_2-axis has become

$$dx_2 + du_2 = \left(1 + \frac{\partial u_2}{\partial x_2}\right) dx_2 = (1 + \varepsilon_{22})\, dx_2$$

which shows that ε_{22} represents the *change in length per unit length* (or *elongation*) in the direction of the x_2-axis. Similarly, ε_{11} and ε_{33} represent elongations in the directions of the x_1- and x_3-axes, respectively. In keeping with the sign convention for stress, elongations are taken as

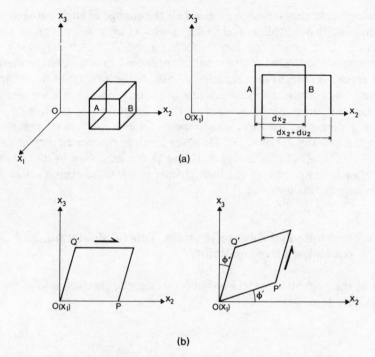

(a)

(b)

Figure 2.10 Examples of strain: (a) elongation, (b) shear strain.

positive when the initial length is increased, and as negative when it is decreased.

The above result is, of course, still valid if the deformation is more complex. So, in the general case, $\varepsilon_{11}, \varepsilon_{22}$, and ε_{33} represent the components of elongation in the three co-ordinate directions.

However, deformation does not consist only of changes of length. Suppose shear stresses act in the positive directions on the faces normal to the x_2- and x_3-axes. The resulting deformation is depicted in Figure 2.10b. Two lines, initially parallel to the x_2- and x_3-axes and consequently forming a right angle before deformation, form an angle $P'\hat{O}Q' = \pi/2 - (\phi' + \phi'')$ after deformation. Since the angular deformation is small, we can write

$$\phi' \simeq \tan \phi' = \partial u_3/\partial x_2 \qquad \phi'' \simeq \tan \phi'' = \partial u_2/\partial x_3$$

and therefore

$$\phi' + \phi'' \simeq \frac{\partial u_3}{\partial x_2} + \frac{\partial u_2}{\partial x_3} = 2\varepsilon_{23}$$

43

i.e. the strain component ε_{23} is one-half the change in angle between two lines initially oriented parallel to the x_2- and x_3-axes. An analogous argument leads to similar conclusions for ε_{12} and ε_{13}. The off-diagonal components of the strain tensor are therefore related to *angular deformation* (or *shear*) of pairs of lines initially oriented in the three co-ordinate directions. Once again, the sign convention is in keeping with the one for stress, the shear shown in Figure 2.10b being taken as positive.

In general, deformation is the result of a combination of changes in lengths and angular changes. However, even in the more complex cases it can be resolved into elongations and shears as shown in this section, with each component of the infinitesimal strain tensor being assigned a precise physical meaning.

2.7 Isotropic and deviatoric strain, cubical dilatation, and equations of compatibility

As in the case of stress, the strain tensor can be resolved into isotropic and deviatoric parts, i.e.

$$\varepsilon_{ij}^0 = \tfrac{1}{3}\varepsilon_{kk}\,\delta_{ij} \tag{2.33}$$

$$\varepsilon_{ij}' = \varepsilon_{ij} - \varepsilon_{ij}^0 \tag{2.34}$$

The term $\varepsilon_{kk} = \partial u_k/\partial x_k$ appearing in Equation 2.33 represents the change in volume per unit volume, or *cubical dilatation*. If an elemental volume is initially $V = dx_1\,dx_2\,dx_3$, and after deformation is $V + dV$, then the change in volume is

$$dV = \left(dx_1 + \frac{\partial u_1}{\partial x_1}\,dx_1\right)\left(dx_2 + \frac{\partial u_2}{\partial x_2}\,dx_2\right)\left(dx_3 + \frac{\partial u_3}{\partial x_3}\,dx_3\right) - dx_1\,dx_2\,dx_3$$

That is, neglecting higher-order terms,

$$dV = \left[dx_1\,dx_2\,dx_3 + \left(\frac{\partial u_1}{\partial x_1} + \frac{\partial u_2}{\partial x_2} + \frac{\partial u_3}{\partial x_3}\right)dx_1\,dx_2\,dx_3\right] - dx_1\,dx_2\,dx_3$$

$$= (\varepsilon_{11} + \varepsilon_{22} + \varepsilon_{33})\,dx_1\,dx_2\,dx_3$$

so that the cubical dilatation (usually denoted by θ) is

$$\theta = \frac{dV}{V} = \varepsilon_{11} + \varepsilon_{22} + \varepsilon_{33} = \varepsilon_{kk} \tag{2.35}$$

($\theta > 0$ for an increase in volume). The isotropic part of the strain tensor therefore represents that part of the deformation resulting purely in a volume change.

The deviatoric strain (Eqn 2.34), on the other hand, represents changes of shape. Obviously, $\varepsilon'_{ij} = \varepsilon_{ij}$, $i \neq j$, i.e. the shear components of the strain deviator and of the total strain are identical – this reflects the fact that off-diagonal strain components are related to angular deformations.

Although in practice deformation may be complex, the strain tensor can always be viewed as the resultant of isotropic and deviatoric strain:

$$\varepsilon_{ij} = \varepsilon^0_{ij} + \varepsilon'_{ij} \qquad (2.36)$$

Volume and shape changes of course can – and in fact do – occur together, although in particular cases it may be that $\varepsilon'_{ij} = 0$ (purely volumetric deformation, i.e. without any change in shape) or $\varepsilon_{kk} = 0$ (no change in volume). Note that the condition $\varepsilon_{ij} = 0$, $i \neq j$, is not sufficient for the deformation to be purely volumetric. A change of shape can occur even if all off-diagonal components of strain in a given co-ordinate system are zero, provided that the three elongations are not all equal. The deformation is purely volumetric – and isotropic – only when $\varepsilon'_{ij} = 0$, which implies $\varepsilon_{11} = \varepsilon_{22} = \varepsilon_{33}$. In other words, the condition $\varepsilon_{kk} = 0$ does not preclude possible shape changes, although it requires the volume change to be zero.

In Section 2.4 we introduced principal stresses and principal directions. The same ideas can be applied to strain. Consequently, a *principal axis of strain* is that direction such that a plane normal to it experiences only normal strains; and the *principal strain tensor* ε^P_{ij} is the strain tensor referred to the principal axes, i.e. with three principal normal strains in the co-ordinate directions and zero off-diagonal components.

The definition of strain (Eqn 2.29) can be regarded as a system of partial differential equations in the displacement $u_i(x_1, x_2, x_3)$. Since the number of equations (in three dimensions) is six and the number of unknowns is three, the system is overdetermined, and continuous and single-valued solutions exist only if ε_{ij} satisfies some additional conditions. (In physical terms, this simply means that strains cannot be prescribed arbitrarily as functions of the co-ordinates if the displacement vector field has to exclude gaps and overlaps of matter in the deformed configuration.) It can be proved that, in order to have single-valued continuous solutions, the components of strain must satisfy the following *condition of compatibility* (see, for example, Fung 1977)

$$\frac{\partial^2 \varepsilon_{ij}}{\partial x_m \partial x_n} + \frac{\partial^2 \varepsilon_{mn}}{\partial x_i \partial x_j} - \frac{\partial^2 \varepsilon_{im}}{\partial x_j \partial x_n} - \frac{\partial^2 \varepsilon_{jn}}{\partial x_i \partial x_m} = 0 \qquad (2.37)$$

Equation 2.37 represents a system of 81 equations in three dimensions; however, because of the symmetry of the strain tensor, they reduce to six. Written in unabridged notation, these are

$$\frac{\partial^2 \varepsilon_{11}}{\partial x_2 \partial x_3} = \frac{\partial}{\partial x_1} \left(-\frac{\partial \varepsilon_{23}}{\partial x_1} + \frac{\partial \varepsilon_{31}}{\partial x_2} + \frac{\partial \varepsilon_{12}}{\partial x_3} \right)$$

$$\frac{\partial^2 \varepsilon_{22}}{\partial x_3 \partial x_1} = \frac{\partial}{\partial x_2} \left(-\frac{\partial \varepsilon_{31}}{\partial x_2} + \frac{\partial \varepsilon_{12}}{\partial x_3} + \frac{\partial \varepsilon_{23}}{\partial x_1} \right)$$

$$\frac{\partial^2 \varepsilon_{33}}{\partial x_1 \partial x_2} = \frac{\partial}{\partial x_3} \left(-\frac{\partial \varepsilon_{12}}{\partial x_3} + \frac{\partial \varepsilon_{23}}{\partial x_1} + \frac{\partial \varepsilon_{31}}{\partial x_2} \right)$$

$$2 \frac{\partial^2 \varepsilon_{12}}{\partial x_1 \partial x_2} = \frac{\partial^2 \varepsilon_{11}}{\partial x_2^2} + \frac{\partial^2 \varepsilon_{22}}{\partial x_1^2}$$

$$2 \frac{\partial^2 \varepsilon_{23}}{\partial x_2 \partial x_3} = \frac{\partial^2 \varepsilon_{22}}{\partial x_3^2} + \frac{\partial^2 \varepsilon_{33}}{\partial x_2^2}$$

$$2 \frac{\partial^2 \varepsilon_{13}}{\partial x_1 \partial x_3} = \frac{\partial^2 \varepsilon_{11}}{\partial x_3^2} + \frac{\partial^2 \varepsilon_{33}}{\partial x_1^2}$$

For two-dimensional plane strain the above equations reduce to one single equation of compatibility:

$$2 \frac{\partial^2 \varepsilon_{12}}{\partial x_1 \partial x_2} = \frac{\partial^2 \varepsilon_{11}}{\partial x_2^2} + \frac{\partial^2 \varepsilon_{22}}{\partial x_1^2}$$

Since strain components do not include rigid-body motions, the solution $u_i(x_1, x_2, x_3)$ to Equation 2.29 is determined up to an arbitrary rigid-body component.

Before we proceed to examine the various kinds of rheological behaviour, something must be said about finite strain, since this plays a very important part in structural geology, which in turn provides many of the fundamental data for the study of the rheology of rocks.

2.8 Finite strain

All of the considerations developed in Sections 2.6 and 2.7 are subject to the assumption that the deformation (the displacement gradient and, consequently, the strain tensor) is *infinitesimal*. This carries us a long way in geodynamics, since seismic wave propagation, to a first approximation, involves only infinitesimal deformation (see Ch. 6), and on the other hand problems involving flow (Chs 7 & 8) are amenable to treatment using the strain rate tensor (introduced in Ch. 4). However, in structural geology one is often faced with situations where the difference between initial and final configurations of a deformed body can be very large, and has to be described in terms of the displacement field. In this case the deformation is no longer infinitesimal, and the concept of *finite strain* has to be brought to bear on the problem (see, for example, Ramsay 1967, Hobbs *et al.* 1976, and Means 1976). We do not go into the details here, but only give the quantitative basis on which the theory of finite strain is built.

The assumption that the components of the displacement gradient tensor $\partial u_i / \partial x_j$ are sufficiently small for their products and second powers to be neglected is dropped. Also, since the difference between Lagrangian and Eulerian descriptions becomes significant, the distinction must be kept between material co-ordinates c_i and spatial co-ordinates x_i (Section 1.6).

Let S_0 be a curve, of any shape, in the body before any displacement occurs and ds_0 be the length of an element of arc of this curve. After displacement, a point $P(c_i)$, originally on S_0, has been mapped into $P(x_i)$, S_0 as a whole has been mapped into S, and ds_0 into ds. Now consider the quantities

$$ds^2 = dx_i \, dx_i \qquad ds_0^2 = dc_i \, dc_i \qquad (2.38)$$

If $ds = ds_0$ for all S, then the length of no line element has been altered, i.e. only rigid-body motion has taken place. The difference $ds - ds_0$ therefore gives a measure of deformation.

We now express $ds - ds_0$ in the Eulerian (spatial) framework, in which the independent variables are the co-ordinates x_i of a particle in the deformed state, $c_i = c_i(x_1, x_2, x_3)$. We can write Equations 2.38 as

$$ds^2 = dx_i \, dx_i = \delta_{jk} \, dx_j \, dx_k$$

$$ds_0^2 = dc_i \, dc_i = \frac{\partial c_i}{\partial x_j} \, dx_j \, \frac{\partial c_i}{\partial x_k} \, dx_k$$

(δ_{jk}, as usual, is the Kronecker delta). It follows that

$$ds^2 - ds_0^2 = \delta_{jk} \, dx_j \, dx_k - \frac{\partial c_i}{\partial x_j} \frac{\partial c_i}{\partial x_k} \, dx_j \, dx_k = 2 \, \varepsilon_{jk}^{(E)} \, dx_j \, dx_k$$

47

where

$$\varepsilon_{jk}^{(E)} = \tfrac{1}{2}\left(\delta_{jk} - \frac{\partial c_i}{\partial x_j}\frac{\partial c_i}{\partial x_k}\right) \tag{2.39}$$

is the finite strain in the Eulerian description.

In order to write Equation 2.39 in terms of the displacement, we recall that the latter is the difference between the final and initial positions of a particle, i.e. $u_i = x_i - c_i$, and therefore, since $\partial x_i/\partial x_j = \delta_{ij}$, and using the properties of the Kronecker delta,

$$\frac{\partial c_i}{\partial x_j}\frac{\partial c_i}{\partial x_k} = \left(\delta_{ij} - \frac{\partial u_i}{\partial x_j}\right)\left(\delta_{ik} - \frac{\partial u_i}{\partial x_k}\right)$$

$$= \delta_{jk} - \frac{\partial u_j}{\partial x_k} - \frac{\partial u_k}{\partial x_j} + \frac{\partial u_i}{\partial x_j}\frac{\partial u_i}{\partial x_k}$$

Consequently, Equation 2.39 for the Eulerian finite strain becomes

$$\varepsilon_{jk}^{(E)} = \tfrac{1}{2}\left(\frac{\partial u_j}{\partial x_k} + \frac{\partial u_k}{\partial x_j} - \frac{\partial u_i}{\partial x_j}\frac{\partial u_i}{\partial x_k}\right) \tag{2.40}$$

which, of course, reduces to the infinitesimal strain (Eqn 2.29) if $\partial u_i/\partial x_j$ is small and the last term can be neglected.

In the Lagrangian (material) framework the independent variables are the co-ordinates c_i of a particle in the initial state, $x_i = x_i(c_1, c_2, c_3)$. Therefore Equations 2.38 become

$$\mathrm{d}s^2 = \mathrm{d}x_i\,\mathrm{d}x_i = \frac{\partial x_i}{\partial c_j}\,\mathrm{d}c_j\,\frac{\partial x_i}{\partial c_k}\,\mathrm{d}c_k$$

$$\mathrm{d}s_0^2 = \mathrm{d}c_i\,\mathrm{d}c_i = \delta_{jk}\,\mathrm{d}c_j\,\mathrm{d}c_k$$

and the measure of strain is

$$\mathrm{d}s^2 - \mathrm{d}s_0^2 = \frac{\partial x_i}{\partial c_j}\frac{\partial x_i}{\partial c_k}\,\mathrm{d}c_j\,\mathrm{d}c_k - \delta_{jk}\,\mathrm{d}c_j\,\mathrm{d}c_k$$

$$= 2\varepsilon_{jk}^{(L)}\,\mathrm{d}c_j\,\mathrm{d}c_k$$

where

$$\varepsilon_{jk}^{(L)} = \tfrac{1}{2}\left(\frac{\partial x_i}{\partial c_j}\frac{\partial x_i}{\partial c_k} - \delta_{jk}\right) \tag{2.41}$$

is the finite strain in the Lagrangian description.

48

As in the previous case, using the fact that $x_i = c_i + u_i$, we can write

$$\frac{\partial x_i}{\partial c_j} \frac{\partial x_i}{\partial c_k} = \left(\delta_{ij} + \frac{\partial u_i}{\partial c_j}\right)\left(\delta_{ik} + \frac{\partial u_i}{\partial c_k}\right)$$

$$= \delta_{jk} + \frac{\partial u_j}{\partial c_k} + \frac{\partial u_k}{\partial c_j} + \frac{\partial u_i}{\partial c_j} \frac{\partial u_i}{\partial c_k}$$

so that Equation 2.41 can be expressed as

$$\varepsilon_{jk}^{(L)} = \tfrac{1}{2}\left(\frac{\partial u_j}{\partial c_k} + \frac{\partial u_k}{\partial c_j} + \frac{\partial u_i}{\partial c_j} \frac{\partial u_i}{\partial c_k}\right) \tag{2.42}$$

The last term of Equation 2.42 can, of course, be dropped when the deformation is infinitesimal; moreover, it is immaterial whether derivatives are calculated at a point before or after deformation. For infinitesimal strain, then, the Eulerian and Lagrangian descriptions become identical.

In structural geology one is not interested in further analytical applications of Equations 2.40 and 2.42 as much as in the determination of the geometrical characteristics of finite inhomogeneous strain. Readers should consult the references given at the beginning of this section if they wish to pursue this matter further.

3 Elasticity

One of the most important rheological classes is elasticity, in which the stress is linearly proportional to the strain and the latter is fully recoverable. The elastic behaviour of a material is described by Hooke's Law and the material coefficients. Rocks are approximately elastic at low pressure and temperature (i.e in the upper lithosphere), and also in the bulk of the Earth for stresses at seismic frequencies.

Some of the fundamentals of the theory of elasticity are developed in this chapter; i.e. the equations of motion, the strain energy function, and the stress compatibility conditions. These lead to the introduction of the Airy stress function, and to a discussion of the possible states of stress in the lithosphere, including Anderson's standard state.

3.1 Hooke's Law

When a continuous deformable body is subjected to a load, stresses and strains are set up within it. The rheological behaviour can be studied experimentally by applying loads generating simple stress systems (uniaxial normal stress, simple shear, etc.) and measuring the corresponding strains. Results can be presented in terms of *strain–time diagrams* (or deformation–time diagrams) by plotting strain as a function of time under constant stress, or in terms of *stress–strain diagrams* (or stress–strain rate diagrams if flow occurs) by plotting the strain observed at different stresses. (In these diagrams and in the experimental literature it is customary to denote stress and strain simply by σ and ε, omitting indices. The stress and strain components are those relevant to the situation, e.g. tensile stress and corresponding elongation in a uniaxial tension experiment, shear stress and shear strain in a shear experiment, and so on.)

In practice, one often observes a behaviour having the following characteristics: instantaneous deformation upon application of a load, instantaneous and total recovery upon removal of the load, and linear

50

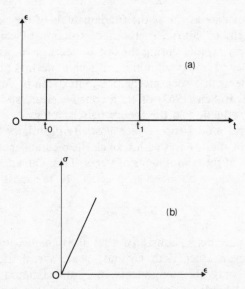

Figure 3.1 Rheological response of an elastic body: (a) strain–time diagram under constant stress applied at time t_0 and removed at t_1; (b) stress–strain diagram for stresses below the elastic limit.

proportionality between stress and strain. In the ideal case the strain–time diagram and the stress–strain diagram are as depicted in Figures 3.1a and b, respectively. Among the materials showing approximately this type of behaviour there are rocks at room temperature and pressure (a brief survey of results on experimental rock deformation is given in Section 5.2). It should be pointed out at the outset that this type of rheological behaviour, termed *elasticity*, is observed only for stresses *below* a threshold stress (itself a function not only of the material, but also of temperature and pressure), above which the material first diverges from linear elasticity and eventually fails, either in a brittle or in a ductile fashion (see Ch. 5). Here we are concerned with behaviour below the threshold stress, or *elastic limit*, which may vary in rocks at low temperature and pressure from 100 to 1000 MPa as orders of magnitude, depending on intrinsic and extrinsic rheological parameters.

Elasticity has wide applications in geodynamics. Rocks in the upper lithosphere can be considered elastic up to loads of duration comparable with the age of the Earth (cf. Section 8.4). Furthermore, the whole Earth behaves elastically – with imperfections – for loads of short duration (seconds to hours), which makes the whole of seismology (Ch. 6) amenable to analysis using elasticity theory.

The theory of elasticity is treated in several classic books. Among these, one may mention Love (1944), which is very comprehensive and not really out of date as far as the fundamentals of the theory are concerned (although, of course, methods of solution of elastic problems have evolved very rapidly during the last 20 years or so, as witnessed by the large number of books on numerical techniques). A good treatment of elastic problems in a rock mechanics context can be found in Jaeger & Cook (1979). Bullen (1963) offers a concise, clear, and elegant summary as an introduction to the theory of seismology.

The theoretical model for *perfect elasticity* postulates that the components of strain at any point within an elastic medium are homogeneous linear functions of the components of stress. This is an adequate description of the behaviour observed in practice. In the most general case, therefore,

$$\sigma_{ij} = C_{ijkl}\varepsilon_{kl} \tag{3.1}$$

and the elastic parameters consist of a set of 81 coefficients, represented by C_{ijkl}. However, since both ε_{kl} and σ_{ij} are symmetric tensors and therefore have only six components, only a maximum of 36 elastic parameters can be independent. Further relations, that can be proved on the basis of the existence of the strain energy function (Section 3.3) reduce this number to 21.

The considerations above apply to a material which is *anisotropic* with respect to elastic properties. A single crystal shows some degree of symmetry in its elastic behaviour (according to the symmetry class to which it belongs), and this further reduces the number of independent elastic parameters. For an *isotropic* body, in which the elastic behaviour does not depend on direction, the number of *independent elastic parameters* reduces to two, which are invariant under a transformation of coordinates, and are therefore scalars (see Bullen 1963 for more details). In this chapter we develop the fundamentals of the theory of perfect (infinitesimal) elasticity for homogeneous, isotropic elastic bodies. The emphasis is not on the solution of particular elastic problems, but on the development of basic equations and ideas that are needed to understand the theory of seismology and other problems in geodynamics where elasticity theory applies.

Elastic behaviour was originally disentangled by Robert Hooke, who in 1676 published the law of proportionality of stress and strain in the form of an anagram of the words '*ut tensio sic vis*'. The law of elasticity, of which Equation 3.1 is the most general form, is called *Hooke's Law* after him. For isotropic bodies it takes the form

$$\sigma_{ij} = \lambda\varepsilon_{kk}\,\delta_{ij} + 2\mu\varepsilon_{ij} \tag{3.2}$$

where δ_{ij} is the Kronecker delta, and λ and μ are two elastic parameters termed *Lamé's constants* (note, however, that they are not 'constants', as they change with pressure and temperature).

Equation 3.2 shows the linear dependence of strain components on stress components, and vice versa. Denoting the cubical dilatation ε_{kk} by θ, we see that the relationships between normal components of stress and strain are of the type (for, say, $i = j = 2$)

$$\sigma_{22} = \lambda\theta + 2\mu\varepsilon_{22}$$

and between shear components (e.g. $i = 1, j = 2$)

$$\sigma_{12} = 2\mu\varepsilon_{12}$$

From the physical viewpoint the effects of isotropic and deviatoric stresses can be conveniently separated. By setting $i = j$ in Equation 3.2 and denoting $\frac{1}{3}\sigma_{ii} = \sigma_0$ (mean normal stress), one has

$$3\sigma_0 = (3\lambda + 2\mu)\theta$$

which, setting

$$\lambda + \tfrac{2}{3}\mu = k \qquad (3.3)$$

becomes

$$\sigma_0 = k\theta \qquad (3.4)$$

Equation 3.4 gives the relation between mean normal stress and volumetric deformation (cubical dilatation). The parameter k, the *incompressibility* or *bulk modulus*, can be seen to be the ratio between mean normal stress and cubical dilatation

$$k = \sigma_0/\theta \qquad (3.5)$$

i.e. the inverse of the compressibility. Elastic moduli of rocks under laboratory conditions are of the order of 10^{10}–10^{11} Pa, i.e. comparable with those of metals (see the compilation by Birch 1966). Elastic parameters as a function of depth in the interior of the Earth are discussed in Chapter 6.

The relation between deviatoric stress and deviatoric strain may be obtained from Equation 3.2 and the definition of deviator:

$$\sigma_{ij} - \tfrac{1}{3}\sigma_{kk}\,\delta_{ij} = \lambda\theta\,\delta_{ij} + 2\mu\varepsilon_{ij} - \sigma_0\,\delta_{ij}$$

53

which, using the result established in Equation 3.4, can be written as

$$\sigma_{ij} - \tfrac{1}{3}\sigma_{kk}\,\delta_{ij} = 2\mu(\varepsilon_{ij} - \tfrac{1}{3}\theta\,\delta_{ij})$$

that is,

$$\sigma'_{ij} = 2\mu\varepsilon'_{ij} \tag{3.6}$$

Taken together, Equations 3.4 and 3.6 show that in a perfectly elastic body isotropic stress is proportional to isotropic strain, and deviatoric stress is proportional to deviatoric strain. Lamé's parameter μ is seen from Equation 3.6 to be given by (no summation implied)

$$\mu = \tfrac{1}{2}(\sigma'_{ij}/\varepsilon'_{ij}) \tag{3.7}$$

i.e. it is simply half of the ratio between shear stress and shear strain, and is called the *rigidity* or *shear modulus*. The order of magnitude of the rigidity of rocks is the same as their incompressibility. Note that k, μ, and λ (which has no special name) all have the dimensions of stress, $[ml^{-1}t^{-2}]$. Since the three elastic parameters are related (Eqn 3.3), only two are independent.

Equation 3.2 can be rewritten (using Eqn 3.4) to give strain in terms of stress:

$$\varepsilon_{ij} = \frac{\sigma_{ij}}{2\mu} - \frac{\lambda\sigma_{kk}\,\delta_{ij}}{2\mu(3\lambda + 2\mu)} \tag{3.8}$$

Consider now the uniform extension of a cylindrical rod with axis coinciding with x_1 ($\sigma_{11} \neq 0$, all other stress components equal to zero). Equation 3.8 becomes ($\varepsilon_{ij} = 0$ for $i \neq j$ in this case)

$$\varepsilon_{11} = \frac{(\lambda + \mu)\sigma_{11}}{\mu(3\lambda + 2\mu)}$$

$$\varepsilon_{22} = \varepsilon_{33} = -\frac{\lambda\sigma_{11}}{2\mu(3\lambda + 2\mu)}$$

and the ratio between transverse and longitudinal elongation is

$$\frac{\varepsilon_{33}}{\varepsilon_{11}} = \frac{\varepsilon_{22}}{\varepsilon_{11}} = -\frac{\lambda}{2(\lambda + \mu)}$$

Introducing two further elastic parameters, *Young's modulus E* and the

Poisson ratio v, defined as

$$E = \sigma_{11}/\varepsilon_{11} \qquad v = -\varepsilon_{33}/\varepsilon_{11} \qquad (3.9)$$

it can immediately be seen that

$$E = \frac{\mu(3\lambda + 2\mu)}{\lambda + \mu} \qquad v = \frac{\lambda}{2(\lambda + \mu)} \qquad (3.10)$$

The definitions of Young's modulus and the Poisson ratio (Eqns 3.9) show how to determine E and v by a simple tension experiment. Young's modulus has dimensions $[ml^{-1}t^{-2}]$, while the Poisson ratio is dimensionless.

Of the five elastic parameters λ, μ, k, E, and v, only two are independent. Other relations that can be derived between them are

$$k = \frac{E}{3(1 - 2v)} \qquad \mu = \frac{E}{2(1 + v)} \qquad \lambda = \frac{Ev}{(1 + v)(1 - 2v)} \qquad (3.11)$$

(Note that, as $k > 0$, they imply $v \leqslant \frac{1}{2}$. For $v = \frac{1}{2}$ the material is incompressible; $v = \frac{1}{4}$ is a reasonable approximation for rocks at seismic frequencies.)

Using Equations 3.11, Equation 3.8 can be rewritten in terms of E and v as follows:

$$\varepsilon_{ij} = (1/E)[(1 + v)\sigma_{ij} - v\sigma_{kk}\,\delta_{ij}] \qquad (3.12)$$

For instance, the shear strains are of the form

$$\varepsilon_{12} = (1 + v)\sigma_{12}/E$$

and the normal strains

$$\varepsilon_{11} = (1/E)[\sigma_{11} - v(\sigma_{22} + \sigma_{33})]$$

and similarly for the other components. Equations 3.2 and 3.12 are the most frequently used forms of Hooke's Law for isotropic elastic media. We finally note that it is a direct consequence of Hooke's Law that the principal axes of strain and the principal axes of stress coincide.

3.2 Equations of motion and the Cauchy–Navier equations

The equations of equilibrium for continuous bodies were derived in Section 2.2 (Eqn 2.7). Using d'Alembert's Principle, they can be converted to the equations of motion for small movements. Denoting acceleration by a_i, these are of the form

$$\frac{\partial \sigma_{ij}}{\partial x_j} + \rho X_i = \rho a_i$$

Since $a_i = \ddot{u}_i = D^2 u_i / Dt^2$ (which reduces to $\partial^2 u_i / \partial t^2$ when displacements and velocities are small, as is the case in elasticity), we can write

$$\frac{\partial \sigma_{ij}}{\partial x_j} + \rho X_i = \rho \ddot{u}_i \tag{3.13}$$

which is the usual form of the *equations of motion* for continuous bodies.

To derive the equations of motion for elastic media, we substitute for σ_{ij} in Equation 3.13 the expression given by Hooke's Law (Eqn 3.2), to obtain

$$\frac{\partial}{\partial x_j} (\lambda \theta \, \delta_{ij} + 2\mu \varepsilon_{ij}) + \rho X_i = \rho \frac{\partial^2 u_i}{\partial t^2}$$

Assuming the body to be homogeneous, recalling the definition of strain (Eqn 2.29), and using the properties of the Kronecker delta, this equation may be written as

$$\lambda \frac{\partial \theta}{\partial x_i} + \mu \frac{\partial}{\partial x_j} \left(\frac{\partial u_i}{\partial x_j} + \frac{\partial u_j}{\partial x_i} \right) + \rho X_i = \rho \frac{\partial^2 u_i}{\partial t^2}$$

Changing the order of derivation when required, recalling that $\theta = \partial u_j / \partial x_j$, and denoting the operator $\partial^2 / \partial x_j \, \partial x_j$ by ∇^2, this becomes

$$(\lambda + \mu) \frac{\partial \theta}{\partial x_i} + \mu \nabla^2 u_i + \rho X_i = \rho \frac{\partial^2 u_i}{\partial t^2} \tag{3.14}$$

Equation 3.14 represents the equations of motion in terms of the displacement for a continuous elastic body, or *Cauchy–Navier equations*. The operator

$$\frac{\partial^2}{\partial x_j \, \partial x_j} \equiv \nabla^2 = \frac{\partial^2}{\partial x_1^2} + \frac{\partial^2}{\partial x_2^2} + \frac{\partial^2}{\partial x_3^2}$$

is called the *Laplacian*. The corresponding equations of equilibrium are obtained from Equation 3.14 by setting $\partial^2 u_i / \partial t^2 = 0$.

The elastic equations of motion form a system of three equations in three unknowns (the components of the displacement). The x_1-component, for instance, is

$$(\lambda + \mu) \frac{\partial}{\partial x_1} \left(\frac{\partial u_1}{\partial x_1} + \frac{\partial u_2}{\partial x_2} + \frac{\partial u_3}{\partial x_3} \right) + \mu \left(\frac{\partial^2 u_1}{\partial x_1^2} + \frac{\partial^2 u_1}{\partial x_2^2} + \frac{\partial^2 u_1}{\partial x_3^2} \right) + \rho X_1 = \rho \frac{\partial^2 u_1}{\partial t^2}$$

and similarly for the x_2- and x_3-components. Most of the classical theory of elasticity consists of devising methods of solution for Equation 3.14 – or the corresponding equations of equilibrium – for bodies subject to given stress and displacement boundary conditions. The equations of motion will be used in Section 6.1 in deriving the equations for seismic waves.

3.3 Strain energy function

In order to strain a body it is necessary to do work on it. The elastic strain energy (per unit volume) is the energy contained in the material as a consequence of its elastic deformation, and it is important in considerations on the failure of materials (Section 5.5).

Consider an infinitesimal volume of matter within an elastic body. At time t_0 let x_i, u_i, v_i, and a_i be the position, displacement, velocity, and acceleration of a material point within the element. During the ensuing small time interval dt, let du_i be the additional displacement, dW be the work done, and dQ, dE, and dU be the heat emitted from the element, the increase in kinetic energy, and the increase in internal energy, respectively (all referred to unit volume). By the First Law of Thermodynamics,

$$dW = dE + dQ + dU \qquad (3.15)$$

The work done on the element can be subdivided into work done by the body force X_i

$$dW_B = \rho X_i \, du_i$$

and work done by the surface forces across the faces of the elemental volume

$$dW_S = \frac{\partial}{\partial x_j} (\sigma_{ij} \, du_i)$$

57

The kinetic energy is

$$dE = d(\tfrac{1}{2}\rho v_i^2)$$

i.e. to a sufficient approximation for small movements,

$$dE = \rho v_i \, dv_i = \rho v_i \, \frac{\partial v_i}{\partial t} \, dt = \rho \, du_i \, \frac{\partial^2 u_i}{\partial t^2}$$

Equation 3.15 then becomes

$$dW - dE = dQ + dU$$

$$= \rho X_i \, du_i + \frac{\partial}{\partial x_j} \, (\sigma_{ij} \, du_i) - \rho \, du_i \, \frac{\partial^2 u_i}{\partial t^2}$$

which, using the equation of motion (Eqn 3.13), can be written as

$$dW - dE = dQ + dU$$

$$= \rho X_i \, du_i + \frac{\partial}{\partial x_j} \, (\sigma_{ij} \, du_i) - \left(\frac{\partial \sigma_{ij}}{\partial x_j} + \rho X_i\right) du_i$$

$$= \sigma_{ij} \, \frac{\partial}{\partial x_j} \, (du_i) = \sigma_{ij} \, d\left(\frac{\partial u_i}{\partial x_j}\right) \tag{3.16}$$

But

$$\frac{\partial u_i}{\partial x_j} = \varepsilon_{ij} - \omega_{ij}$$

where ε_{ij} and ω_{ij} are the strain and the rotation tensors, respectively. Consequently, Equation 3.16 takes the form

$$\sigma_{ij}(d\varepsilon_{ij} - d\omega_{ij}) = \sigma_{ij} \, d\varepsilon_{ij} = dQ + dU \tag{3.17}$$

as $\sigma_{ij} \, d\omega_{ij} = 0$, being the product of a symmetric and an antisymmetric tensor. For both isothermal (T = constant) and adiabatic ($dQ = 0$) processes, the right-hand side of Equation 3.17 is an exact differential, and therefore

$$\sigma_{ij} \, d\varepsilon_{ij} = d\Omega \tag{3.18}$$

where Ω is the *strain energy function*, i.e. the stored elastic energy per unit volume.

We now recall Euler's Theorem, which states that, if $f(x_1, x_2, \ldots)$ is a homogeneous function of degree n, then the following relation holds:

$$x_i \frac{\partial f}{\partial x_i} = nf$$

Therefore, since Ω is a homogeneous quadratic function of ε_{ij},

$$\varepsilon_{ij} \frac{\partial \Omega}{\partial \varepsilon_{ij}} = 2\Omega$$

i.e.

$$\tfrac{1}{2} \varepsilon_{ij} \sigma_{ij} = \Omega \tag{3.19}$$

Equations 3.18 and 3.19 have been derived under no restrictive assumptions as to the kind of elastic behaviour involved (except for the strain to be infinitesimal). If the material is elastically isotropic, then Hooke's Law can be substituted into Equation 3.19, which allows separation of the strain energy function into isotropic and deviatoric parts:

$$\Omega = \tfrac{1}{2}\varepsilon_{ij}[\lambda\theta \ \delta_{ij} + 2\mu(\varepsilon'_{ij} + \tfrac{1}{3}\theta \ \delta_{ij})]$$

$$= \tfrac{1}{2}k\theta^2 + \mu\varepsilon'_{ij}\varepsilon'_{ij} \tag{3.20}$$

where use has been made of the definition of bulk modulus (Eqn 3.3). The isotropic and deviatoric strain energy functions are therefore

$$\Omega_0 = \tfrac{1}{2}k\theta^2 \qquad \Omega' = \mu\varepsilon'_{ij}\varepsilon'_{ij} \tag{3.21}$$

They can, of course, be expressed in terms of stress on the basis of the relations between isotropic and deviatoric stresses and strains (Eqns 3.4 & 6)

$$\Omega_0 = \tfrac{1}{2}\sigma_0^2/k \qquad \Omega' = \sigma''_{ij}\sigma''_{ij}/4\mu \tag{3.22}$$

Equations 3.18–3.22 are valid for both isothermal and adiabatic processes, provided that the relevant elastic parameters are used. (An *isothermal* process occurs at constant temperature, an *adiabatic* process occurs without exchange of heat with the surroundings.) Thermodynamic relationships between isothermal and adiabatic elastic parameters (see Section 7.2) can be used to show that the difference between them does not exceed a few per cent for most solids at usual temperatures.

In purely elastic processes the total energy is conserved. In non-elastic processes, $dQ + dU$ may exceed $d\Omega$; consequently, energy dissipation (per unit volume) occurs which, on the basis of the previous considerations, is proportional to $\sigma_{ij}\, d\varepsilon_{ij} - d\Omega$. As dissipation generates a local increase in temperature and therefore affects the rheology, the processes of *shear heating* and *thermomechanical coupling* are of considerable importance in geodynamics (cf. Sections 4.3 & 12.2 for a further discussion of energy dissipation).

3.4 Stress compatibility equations

In Section 2.7 we derived the strain compatibility equations. We now express the conditions of compatibility for a homogeneous elastic medium in terms of stress. This leads to the definition of a special function of stress, the Airy stress function, which is useful in the solution of some two-dimensional elastic problems.

To derive elastic compatibility equations in terms of stress we substitute Hooke's Law (Eqn 3.12) into the strain compatibility condition (Eqn 2.37) to obtain

$$\frac{1+\nu}{E}\left(\frac{\partial^2 \sigma_{ij}}{\partial x_m\, \partial x_n} + \frac{\partial^2 \sigma_{mn}}{\partial x_i\, \partial x_j} - \frac{\partial^2 \sigma_{im}}{\partial x_j\, \partial x_n} - \frac{\partial^2 \sigma_{jn}}{\partial x_i\, \partial x_m}\right)$$

$$= \frac{\nu}{E}\left(\frac{\partial^2 \sigma_{kk}}{\partial x_m\, \partial x_n}\delta_{ij} + \frac{\partial^2 \sigma_{kk}}{\partial x_i\, \partial x_j}\delta_{mn} - \frac{\partial^2 \sigma_{kk}}{\partial x_j\, \partial x_n}\delta_{im} - \frac{\partial^2 \sigma_{kk}}{\partial x_i\, \partial x_m}\delta_{jn}\right) \qquad (3.23)$$

These 81 equations can be reduced to nine (of which only six are independent) by setting $m = n$ and contracting:

$$\frac{\partial^2 \sigma_{ij}}{\partial x_m\, \partial x_m} + \frac{\partial^2 \sigma_{mm}}{\partial x_i\, \partial x_j} - \frac{\partial^2 \sigma_{im}}{\partial x_j\, \partial x_m} - \frac{\partial^2 \sigma_{jm}}{\partial x_i\, \partial x_m}$$

$$= \frac{\nu}{1+\nu}\left(\frac{\partial^2 \sigma_{kk}}{\partial x_m\, \partial x_m}\delta_{ij} + \frac{\partial^2 \sigma_{kk}}{\partial x_i\, \partial x_j}\delta_{mm} - \frac{\partial^2 \sigma_{kk}}{\partial x_j\, \partial x_m}\delta_{im} - \frac{\partial^2 \sigma_{kk}}{\partial x_i\, \partial x_m}\delta_{jm}\right)$$

that is, using the properties of the Kronecker delta ($\delta_{mm} = 3$ and

$$\partial^2 \sigma_{kk} \; \delta_{im}/\partial x_j \; \partial x_m = \partial^2 \sigma_{kk}/\partial x_j \; \partial x_i),$$

$$\frac{\partial^2 \sigma_{ij}}{\partial x_m \; \partial x_m} + \frac{\partial^2 \sigma_{kk}}{\partial x_i \; \partial x_j} - \frac{\partial^2 \sigma_{im}}{\partial x_j \; \partial x_m} - \frac{\partial^2 \sigma_{jm}}{\partial x_i \; \partial x_m} = \frac{\nu}{1+\nu} \left(\frac{\partial^2 \sigma_{kk}}{\partial x_m \; \partial x_m} \delta_{ij} + \frac{\partial^2 \sigma_{kk}}{\partial x_i \; \partial x_j} \right)$$

This equation becomes, by the equilibrium condition (Eqn 2.7) and after some algebra,

$$\frac{\partial^2 \sigma_{ij}}{\partial x_m \; \partial x_m} + \left(\frac{1}{1+\nu} \right) \frac{\partial^2 \sigma_{kk}}{\partial x_i \; \partial x_j} - \left(\frac{\nu}{1+\nu} \right) \frac{\partial^2 \sigma_{kk}}{\partial x_m \; \partial x_m} \delta_{ij} = -\rho \left(\frac{\partial X_i}{\partial x_j} + \frac{\partial X_j}{\partial x_i} \right)$$

$$(3.24)$$

Equation 3.23 can also be contracted by setting $m = i$ and $n = j$ to yield

$$2 \frac{\partial^2 \sigma_{ij}}{\partial x_i \; \partial x_j} - 2 \frac{\partial^2 \sigma_{ii}}{\partial x_j \; \partial x_j} = \frac{\nu}{1+\nu} \left(2 \frac{\partial^2 \sigma_{kk}}{\partial x_i \; \partial x_j} \delta_{ij} - 2 \frac{\partial^2 \sigma_{kk}}{\partial x_j \; \partial x_j} \delta_{ii} \right)$$

that is, since $\partial^2 \sigma_{kk} \; \delta_{ij}/\delta x_i \; \delta x_j = \partial^2 \sigma_{kk}/\partial x_i \; \partial x_i$,

$$\frac{\partial^2 \sigma_{ij}}{\partial x_i \; \partial x_j} = \left(\frac{1-\nu}{1+\nu} \right) \frac{\partial^2 \sigma_{kk}}{\partial x_i \; \partial x_i}$$

which, again using the equation of equilibrium, becomes

$$\frac{\partial^2 \sigma_{kk}}{\partial x_i \; \partial x_i} = -\rho \left(\frac{1+\nu}{1-\nu} \right) \frac{\partial X_i}{\partial x_i} \qquad (3.25)$$

Substituting Equation 3.25 into Equation 3.24, we obtain

$$\frac{\partial^2 \sigma_{ij}}{\partial x_k \; \partial x_k} + \left(\frac{1}{1+\nu} \right) \frac{\partial^2 \sigma_{kk}}{\partial x_i \; \partial x_j} = -\rho \left(\frac{\nu}{1-\nu} \right) \frac{\partial X_k}{\partial x_k} \delta_{ij} - \rho \left(\frac{\partial X_i}{\partial x_j} + \frac{\partial X_j}{\partial x_i} \right) \quad (3.26)$$

Equation 3.26 represents the six independent equations giving the compatibility conditions in terms of stress in elastic bodies, also known as the *Beltrami–Michell equations*.

If the body force is constant, then Equation 3.26 simplifies to

$$\frac{\partial^2 \sigma_{ij}}{\partial x_k \; \partial x_k} + \left(\frac{1}{1+\nu} \right) \frac{\partial^2 \sigma_{kk}}{\partial x_i \; \partial x_j} = 0 \qquad (3.27)$$

and Equation 3.25 becomes

$$\frac{\partial^2 \sigma_{kk}}{\partial x_i \; \partial x_i} \equiv \nabla^2 \sigma_{kk} = 0 \qquad (3.28a)$$

where ∇^2 is again the Laplacian operator, defined in Section 3.2. It follows by Hooke's Law (Eqn 3.4) that, since $\sigma_{kk} = 3k\theta$,

$$\frac{\partial^2 \theta}{\partial x_i \, \partial x_i} \equiv \nabla^2 \theta = 0 \qquad (3.28b)$$

We have thus shown that, if the body force is not a function of position, the mean normal stress and the cubical dilatation satisfy similar conditions (Eqns 3.28a & b). An equation of the type

$$\nabla^2 f(x_1, x_2, x_3) \equiv \frac{\partial^2 f}{\partial x_1^2} + \frac{\partial^2 f}{\partial x_2^2} + \frac{\partial^2 f}{\partial x_3^2} = 0$$

is called the *Laplace equation*, and a function that satisfies it is termed *harmonic*. Both mean normal stress and cubical dilatation are therefore harmonic functions in elastic bodies under constant body force.

The Laplace equation is one of the most important in physics, and plays an important role in the Earth sciences, especially in the study of the gravitational and the geomagnetic fields (see Section 8.1 for a brief treatment of the former). It may also be mentioned here that two other fundamental equations of physics that are of great importance in geodynamics are the *wave equation*

$$\nabla^2 f(x_1, x_2, x_3, t) = a \frac{\partial^2 f}{\partial t^2}$$

and the *diffusion equation*

$$\nabla^2 f(x_1, x_2, x_3, t) = b \frac{\partial f}{\partial t}$$

which, as their names imply, govern vibrational processes (e.g. seismic waves, Ch. 6) and diffusion process (e.g heat conduction, Ch. 7).

A further interesting property of stress in elastic media becomes apparent when the Laplacian of Equation 3.27 is taken. Assuming that derivatives of σ_{ij} up to the fourth order are continuous, we obtain

$$\frac{\partial^2}{\partial x_m \, \partial x_m} \left[\frac{\partial^2 \sigma_{ij}}{\partial x_k \, \partial x_k} + \left(\frac{1}{1+\nu} \right) \frac{\partial^2 \sigma_{kk}}{\partial x_i \, \partial x_j} \right]$$

$$= \nabla^2 (\nabla^2 \sigma_{ij}) + \left(\frac{1}{1+\nu} \right) \frac{\partial^2}{\partial x_i \, \partial x_j} (\nabla^2 \sigma_{kk}) = 0$$

where we have written $\nabla^2 \equiv \partial^2/\partial x_k\, \partial x_k$ and have inverted the order of differentiation in the second term. By Equation 3.28a this second term is zero, and therefore

$$\nabla^2(\nabla^2\sigma_{ij}) \equiv \nabla^4\sigma_{ij} = 0 \qquad (3.29a)$$

where the operator ∇^4 is, when written in full

$$\nabla^4 \equiv \frac{\partial^2}{\partial x_i\, \partial x_i}\, \frac{\partial^2}{\partial x_j\, \partial x_j}$$

$$= \frac{\partial^4}{\partial x_1^4} + 2\,\frac{\partial^4}{\partial x_1^2\, \partial x_2^2} + \frac{\partial^4}{\partial x_2^4} + 2\,\frac{\partial^4}{\partial x_2^2\, \partial x_3^2} + \frac{\partial^4}{\partial x_3^4} + 2\,\frac{\partial^4}{\partial x_1^2\, \partial x_3^2}$$

From the linear correspondence between stress and strain components it follows from Equation 3.29a that

$$\nabla^4\varepsilon_{ij} = 0 \qquad (3.29b)$$

A function satisfying a relation such as Equation 3.29a or b is called *biharmonic*. Both stress and strain in elastic bodies under constant body force are biharmonic functions.

The Beltrami–Michell equations and their corollaries establish important results in the theory of elasticity. In the next section we study a simple two-dimensional application which is of some relevance to geodynamics.

3.5 The Airy stress function

For the sake of simplicity, and because the approach lends itself naturally to the study of stresses in tectonic blocks, we consider here a two-dimensional problem in Cartesian co-ordinates x_1, x_2. Reduction to two dimensions, which achieves a considerable simplification, is justified whenever the geological structure to be considered is much longer than it is wide and can be taken not to vary significantly along its strike; then a particular problem (e.g. the stress distribution) may be studied in a typical cross-section perpendicular to the strike.

The assumption that body force is independent of position is sufficiently good for problems dealing with geodynamic processes, if buoyancy does not play a significant role. However, we can introduce a different assumption; namely, that the body force has a potential U that is itself harmonic, i.e.

$$X_i = \partial U/\partial x_i \qquad \nabla^2 U = 0$$

which holds for the force of gravity (Section 8.1). Then, since in Equation 3.25 $\partial X_i/\partial x_i = \nabla^2 U = 0$, Equation 3.28a is still valid. Written in full in two dimensions, it is

$$\nabla^2(\sigma_{11} + \sigma_{22}) \equiv \frac{\partial^2(\sigma_{11} + \sigma_{22})}{\partial x_1^2} + \frac{\partial^2(\sigma_{11} + \sigma_{22})}{\partial x_2^2} = 0 \qquad (3.30)$$

Now assume there exists a function $\Phi(x_1, x_2)$, called the *Airy stress function*, such that

$$\sigma_{11} = \frac{\partial^2\Phi}{\partial x_2^2} - \rho U \qquad \sigma_{22} = \frac{\partial^2\Phi}{\partial x_1^2} - \rho U \qquad \sigma_{12} = -\frac{\partial^2\Phi}{\partial x_1\,\partial x_2} \qquad (3.31)$$

which naturally satisfies the equations of equilibrium, as can be verified by substitution. Using Equations 3.31 in Equation 3.30 gives

$$\nabla^2(\sigma_{11} + \sigma_{22}) = \nabla^2\left(\frac{\partial^2\Phi}{\partial x_2^2} + \frac{\partial^2\Phi}{\partial x_1^2} - 2\rho U\right)$$

$$= \nabla^2(\nabla^2\Phi) - 2\rho\nabla^2 U = 0$$

i.e. since $\nabla^2 U = 0$,

$$\nabla^4\Phi \equiv \frac{\partial^4\Phi}{\partial x_1^4} + 2\frac{\partial^4\Phi}{\partial x_1^2\,\partial x_2^2} + \frac{\partial^4\Phi}{\partial x_2^4} = 0 \qquad (3.32)$$

The function Φ is therefore biharmonic, and any solution to the biharmonic equation satisfies the conditions of equilibrium and compatibility.

If the Airy stress function is known, then the stress components as a function of position can be found by differentiation (Eqns 3.31). The Airy stress function therefore has a role in two-dimensional elasticity somewhat analogous to that of the potential in conservative force fields. The problem is reduced to finding only one unknown function (the solution to Eqn 3.32 under given boundary conditions), from which the stress components can be found. The biharmonic equation is consequently of great importance in the theory of elasticity, although the methods for its solution are beyond the scope of this book.

It is often important in geodynamics to know the stress state in a given tectonic feature. Of all the possible stress systems the so-called *standard state* (Anderson 1951) is often used as a reference state. Considering a two-dimensional elastic half-space referred to a Cartesian co-ordinate system (origin on the surface, x_2-axis vertical and positive downward), the standard state is defined as

$$\sigma_{11} = \sigma_{22} = -\rho g x_2 \qquad \sigma_{12} = 0 \qquad\qquad (3.33)$$

i.e. it is everywhere hydrostatic with normal stresses at depth x_2 equal to the *overburden* – or *lithostatic* – pressure (ρ and g are density of the material and acceleration of gravity, respectively).

The standard state is an idealization that does not occur in nature except in special cases. (The *in situ* stresses in the lithosphere are discussed in Section 8.5.) However, it is often useful to consider the actual state of stress in a tectonic block as the sum of the standard condition (Eqn 3.33) and some supplementary stress system. This approach is not dissimilar from the resolution of the stress into isotropic and deviatoric components (Section 2.4). As an example, consider a supplementary stress system whose only non-zero component is $\sigma_{11} = c$ (uniform horizontal stress). The total stress system is then

$$\sigma_{11} = -\rho g x_2 + c \qquad \sigma_{22} = -\rho g x_2 \qquad \sigma_{12} = 0$$

(the co-ordinate axes are principal directions). If $c < 0$, then the horizontal compression is larger than the overburden pressure and we speak of a compressive state of stress; if $c > 0$, then the horizontal stress, while in general still a compression except very near the surface, is less (in magnitude) than the overburden pressure at all depths, and we speak of a state of relative horizontal tension. The same type of discussion can be carried out taking as reference not the standard state but the mean normal stress, in which case the horizontal relative stress (compression or tension) is simply the horizontal component of the stress deviator.

4 Flow, strain rate, and viscosity

Fluids, i.e. bodies that show steady-state flow under constant deviatoric stress, are at the other end of the rheological spectrum from elastic solids. Their kinematic state is described by the strain rate tensor, and their characteristic material parameter is the viscosity. The mantle of the Earth is a fluid at geological timescales.

The two classes of fluids that are most important in geodynamics are Newtonian bodies – where strain rate is linearly proportional to stress – and non-Newtonian bodies – where strain rate is proportional to the nth power of the stress ($n > 1$). We discuss the equation of continuity (conservation of mass), the equations of motion for Newtonian bodies (Navier–Stokes equations), and the tensor form of the non-Newtonian power-law equation.

Many materials show time effects with a stage of transient creep between instantaneous (elastic) and steady-state (viscous) deformation. This can be modelled by linear rheological bodies (Kelvin, Maxwell, and Burgers), which can account for elastic afterworking and stress relaxation.

4.1 Strain rate tensor

A perfectly elastic body shows only instantaneous deformation, and time effects are absent; consequently, it is natural to use displacement, displacement gradient, and strain as kinematic quantities. At the opposite end of the rheological spectrum one may envisage a medium that flows under any stress, however small, i.e. a medium in which the deformation under constant load changes continuously with time. In this case the basic kinematic quantity is the velocity. On the basis of the considerations developed in Section 1.6, the velocity v_i in the Eulerian description is the total time derivative of the displacement, i.e. $v_i = Du_i/Dt$ (cf. Eqn 1.14). If, following a line of thought analogous to that used in elasticity, we define a velocity gradient tensor $\partial v_i/\partial x_j$, it follows that (neglecting second-order terms)

$$\frac{\partial v_i}{\partial x_j} = \frac{\partial}{\partial x_j} \left(\frac{\partial u_i}{\partial t} + v_k \frac{\partial u_i}{\partial x_k} \right) = \frac{\partial}{\partial t} \left(\frac{\partial u_i}{\partial x_j} \right) + v_k \frac{\partial}{\partial x_k} \left(\frac{\partial u_i}{\partial x_j} \right) = \frac{\mathrm{D}}{\mathrm{D}t} \left(\frac{\partial u_j}{\partial x_j} \right)$$

The *strain rate tensor* $\dot{\varepsilon}_{ij}$ is then defined as

$$\dot{\varepsilon}_{ij} \equiv \frac{\mathrm{D}\varepsilon_{ij}}{\mathrm{D}t} = \frac{1}{2} \left[\frac{\mathrm{D}}{\mathrm{D}t} \left(\frac{\partial u_i}{\partial x_j} \right) + \frac{\mathrm{D}}{\mathrm{D}t} \left(\frac{\partial u_j}{\partial x_i} \right) \right]$$

$$= \frac{1}{2} \left(\frac{\partial v_i}{\partial x_j} + \frac{\partial v_j}{\partial x_i} \right) . \tag{4.1}$$

where, as usual, an overdot denotes the time derivative.

The properties of the strain rate tensor are formally analogous to those of the strain tensor, with the important difference that strain rate is related to variations of *velocity* from point to point, rather than to variations of displacement. Since strain is dimensionless, the dimensions of the strain rate tensor are $[\dot{\varepsilon}] = [t^{-1}]$, and its unit in both the SI and the CGS systems is the reciprocal second (s^{-1}).

Before discussing the constitutive equations relating stress and strain rate, it is perhaps useful to clarify the meaning of 'solid' and 'fluid' in the rheological sense. We agree to call 'solid' any material that does not yield unless the stress reaches a given threshold (thus, an elastic body would be a solid), and to call 'fluid' any material that exhibits steady-state flow under constant stress, no matter how small. As noted at the outset of this book, this definition differs from the notions of 'solidity' and 'fluidity' based on the lattice structure. Also, the provision of a constant rate of flow under constant stress (*steady-state flow*) is of considerable importance: *transient flow*, where the strain rate varies with time under constant stress, can also occur (see Sections 4.5 & 4.6).

A further word on terminology: very often, the term *creep* is used to denote slow flow under constant stress (thus one can have transient creep and steady-state creep). As such, this term is used simply to emphasize the slowness of the flow; in this sense we may say that water flows, while metals and rocks at high temperature creep. However, it must be understood that any material exhibiting steady-state creep is a 'fluid' in the rheological sense.

The topics treated in this chapter are dealt with in several books (cf especially Reiner 1960 and Fung 1977). Lamb (1945) is a classic text among the larger and more-advanced treatises on fluid mechanics; a more modern coverage is given in Batchelor (1967).

4.2 Constitutive equations for Newtonian viscosity

The simplest type of relation between stress and strain rate (excluding an inviscid fluid, in which no shear stress exists) is depicted in Figures 4.1a and b by means of a strain–time diagram and a stress–strain rate diagram, respectively. Steady-state flow occurs under constant stress (the strain rate $\dot{\varepsilon}$ is the slope of the $\varepsilon(t)$ curve), the deformation is non-recoverable, and the relation between stress and strain rate is linear.

Linear dependence between stress and strain rate components may be expressed as

$$\sigma_{ij} = -p\,\delta_{ij} + C'_{ijkl}\dot{\varepsilon}_{kl} \tag{4.2}$$

where p is the fluid pressure, i.e. $-p\,\delta_{ij}$ is the (isotropic) state of stress present when the fluid is at rest ($\dot{\varepsilon}_{kl} = 0$), and the components of the tensor C'_{ijkl} do not depend on σ_{ij} or $\dot{\varepsilon}_{kl}$. On the basis of symmetry and isotropy considerations formally analogous to those used to derive Hooke's Law, Equation 4.2 for isotropic, homogeneous bodies reduces to (Fung 1977)

$$\sigma_{ij} = -p\,\delta_{ij} + \lambda'\dot{\theta}\,\delta_{ij} + 2\eta\dot{\varepsilon}_{ij} \tag{4.3}$$

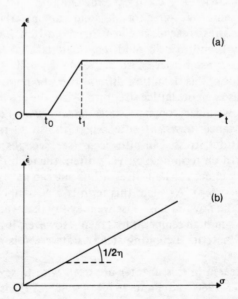

Figure 4.1 (a) Strain–time diagram and (b) stress–strain rate diagram for a linearly viscous body. For shear stress and strain rate the slope of (b) is related to viscosity as $1/2\eta$.

where $\dot{\theta} = \dot{\varepsilon}_{kk}$, and λ' and η are material parameters, the latter being termed the *coefficient of linear viscosity* (or, briefly, *viscosity*) of the fluid. The viscosity, like any other material parameter, is a function of temperature and pressure, but in the linear case does not depend on σ_{ij} and $\dot{\varepsilon}_{ij}$.

Contracting Equation 4.3 by setting $i = j$, we obtain

$$\sigma_{ii} = -3p + (3\lambda' + 2\eta)\dot{\theta}$$

which implies that, assuming σ_{ii} to be independent of $\dot{\theta}$,

$$3\lambda' + 2\eta = 0 \quad \text{i.e.} \quad \lambda' = -\tfrac{2}{3}\eta$$

Equation 4.3 therefore becomes

$$\sigma_{ij} = -p\,\delta_{ij} - \tfrac{2}{3}\eta\dot{\theta}\,\delta_{ij} + 2\eta\dot{\varepsilon}_{ij} \tag{4.4}$$

A body with a rheological equation such as the above is sometimes called a *Stokes fluid*.

If the fluid is non-viscous ($\eta = 0$), then its rheological equation becomes

$$\sigma_{ij} = -p\,\delta_{ij} \tag{4.5}$$

i.e. the state of stress is everywhere hydrostatic. The term *Pascalian fluid* denotes a non-viscous, incompressible body. The pressure p in a Pascalian fluid is an arbitrary variable determined only by the equations of motion and the boundary conditions.

The assumption of *incompressibility* is used fairly often in geodynamic problems. A change in pressure generates no change in volume or density in an incompressible body. Compressibility – the inverse of the bulk modulus k – is defined as

$$\beta = -\frac{1}{V}\left(\frac{\partial V}{\partial p}\right) = \frac{1}{\rho}\left(\frac{\partial \rho}{\partial p}\right)$$

where V and ρ are the volume and density, respectively. It follows that, for an incompressible body, $\theta = \dot{\theta} = 0$, and Equation 4.4 therefore reduces to

$$\sigma_{ij} = -p\,\delta_{ij} + 2\eta\dot{\varepsilon}_{ij} \tag{4.6}$$

A relation such as Equation 4.4 or Equation 4.6 can also be derived by assuming that the elastic relation between mean normal stress and cubical dilatation (Eqn 3.4) continues to hold (i.e. a purely isotropic stress does not induce flow – an observation supported by experiment), and that the relation between deviatoric stress and deviatoric strain rate is

$$\sigma'_{ij} = 2\eta\dot{\varepsilon}'_{ij} \tag{4.7}$$

Equation 4.7, recalling the definition of stress deviator and considering that $\dot{\varepsilon}'_{ij} = \dot{\varepsilon}_{ij} - \frac{1}{3}\dot{\theta}\,\delta_{ij}$, immediately yields Equation 4.4.

A body whose rheological equation is such as Equation 4.3, 4.4, 4.6, or 4.7, i.e. in which the relation between deviatoric stress and deviatoric strain rate is linear with coefficient of proportionality η, is called a *Newtonian fluid*. The *Newtonian* (or *linear*) *viscosity* is defined as the ratio between deviatoric stress and twice the deviatoric strain rate (or between the shear stress and twice the corresponding strain rate):

$$\eta = \sigma_{ij}/2\dot{\varepsilon}_{ij} \qquad i \neq j \tag{4.8}$$

where, of course, no summation is implied.

In fluid dynamics one sometimes uses for shear strain rate the quantity

$$\dot{\gamma}_{ij} = \frac{\partial v_i}{\partial x_j} + \frac{\partial v_j}{\partial x_i} = 2\dot{\varepsilon}_{ij} \qquad i \neq j \tag{4.9}$$

in which case the viscosity is simply $\eta = \sigma_{ij}/\dot{\gamma}_{ij}$. In either case the viscosity is related to the slope of the stress–strain rate curve (Fig. 4.1b).

The viscosity is a measure of the resistance of a fluid to flow, i.e. of the internal friction between contiguous layers of fluid in motion relative to one another. Suppose that the flow is *laminar* (i.e. the fluid moves in parallel laminae, each with constant velocity) and parallel to the x_2-axis (Fig. 4.2). The two planes A and B are separated by the distance dx_3, and they move with velocities v_2 and $v_2 + (\partial v_2/\partial x_3)\,dx_3$, respectively; $\partial v_2/\partial x_3 = \dot{\gamma}_{23}$ is therefore the shear strain rate. If the fluid is Newtonian, then one has $\eta = \sigma_{23}/\dot{\gamma}_{23}$, where σ_{23} is the shear stress necessary to maintain the relative motion between the two planes; that is, to overcome the internal friction between contiguous layers of fluid.

The dimensions of viscosity are those of stress divided by strain rate, i.e. $[ml^{-1}t^{-1}]$. Its units are the pascal-second, $(1\ \text{Pa s} = 1\ \text{N s m}^{-2})$ in the SI system, and the poise (1 P = 1 dyn s cm^{-2}) in the CGS system, the conversion factor being 1 Pa s = 10 P. Sometimes it is convenient to use as a parameter the *kinematic viscosity* ν, defined as the viscosity divided by the density, $\nu = \eta/\rho$ (the term *dynamic viscosity* is used for η in con-

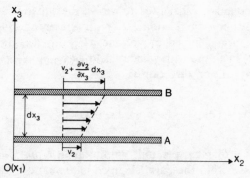

Figure 4.2 Laminar viscous flow in the x_2-direction between parallel planes normal to the x_3-axis.

texts where confusion may arise). The dimensions of kinematic viscosity are $[l^2 t^{-1}]$, and the units are therefore the square metre per second $(\text{m}^2\,\text{s}^{-1})$ in the SI system, and the stokes $(1\,\text{St} = 1\,\text{cm}^2\,\text{s}^{-1})$ in the CGS system.

The viscosity is a parameter of fundamental importance in the mantle of the Earth, for it affects long-term flow processes that are related to plate tectonics and the thermal evolution of the planet. These aspects of the rheology of the Earth are discussed in Chapters 7, 8 and 12. Here we wish to emphasize only that the inferred viscosities of the mantle $(\eta \simeq 10^{20} - 10^{22}\,\text{Pa s})$ are extremely large when compared not only with the viscosity of water $(10^{-3}\,\text{Pa s}$ at $T = 20°\text{C})$, but also with the viscosities of other materials (for instance $\eta \simeq 10^2 - 10^4\,\text{Pa s}$ for molten basalt at $1100-1400°\text{C}$, $\eta \simeq 10^9\,\text{Pa s}$ for pitch at $15°\text{C}$, and $\eta = 1.2 \times 10^{13}\,\text{Pa s}$ for glacier ice). Thus, a shear stress of 10 MPa would generate a shear strain rate of about $10^{-6}\,\text{s}^{-1}$ in ice, but of only $10^{-14}\,\text{s}^{-1}$ in the mantle. To achieve strain rates in rocks of the order of those observable in the laboratory ($\dot{\varepsilon} \geqslant 10^{-8}\,\text{s}^{-1}$; see Section 10.4), it is necessary to perform experiments at very high temperature, sometimes not far below the solidus temperature.

4.3 Equation of continuity and the Navier–Stokes equations

Before setting up the equations of motion in terms of the velocity (the equivalent for viscous bodies of the equations of motion in terms of the displacement for an elastic body, Eqn 3.14), it is necessary to express conservation of mass in quantitative form. Mass must be conserved in non-relativistic mechanics (i.e. there cannot be sources or sinks of matter). The *conservation equation* (or *equation of continuity*) is derived in a manner analogous to that used to derive the equation of equilibrium for a continuous body in Section 2.2. With reference to Figure 4.3, consider a small parallelepiped of volume $dV = dx_1\ dx_2\ dx_3$ and density $\rho(x_1, x_2, x_3, t)$ in a fluid with velocity field $v_i(x_1, x_2, x_3, t)$. The net flow of mass in, say, the x_2-direction is

$$\rho v_2\ dx_1\ dx_3 - \left(\rho v_2 + \frac{\partial}{\partial x_2}\ (\rho v_2)\ dx_2\right)\ dx_1\ dx_3 = - \frac{\partial}{\partial x_2}\ (\rho v_2)\ dV$$

i.e. the net flow of mass per unit volume is $-\partial(\rho v_i)/\partial x_i$. The rate of change of mass within the parallelepiped is $(\partial \rho/\partial t)\ dV$. For mass to be conserved, the net flow of matter from the elemental volume must be equal to the change within it, that is

$$\frac{\partial \rho}{\partial t} + \frac{\partial}{\partial x_i}\ (\rho v_i) = 0$$

which becomes, using the definition of total derivative derived in Section 1.6,

$$\frac{\partial \rho}{\partial t} + v_i\ \frac{\partial \rho}{\partial x_i} + \rho\ \frac{\partial v_i}{\partial x_i} = \frac{D\rho}{Dt} + \rho\ \frac{\partial v_i}{\partial x_i} = 0 \tag{4.10}$$

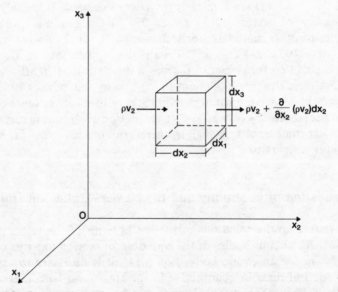

Figure 4.3 Conservation of mass in a moving fluid. For simplicity, only the x_2-components of the velocity field are shown.

72

If the fluid is incompressible, then $\rho = $ constant and the equation of continuity becomes

$$\frac{\partial v_i}{\partial x_i} = 0 \qquad (4.11)$$

Having thus established the condition for conservation of mass, we derive the equation of motion for viscous fluids, in a manner analogous to the Cauchy–Navier equations for elastic bodies (Section 3.2). The equation of motion is written for the general case as

$$\frac{\partial \sigma_{ij}}{\partial x_j} + \rho X_i = \rho \frac{Dv_i}{Dt}$$

Substituting Equation 4.4 for the stress, we obtain

$$\frac{\partial}{\partial x_j} \left(-p\,\delta_{ij} - \tfrac{2}{3}\eta\dot{\theta}\,\delta_{ij} + 2\eta\dot{\varepsilon}_{ij} \right) + \rho X_i = \rho \frac{Dv_i}{Dt}$$

i.e. for a homogeneous fluid, since $\dot{\theta} = \partial v_j / \delta x_j$,

$$-\frac{\partial p}{\partial x_i} - \tfrac{2}{3}\eta \frac{\partial}{\partial x_i}\left(\frac{\partial v_j}{\partial x_j}\right) + \eta \frac{\partial}{\partial x_j}\left(\frac{\partial v_i}{\partial x_j} + \frac{\partial v_j}{\partial x_i}\right) + \rho X_i = \rho \frac{Dv_i}{Dt}$$

Inverting the order of differentiation where needed, and writing $\partial^2 v_i / \partial x_j\,\partial x_j \equiv \nabla^2 v_i$, this equation becomes

$$-\frac{\partial p}{\partial x_i} + \tfrac{1}{3}\eta \frac{\partial}{\partial x_i}\left(\frac{\partial v_j}{\partial x_j}\right) + \eta\nabla^2 v_i + \rho X_i = \rho \frac{Dv_i}{Dt} \qquad (4.12a)$$

or, in terms of the kinematic viscosity $\nu = \eta/\rho$,

$$-\frac{1}{\rho}\frac{\partial p}{\partial x_i} + \tfrac{1}{3}\nu \frac{\partial}{\partial x_i}\left(\frac{\partial v_j}{\partial x_j}\right) + \nu\nabla^2 v_i + X_i = \frac{Dv_i}{Dt} \qquad (4.12b)$$

Equations 4.12a or b are the *equations of motion* (also known as the *Navier–Stokes equations*) for a fluid with Newtonian viscosity. If the fluid is incompressible, then they reduce to

$$-\frac{1}{\rho}\frac{\partial p}{\partial x_i} + \nu\nabla^2 v_i + X_i = \frac{Dv_i}{Dt} \qquad (4.13)$$

If the material acceleration is negligible, $Dv_i/Dt = 0$, then the

73

Navier–Stokes equations become

$$-\frac{1}{\rho}\frac{\partial p}{\partial x_i} + \frac{1}{3}\nu\frac{\partial}{\partial x_i}\left(\frac{\partial v_j}{\partial x_j}\right) + \nu\nabla^2 v_i + X_i = 0 \qquad (4.14)$$

$$-\frac{1}{\rho}\frac{\partial p}{\partial x_i} + \nu\nabla^2 v_i + X_i = 0 \qquad (4.15)$$

for the compressible and incompressible case, respectively. Equations 4.14 and 4.15 are often used in geodynamic applications.

The fluid pressure p marks an important difference between elasticity and fluid dynamics. The equations of motion (Eqns 4.12–15), representing force balance, and the equations of continuity (Eqns 4.10 & 11), representing mass balance, which together impose conditions on the velocity field, must be complemented by some *equation of state* relating density, pressure and temperature

$$f(\rho,\ p,\ T) = 0 \qquad (4.16)$$

The presence of non-zero time derivatives in the constitutive equation for viscous fluids is associated with dissipation of energy. Using the terminology developed in Section 3.3 for elastic solids, the energy dissipated per unit volume in time dt is $\sigma_{ij}\,d\varepsilon_{ij} - d\Omega$, and therefore the rate of energy dissipation is

$$\dot{W} = \sigma_{ij}\dot{\varepsilon}_{ij} - \dot{\Omega} \qquad (4.17)$$

The strain energy function is given by Equation 3.20, and $\mu = 0$ in fluids. This, together with the definition of deviator, yields

$$\dot{W} = (\sigma'_{ij} + \tfrac{1}{3}\sigma_{kk}\ \delta_{ij})(\dot{\varepsilon}'_{ij} + \tfrac{1}{3}\dot{\theta}\ \delta_{ij}) - k\theta\dot{\theta} = \sigma'_{ij}\dot{\varepsilon}'_{ij} + \tfrac{1}{3}\sigma_{kk}\dot{\theta} - k\theta\dot{\theta}$$

since $\delta_{ij}\dot{\varepsilon}'_{ij} = \dot{\varepsilon}'_{ii} = 0$ and $\delta_{ij}\sigma'_{ij} = \sigma'_{ii} = 0$. Recalling that $\tfrac{1}{3}\sigma_{kk} = k\theta$ and Equation 4.7, we obtain

$$\tfrac{1}{2}\dot{W} = \eta\dot{\varepsilon}'_{ij}\dot{\varepsilon}'_{ij} \qquad (4.18)$$

This equation states that the rate of energy dissipation per unit volume is proportional to the square of the shear strain rate. More precisely, it can be proved that the quantity $\dot{\varepsilon}'_{ij}\dot{\varepsilon}'_{ij}$ is related to the second invariant of the strain rate deviator (Section 4.4). Therefore, the larger the velocity gradient is, the higher the rate of heat generation by energy dissipation – this is the phenomenon of 'shear heating' referred to at the end of Section 3.3. As an increase in temperature has a softening effect (i.e. it decreases the viscosity), a positive feedback exists between rate of deformation and ease of deformation. This thermomechanical coupling affects the resistance to deformation in many situations of geodynamic significance (see Section 12.2).

4.4 Power-law creep and non-Newtonian viscosity

In Newtonian fluids the relation between shear stress and shear strain rate is linear. There are other types of fluid behaviour where the material exhibits steady-state flow even at small stresses, but the relation between the variables is non-linear. Depending on the extrinsic rheological parameters and the level of stress, the rheological equation may vary even for the same material. By far the most important non-linear stress—strain rate relation in geodynamics is the so-called *power-law creep* equation, where strain rate is related to the nth power of the stress ($n > 1$). This non-Newtonian flow behaviour is very common in silicate polycrystals at high temperature and low stresses ($\sigma \simeq 10$–100 MPa; see Section 10.4). The physical basis for such a creep law will be studied in detail in Chapter 10, as will be that for Newtonian creep. Here, we establish the phenomenological law from a continuum-mechanics standpoint, and discuss some of the implications of non-linear viscosity.

Most treatises on fluid mechanics focus almost exclusively on Newtonian viscosity. Non-Newtonian viscosity, being often a property of materials that are 'solids' in the everyday sense of the word, is discussed elsewhere; for rocks, ice, and metals, respectively, reference may be made to Nicolas & Poirier (1976), Paterson (1981), and Poirier (1985).

The general tensor form of power-law creep can be established starting from laboratory experiments. Without going into the details of experimental arrangements, a specimen can be deformed in *simple shear*, where the only non-zero stress is a shear stress (denoted by σ_s in the following discussion), or under *uniaxial* compression or tension (denoted by σ_l). The corresponding components of strain rate are $\dot{\varepsilon}_s$ and $\dot{\varepsilon}_l$, respectively. Suppose, then, that shear experiments are performed on a homogeneous, isotropic material, and that the resulting steady-state flow is of the form

$$\dot{\varepsilon}_s = A\sigma_s^n \qquad (4.19)$$

where A is, in general, a function of pressure, temperature, and material parameters, while the stress exponent is a constant ($n > 1$). The problem is to extend the experimental results to a general state of stress, which will enable us to write down the power-law creep equation in tensor form.

Since a flow law reflects the physical properties of the material, it must be valid under any co-ordinate transformation (incidentally, this is true for both elastic and linearly viscous rheological equations). It is therefore natural to express Equation 4.19 in terms of invariants. However, as the flow law is not affected by hydrostatic pressure, stress and strain rate deviators can be used. The first and second invariants of the stress deviator and of the strain rate deviator are, respectively (cf. Section 2.4),

$$I_1' = \sigma_{11}' + \sigma_{22}' + \sigma_{33}' = 0$$

$$\dot{E}_1' = \dot{\varepsilon}_{11}' + \dot{\varepsilon}_{22}' + \dot{\varepsilon}_{33}' = 0$$

$$I_2' = -(\sigma_{11}'\sigma_{22}' + \sigma_{11}'\sigma_{33}' + \sigma_{22}'\sigma_{33}') + \sigma_{12}'^2 + \sigma_{13}'^2 + \sigma_{23}'^2$$

$$\dot{E}_2' = -(\dot{\varepsilon}_{11}'\dot{\varepsilon}_{22}' + \dot{\varepsilon}_{11}'\dot{\varepsilon}_{33}' + \dot{\varepsilon}_{22}'\dot{\varepsilon}_{33}') + \dot{\varepsilon}_{12}'^2 + \dot{\varepsilon}_{13}'^2 + \dot{\varepsilon}_{23}'^2$$

where the equalities $I_1' = 0$, $\dot{E}_1' = 0$ are a direct consequence of the definition of deviator. Multiplying the second invariants by two and adding $I_1'^2$ and $\dot{E}_1'^2$, respectively (a permissible operation since $I_1'^2 = \dot{E}_1'^2 = 0$), we obtain

$$I_2' = \tfrac{1}{2}(\sigma_{11}'^2 + \sigma_{22}'^2 + \sigma_{33}'^2) + \sigma_{12}'^2 + \sigma_{13}'^2 + \sigma_{23}'^2 = \tfrac{1}{2}\sigma_{ij}'\sigma_{ij}'$$

$$\dot{E}_2' = \tfrac{1}{2}(\dot{\varepsilon}_{11}'^2 + \dot{\varepsilon}_{22}'^2 + \dot{\varepsilon}_{33}'^2) + \dot{\varepsilon}_{12}'^2 + \dot{\varepsilon}_{13}'^2 + \dot{\varepsilon}_{23}'^2 = \tfrac{1}{2}\dot{\varepsilon}_{ij}'\dot{\varepsilon}_{ij}'$$

The square roots of the quantities defined in the equations above are sometimes referred to as *effective shear stress* and *effective strain rate*, respectively. If the material is incompressible ($\dot{\theta} = 0$, usually a fair approximation in flow problems), the second invariant $\dot{E}_2'$ becomes

$$\dot{E}_2' = \tfrac{1}{2}(\dot{\varepsilon}_{ij} - \tfrac{1}{3}\dot{\theta}\ \delta_{ij})(\dot{\varepsilon}_{ij} - \tfrac{1}{3}\dot{\theta}\ \delta_{ij}) = \tfrac{1}{2}\dot{\varepsilon}_{ij}\dot{\varepsilon}_{ij} = \dot{E}_2$$

i.e. the strain rate tensor rather than its deviator may be used.

For a homogeneous, isotropic, incompressible medium with steady-state flow in simple shear expressed by Equation 4.19, we can postulate that the general flow law expresses a relation between effective shear stress and effective strain rate. Denoting these as

$$\sigma_E' = (\tfrac{1}{2}\sigma_{ij}'\sigma_{ij}')^{1/2} \qquad \dot{\varepsilon}_E = (\tfrac{1}{2}\dot{\varepsilon}_{ij}\dot{\varepsilon}_{ij})^{1/2} \tag{4.20}$$

we can rewrite the flow law in the form

$$\dot{\varepsilon}_E = A\sigma_E'^{\,n} \tag{4.21}$$

76

Equation 4.21 gives the power-law creep equation in terms of invariants.

We now assume that the components of strain rate at any point are proportional to the corresponding components of the stress deviator, i.e.

$$\dot{\varepsilon}_{ij} = \lambda \sigma'_{ij} \qquad (4.22)$$

where λ is not a constant, but a function of the state of stress and therefore of position. This is a reasonable assumption for isotropic materials. Then Equations 4.20 and 4.22 imply that

$$\dot{\varepsilon}_E = \lambda \sigma'_E \qquad (4.23)$$

Combining Equations 4.21 and 4.23,

$$\lambda = A \sigma_E'^{(n-1)}$$

and Equation 4.22 becomes

$$\dot{\varepsilon}_{ij} = A \sigma_E'^{(n-1)} \sigma'_{ij} \qquad (4.24)$$

Equation 4.24 is the required power-law creep equation in tensor form, based on the assumptions of (a) power-law dependence of effective shear strain rate on effective shear stress and (b) linear dependence of components of strain rate and stress deviator. As this linear dependence is a function of the state of stress, the resulting relation between $\dot{\varepsilon}_{ij}$ and σ'_{ij} is not linear.

As an example, consider the case of simple shear, where σ_s and $\dot{\varepsilon}_s$ are the only non-zero components of stress and strain rate. Then, by Equation 4.20, $\sigma'_E = \sigma_s$, $\dot{\varepsilon}_E = \dot{\varepsilon}_s$ and therefore

$$\dot{\varepsilon}_s = A \sigma_s^n$$

which is, of course, identical to Equation 4.19 which was derived experimentally.

If we now take the case of uniaxial stress σ_1, the only non-zero components of the stress deviator are $\sigma'_{11} = \frac{2}{3}\sigma_1$, $\sigma'_{22} = \sigma'_{33} = -\frac{1}{3}\sigma_1$, and the corresponding strain rates are $\dot{\varepsilon}_{11} = \dot{\varepsilon}_1$, $\dot{\varepsilon}_{22} = \dot{\varepsilon}_{33} = -\frac{1}{2}\dot{\varepsilon}_1$ (because of incompressibility). Consequently,

$$\sigma'_E = \frac{1}{3^{1/2}}\sigma_{1'} \qquad \dot{\varepsilon}_E = \frac{3^{1/2}}{2}\dot{\varepsilon}_1$$

77

and the creep law (Eqns 4.21 or 24) becomes

$$\dot{\varepsilon}_1 = \frac{2}{3^{(n+1)/2}} \, A\sigma_1^n$$

Consequently, for a stress of the same magnitude ($\sigma_1 = \sigma_s$), the elongation rate under uniaxial stress is $\dot{\varepsilon}_1 = 2\dot{\varepsilon}_s/3^{(n+1)/2}$. For instance, for $n = 3$, $\dot{\varepsilon}_1 \simeq 0.2\dot{\varepsilon}_s$. This is confirmed experimentally, which proves that the interpretation of the power-law creep equation in terms of second invariants is justified.

By the same token it follows that, if the parameter A used in the power-law creep equation is not that determined in simple shear but that determined in a uniaxial stress experiment (say, A'), then the tensor equations should be multiplied by a factor $3^{(n+1)/2}/2$. Equation 4.24, for instance, becomes

$$\dot{\varepsilon}_{ij} = \frac{3^{(n+1)/2}}{2} \, A' \sigma_E^{(n-1)} \sigma_{ij}' \qquad (4.25)$$

A consequence of a flow law such as Equation 4.24 or 4.25, which could be relevant to the analysis of the rheology of the mantle (see Section 8.3), is that under some stress systems the relation between a component of $\dot{\varepsilon}_{ij}$ and the corresponding component of σ_{ij}' may be apparently linear, even if the material has a non-linear rheology. In experiments on metals exhibiting power-law creep it has been noted that, if for instance a metal wire is subject at the same time to tension (normal

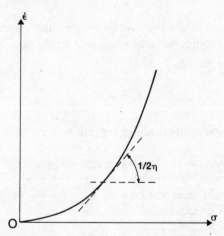

Figure 4.4 Stress–strain rate diagram for power-law rheology. The non-Newtonian viscosity η varies with stress (or strain rate).

stress σ_1) and torsion (shear stress σ_s) and $\sigma_1 \gg \sigma_s$, then the relation between $\dot{\varepsilon}_s$ and σ_s is approximately linear. The reason is that to a first approximation in this case $\sigma'_E \simeq \sigma'_1$ (see Eqn 4.20), and consequently, on the basis of Equation 4.24,

$$\dot{\varepsilon}_s \simeq A\sigma_1'^{(n-1)}\sigma_s$$

Therefore, if a component of stress is sufficiently small when compared with the other components, then its relation with the corresponding component of strain rate is approximately linear. This illustrates one of the possible complications of power-law rheology.

The stress–strain rate diagram for a body with power-law rheology is shown in Figure 4.4, which also illustrates the meaning of viscosity in non-Newtonian media. Formally, *non-linear* (or *non-Newtonian*, or *effective*) *viscosity* is defined in a manner analogous to that of Section 4.2, i.e.

$$\eta = \frac{\sigma_s}{2\dot{\varepsilon}_s} = \frac{\sigma_s}{\dot{\gamma}_s} \tag{4.26}$$

However, while η in Newtonian bodies is not a function of stress, it is a function of stress in non-Newtonian bodies (besides being, of course, a function of temperature, pressure, and material parameters). The slope of the stress–strain rate curve in Figure 4.4 is $1/2\eta$ (or $1/\eta$ if $\dot{\gamma}_s$ is used). The effective viscosity decreases with increasing stress (or strain rate). It is not, therefore, a material coefficient in the same sense as linear viscosity is. For non-Newtonian materials, even when all other parameters (p, T, etc.) are fixed, one cannot speak of *the* viscosity of the material, but only of the viscosity *at a given stress (or strain rate)*. As silicate polycrystals can show either Newtonian or non-Newtonian viscosity according to the pressure and temperature (p, T)-conditions, stress level, and mean grain size (cf. Ch. 10), these considerations are of considerable importance in the discussion of the rheology of the lithosphere and mantle.

Finally, it should be noted that definitions for the viscosity of both Newtonian and non-Newtonian fluids (Eqns 4.8 & 26) are based on *shear* stresses and strain rates. If the viscosity of a material is to be determined from normal stresses and elongation rates (as, for instance, in a uniaxial stress experiment), the numerical factor ($\frac{1}{2}$ in the case of shear components) changes. For a Newtonian material it can be proved from Equation 4.6 or Equation 4.7 that $\eta = \sigma_1/3\dot{\varepsilon}_1$; for a non-Newtonian material the numerical factor depends on the ratio between the elongation rate under uniaxial stress and the shear strain rate in simple shear, which is in turn a function of the stress exponent.

As stated at the beginning of this section, linear creep ($\dot{\varepsilon} \propto \sigma$) and power-law creep ($\dot{\varepsilon} \propto \sigma^n$, $n > 1$) are not the only two types of relation between strain rate and stress in fluids. Other types of rheological equations (i.e. $\dot{\varepsilon} \propto \exp \sigma$, $\dot{\varepsilon} \propto \sinh \sigma$) have been proposed, but they are more apt to describe the behaviour of metals (or sometimes of rocks at high stress, $\sigma \geqslant 100$ MPa) rather than that of rocks at pressures, temperatures, and stress conditions of geodynamic significance.

4.5 Time effects

As we have repeatedly stressed, the rheology of rocks depends on several intrinsic (material) and extrinsic (environmental) parameters. Their behaviour is elastic at low temperature and pressure, as in the upper lithosphere, even under loads of long duration; it is still elastic at high temperature and pressure, as in the bulk of the Earth, provided that the load is of short duration; and it is fluid (Newtonian or non-Newtonian) at high pressure and temperature and under loads of long duration.

Between perfect elasticity and steady-state flow there is usually an intermediate type of behaviour in which there is some time-dependent deformation but the strain rate under constant stress is a decreasing function of time. Figure 4.5 illustrates the strain–time relation in the general case of a polycrystal. When a constant load is applied at time t_0, there is at first *instantaneous elastic deformation* ε_e. This is followed by a stage of *transient creep* ε_t, where the strain increases with time at a decreasing rate. If conditions are right (for instance, if the temperature is sufficiently high) this stage may, in turn, be followed by a stage of *steady-state creep*, where the strain rate $\dot{\varepsilon}$ is constant. (When the load is applied for a sufficiently long time and other conditions are favourable – for example, no

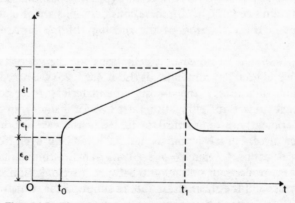

Figure 4.5 Time-dependent deformation under constant stress.

confining pressure or uniaxial stress – fracture eventually occurs, preceded by a stage of accelerating creep.) If the load is removed at time t_1, then there is instantaneous recovery of the elastic strain, followed by time-dependent recovery of the transient strain; the deformation acquired during steady-state creep is permanent. The total creep strain as a function of time under constant stress can therefore be written

$$\varepsilon(t) = \varepsilon_e + \varepsilon_t(t) + \dot{\varepsilon}t \qquad (4.27)$$

where indices have been omitted and each term is to be interpreted as a given component of strain (for instance, elongation in a uniaxial stress experiment).

The last term of Equation 4.27, the steady-state creep strain, is present only if the temperature is sufficiently high (for silicate rocks, $T \gtrsim \frac{1}{2} T_m$, where T_m is the absolute solidus temperature). When this is the case, and for long timescales, this term is much more important than elastic and transient strain combined, and the material can be treated as a viscous fluid (Newtonian or non-Newtonian according to the $\sigma-\dot{\varepsilon}$ relation). When steady-state creep does not occur, and $\dot{\varepsilon}_t$ is a decreasing function of t, the material may be considered to be basically elastic with time-dependent afterworking.

It is worth emphasizing that the occurrence of steady-state creep depends not only on the temperature, but also on the attainment of a sufficiently large total strain or – which amounts to the same thing – the availability of a sufficiently long time. With reference to Figure 4.5, if the process under consideration involves a total strain less than $(\varepsilon_e + \varepsilon_t)$, or a time shorter than that required for the onset of steady-state creep, then the steady-state regime will not be reached. The critical strains and times are functions of intrinsic and extrinsic rheological parameters, and are not well-determined for polycrystals at high temperature and pressure. These considerations may be relevant to the study of the rheology of the mantle from glacio-isostatic rebound, which is usually based on the assumption of steady-state creep, although the relevant strains and times are at the low end of the spectrum (see Sections 8.2 & 3 for a discussion).

The time-dependence of the transient creep function $\varepsilon_t(t)$ varies. Linear rheological models, discussed in Section 4.6, predict a relation of the type

$$\varepsilon_t(t) = A\sigma[1 - \exp(-t/\tau)] \qquad (4.28)$$

where A is some coefficient of proportionality and τ is a material parameter known as the *relaxation time*. In a material exhibiting transient creep according to Equation 4.28, the strain – if steady-state creep does not occur – tends to an asymptotic value as time increases. Other proposed transient creep laws are of the form

$$\varepsilon_t(t) = A\sigma t^{1/r} \qquad r > 1 \tag{4.29a}$$

or

$$\varepsilon_t(t) = A\sigma \ln(1 + at) \tag{4.29b}$$

where A, r, and a are material parameters. In a body following Equation 4.29a or b the transient strain − although increasing at a decreasing rate as time passes − reaches no asymptotic value.

It is difficult to assess from experimental results which transient creep law applies. Although some rocks at low temperature appear to follow a law such as Equation 4.28, experiments more frequently point to a law such as Equation 4.29a or b. The former is derivable from a macroscopic rheological model (Section 4.6), while the latter is primarily a generalization of experimental results, although microphysical models accounting for it are available (see Section 10.1 for a discussion).

Although the logarithmic transient creep law (Equation 4.29b) is observed experimentally in igneous rocks only at room temperature and pressure, it has been claimed (Jeffreys 1976) that a closely related law, the *Modified Lomnitz Law*, is of general applicability to the Earth. The Modified Lomnitz Law is

$$\varepsilon = \frac{\sigma}{\mu} \left(1 + \frac{q}{\alpha} \left[(1 + at)^\alpha - 1 \right] \right) \tag{4.30}$$

where μ, q, a, and α are parameters of the material. (Note that ε denotes the total − i.e. elastic plus transient − strain.)

A law such as Equation 4.30 with $\alpha \simeq 0.25$, or some closely related law, accounts satisfactorily for a range of geophysical phenomena, including damping of seismic waves and of free oscillations, and variations of latitude (see Section 6.5). However, the claim that the Modified Lomnitz Law, being essentially a law of imperfect elasticity with strain rate decreasing with time, precludes the possibility of steady-state flow in the mantle is discounted by most geophysicists, as it goes directly against the theoretical, experimental, and observational evidence, and appears to ignore the fact that the rheological behaviour of rocks varies under different timescales.

4.6 Linear rheological models

In this section we discuss some rheological models which describe phenomenologically the properties of large classes of bodies, including elastic and linearly viscous bodies, and some types of intermediate behaviour. Admittedly, these *analogue models* tell nothing of the microphysics of deformation; however, they make possible a continuum-mechanics description of the rheology. A classical treatment may be found in Reiner (1960).

Since rheological models of this kind are uni-dimensional, we omit indices and refer to shear or deviatoric stress components. Generalization to complex stress states is possible, and leads to the corresponding tensor form of the relevant equations (see Section 8.3 for some examples in the context of mantle rheology).

We consider *linear rheological bodies*, i.e. materials having a rheological equation that can be written as

$$f(D)\sigma = g(D)\varepsilon \tag{4.31}$$

where D denotes differentiation with respect to time and $f(D)$ and $g(D)$ are polynomials in D with coefficients – material parameters – that do not depend on σ and ε.

The simplest linear model is that of an *elastic* (or *Hooke*) *body* in which stress is proportional to strain; i.e. for shear components

$$\sigma = 2\mu\varepsilon \tag{4.32}$$

where μ is the rigidity. The strain–time and stress–strain diagrams for an elastic body were given in Figure 3.1; its analogue model is the spring (Fig. 4.6a).

At the other end of the spectrum of linear rheological bodies there is the *linearly viscous* (or *Newton*) *body* in which stress is proportional to strain rate:

$$\sigma = 2\eta\dot{\varepsilon} \tag{4.33}$$

where η is the linear, or Newtonian, viscosity. The strain–time and stress–strain rate diagrams were given in Figure 4.1. The analogue model of a viscous body is the dashpot (Fig. 4.6b).

We saw in the previous section that in many materials the effects of

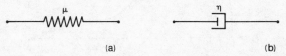

(a) (b)

Figure 4.6 Rheological models for (a) elastic (Hooke) and (b) linearly viscous (Newton) bodies.

loading and unloading are not instantaneous. *Elastic afterworking* – that is, time-dependent strain tending asymptotically to a limiting value – can be modelled by a *firmoviscous (Kelvin) body,* consisting of an elastic and a viscous element disposed in parallel (Fig. 4.7a). Upon loading, the elastic response of the spring is delayed by the viscous response of the dashpot, and the rheological equation is obtained by superposition of Equations 4.32 and 4.33 as

$$\sigma = 2\mu_K \varepsilon + 2\eta_K \dot{\varepsilon} \qquad (4.34)$$

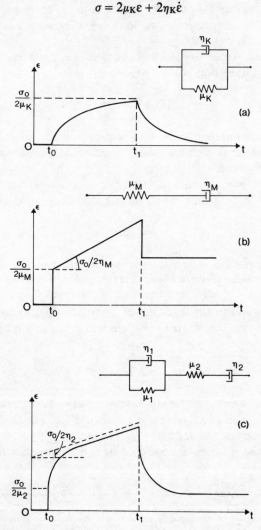

Figure 4.7 Rheological models and strain–time diagrams for (a) firmoviscous (Kelvin), (b) viscoelastic (Maxwell), and (c) general linear (Burgers) bodies.

84

where μ_K and η_K are the Kelvin-rigidity and Kelvin-viscosity, respectively.

In order to analyse the time-dependence of strain in a Kelvin body, Equation 4.34 may be rearranged and multiplied by the integrating factor $\exp(\mu_K t/\eta_K)$ to obtain

$$\dot{\varepsilon} \exp\left(\frac{\mu_K}{\eta_K} t\right) + \frac{\mu_K}{\eta_K} \varepsilon \exp\left(\frac{\mu_K}{\eta_K} t\right) = \frac{\sigma}{2\eta_K} \exp\left(\frac{\mu_K}{\eta_K} t\right)$$

i.e.

$$\frac{D}{Dt}\left[\varepsilon \exp\left(\frac{\mu_K}{\eta_K} t\right)\right] = \frac{\sigma}{2\eta_K} \exp\left(\frac{\mu_K}{\eta_K} t\right)$$

which yields, by integration,

$$\varepsilon = \frac{\sigma}{2\mu_K} \exp\left(-\frac{\mu_K}{\eta_k} t\right)\left[\exp\left(\frac{\mu_K}{\eta_K} t\right) + C'\right]$$

With the initial condition $\varepsilon = 0$ at $t = 0$, and under constant stress σ_0, the constant of integration becomes $C' = -1$ and the above equation reduces to

$$\varepsilon = \frac{\sigma_0}{2\mu_K}\left[1 - \exp\left(-\frac{\mu_K}{\eta_K} t\right)\right] \tag{4.35}$$

which shows that the asymptotic value of the strain is approached exponentially as $t \to \infty$. When the load is removed, $\sigma = 0$ in Equation 4.34, and integration yields

$$\varepsilon = \exp\left(-\frac{\mu_K}{\eta_K} t + C''\right)$$

which becomes, since $\exp C'' = \varepsilon_0$ is the strain when the load is removed,

$$\varepsilon = \varepsilon_0 \exp\left(-\frac{\mu_K}{\eta_K} t\right) \tag{4.36}$$

Hence, the strain goes exponentially to zero after the body is unloaded (Fig. 4.7a). The Kelvin body is therefore essentially a solid showing elastic after-effects. The instantaneous part of the strain, if necessary, can be modelled by a Hooke element in series with the Kelvin element.

The quantity η_K/μ_K, which appears in Equations 4.35 and 4.36, has dimensions $[t]$ and is termed the *Kelvin relaxation time* τ_K. It is the time required for the strain to change by a factor of $\exp(-1)$ under constant load.

To combine elastic and viscous effects, a Hooke element and a Newton element may be combined in series (Fig. 4.7b) to give what is known as a *viscoelastic* or *elasticoviscous* (*Maxwell*) *body*, in which the instantaneous response of the spring and the viscous response of the dashpot are coupled. Differentiating Equation 4.32 with respect to time, and combining it with Equation 4.33, we obtain the rheological equation for a Maxwell body:

$$\dot{\varepsilon} = \frac{\dot{\sigma}}{2\mu_M} + \frac{\sigma}{2\eta_M} \tag{4.37}$$

where μ_M and η_M are the Maxwell-rigidity and Maxwell-viscosity, respectively. Obviously, the total strain under constant stress σ_0 is

$$\varepsilon = \frac{\sigma_0}{2\mu_M} + \frac{\sigma_0}{2\eta_M} t \tag{4.38}$$

so that the instantaneous elastic strain is followed by steady-state linearly viscous deformation. When the load is removed, only the elastic strain is recovered. The strain–time curve of a Maxwell body is shown in Figure 4.7b. It follows from Equations 4.37 and 4.38 that the strain rate in a Maxwell body is linearly related to the stress, and that the elastic part of the deformation can be neglected when it is much smaller than the viscous deformation (as, for instance, when the latter lasts a very long time).

An interesting property of the Maxwell body becomes apparent when the strain is kept constant. For $\dot{\varepsilon} = 0$ in Equation 4.37, the solution for the stress is

$$\sigma = \sigma_0 \exp\left(-\frac{\mu_M}{\eta_M} t\right) \tag{4.39}$$

where σ_0 is the initial stress. Equation 4.39 shows that a Maxwell body exhibits exponential stress relaxation, with *Maxwell relaxation time* $\tau_M = \eta_M/\mu_M$. Consequently, under constant strain (e.g. when laterally confined) the stress in rocks should tend to become hydrostatic over times larger than the relaxation time if they are Maxwell bodies. However, very old stress systems (say, 1 Ga as an order of magnitude) are sometimes present in rocks of the upper lithosphere: the presence of these *residual stresses* shows that the upper lithosphere either has a finite yield strength or a Maxwell time on the geological timescale. (Stresses in the lithosphere are further discussed in Section 8.5.)

A more general type of linear rheology can be obtained by combining a Maxwell element in series with a Kelvin element, resulting in a *general linear (Burgers) body* (Fig. 4.7c). The Burgers body requires four material parameters for its description (μ_1, η_1, μ_2, and η_2; i.e. the rigidities and viscosities of the Kelvin and Maxwell elements, respectively). Its rheological equation can be derived by noting that the total strain is $\varepsilon = \varepsilon_1 + \varepsilon_2$, where ε_1 and ε_2 are the strains related to the Kelvin and Maxwell elements, respectively). Time-differentiation of Equation 4.34, considering that $\varepsilon_1 = \varepsilon - \varepsilon_2$, gives (a double overdot denotes D^2/Dt^2)

$$\dot{\sigma} = 2\mu_1(\dot{\varepsilon} - \dot{\varepsilon}_2) + 2\eta_1(\ddot{\varepsilon} - \ddot{\varepsilon}_2)$$

and, substituting Equation 4.37 for $\dot{\varepsilon}_2$,

$$\dot{\sigma} = 2\mu_1\left(\dot{\varepsilon} - \frac{\dot{\sigma}}{2\mu_2} - \frac{\sigma}{2\eta_2}\right) + 2\eta_1\left(\ddot{\varepsilon} - \frac{\ddot{\sigma}}{2\mu_2} - \frac{\dot{\sigma}}{2\eta_2}\right)$$

Rearranging, we obtain the rheological equation for a Burgers body in its usual form

$$2\eta_1\ddot{\varepsilon} + 2\mu_1\dot{\varepsilon} = \frac{\eta_1}{\mu_2}\ddot{\sigma} + \left(\frac{\eta_1}{\eta_2} + \frac{\mu_1}{\mu_2} + 1\right)\dot{\sigma} + \frac{\mu_1}{\eta_2}\sigma \qquad (4.40)$$

The strain–time diagram for a Burgers body under constant stress σ_0 is shown in Figure 4.7c, and is given by

$$\varepsilon = \frac{\sigma_0}{2\mu_2} + \frac{\sigma_0}{2\mu_1}\left[1 - \exp\left(-\frac{\mu_1}{\eta_1}t\right)\right] + \frac{\sigma_0}{2\eta_2}t \qquad (4.41)$$

When the load is removed, there is instantaneous recovery of the elastic strain followed by time-dependent recovery of the transient creep strain; the viscous contribution to the deformation is permanent.

The Burgers body exhibits instantaneous elastic behaviour, followed by transient creep and damping of vibrations over the short timescale, and by linearly viscous behaviour over the long timescale. It has been proposed that it provides a unifying model for the rheology of the mantle of the Earth over the whole frequency spectrum; this idea will be discussed in Section 8.3.

Finally, it should be mentioned that the terminology for linear rheological bodies is not uniform. *Firmoviscosity* and *elasticoviscosity* are used here in the sense of Jeffreys (1976); the term *viscoelasticity*, that (following common usage) we apply to Maxwellian rheology, is sometimes used in a more general sense (to include general linear bodies). The terminology adopted in this book is that most commonly used in the Earth sciences.

87

5 Strength, fracture, and plasticity

> The rheological behaviour of solids depends on the level of deviatoric stress. When a critical stress is reached, the material fails, in either a brittle or a ductile manner. The critical stress (yield strength) and the mode of failure are functions of intrinsic and extrinsic rheological parameters.
>
> In this chapter we first survey experimental results on rock deformation, and discuss the various empirical criteria for brittle failure. We then examine the phenomenon of ductile failure with attendant plastic flow and present some rheological models for plasticity, together with the theory of plane strain slip-line fields. Both brittle and ductile failure occur in the lithosphere (e.g. earthquakes, faulting, folding, and orogenesis).

5.1 Failure of materials

The two main classes of rheological bodies considered in the previous chapters (elastic and viscous), together with their linear combinations and attendant time-effects, account satisfactorily for the rheological behaviour of the Earth over a very large spectrum of frequencies, from seismic waves to large-scale creep in the mantle. However, there is another important group of phenomena that has not yet been considered – phenomena involving *failure* of materials.

It was pointed out in Section 3.1 that linear elastic behaviour applies only below a critical stress – the elastic limit of the material. Above the elastic limit deformation is at first non-linear, and then, when the (deviatoric) stress reaches a value σ_Y termed the *yield strength* or *yield stress*, the material fails. The yield stress is a function of the material, temperature, pressure, chemical environment, and conditions of loading. As the difference between elastic limit and strength is usually small, it is neglected here. Failure can result either in discontinuous deformation (*fracture*), when the material shows loss of continuity along an approximately well-defined fracture surface, or in irrecoverable continuous deformation (*plastic flow*), when the material yields without any apparent loss of continuity. In both cases only deviatoric stresses can cause failure. Phenomenologically, yielding by flow may be similar to fluid flow, but with the substantial difference that it occurs only when the critical stress is reached, while fluids flow at any stress (in other words, have vanishingly small strength).

As the strength itself is, so the mode of failure is a function not only of the material but also of the physical and chemical environment. Under given conditions a material is termed *brittle* if it fails by fracture when its strength is reached, with very little or no permanent deformation preceding it; it is termed *ductile* if it fails by plastic flow. A stage of non-elastic deformation usually precedes the onset of failure. (However, fracture is not a phenomenon restricted to the brittle field: hot creep fractures, occurring during plastic flow, have been observed experimentally and may have some relevance to deep-focus earthquakes.) Figure 5.1 shows schematically the stress–strain curves for brittle and ductile failure. It also shows *strain* (or *work*) *hardening*, i.e. an increase in yield stress with increasing deformation. Strain hardening disappears at high temperature, in which case the body can sustain no stress larger than the yield stress.

Processes of fracture and plastic flow play an important role in the mechanics of the upper lithosphere (earthquakes, faulting, and folding). At temperatures $T \gtrsim \frac{1}{2} T_m$, i.e. below the upper lithosphere, silicate polycrystals have vanishing yield stress over the long timescale. In this chapter we survey the phenomenology of failure, and some characteristics of plastic flow. Microscopic theories of brittle fracture are mentioned only briefly; although they are of the greatest importance in rock mechanics, their role in the study of the rheology of the Earth is relatively minor. The books by Paterson (1978) and Jaeger & Cook (1979) cover the brittle field exhaustively. First we give a brief account of the experimental results on rock deformation.

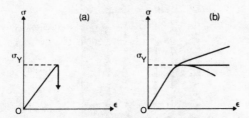

Figure 5.1 Stress–strain curves for (a) brittle failure and (b) ductile failure (σ_Y is the yield strength). In theory, the stress goes to zero at brittle failure (downward arrow in (a)); in ductile flow the stress is ideally constant with increasing strain, but it can either increase (strain hardening) or decrease (strain softening), as shown in (b).

5.2 Experimental rock deformation and the brittle–ductile transition

An account of laboratory techniques and procedures for rock deformation tests is given by Paterson (1978). The geologically most significant experiment is the so-called *triaxial test,* in which a cylindrical rock specimen is subjected to an axial compression σ_1 and to a lateral confining pressure $p = \sigma_3$, while being kept, if desired, at a temperature T, greater than the room temperature (Fig. 5.2). As is usual in the experimental literature, compressive stress is considered positive, and $\sigma_1 > \sigma_3$. The results can conveniently be plotted as a stress–strain diagram, with the stress difference $\sigma_1 - \sigma_3$ on the vertical axis, and strain as percentage axial shortening on the horizontal axis.

Figure 5.3 shows schematically how the strength and the mode of failure are affected by temperature and confining pressure. Increasing temperature promotes ductility and decreases the strength; increasing pressure promotes ductility but increases the strength. Table 5.1 gives a synopsis of the results on the strength of various rocks at various (p, T)-conditions. Results can also be presented by plotting strength as a function of temperature or pressure.

The various modes of failure are depicted in Figure 5.4. Extension fracture and axial splitting fracture require tensional axial stress and no confining pressure, respectively, and are therefore of relatively minor importance in geodynamics. Under triaxial stress the most common mode of failure in the brittle field is *shear fracture*, in either one or two conjugate planes forming an angle of less than 45° with the direction of max-

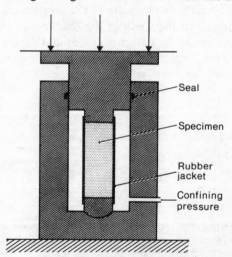

Figure 5.2 Triaxial test deformation apparatus (from Paterson 1978).

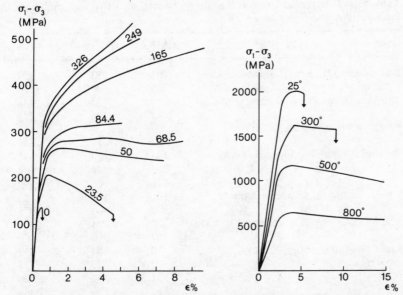

Figure 5.3 Effects of pressure (left) and temperature (right) on the stress–strain behaviour of rocks. Numbers on curves give confining pressure (in MPa) and temperature (in °C), respectively (from Jaeger & Cook 1979).

imum compression. As pressure (or temperature, or both) is increased, the relative importance of continuous deformation – as opposed to well-defined planes of fracture – increases until, fully in the ductile field, the specimen shows (at least macroscopically) no recognizable loss of cohesion.

Pressure and temperature are not the only parameters that affect the strength and mode of failure of a rock. If experiments are carried out at a constant rate of deformation in the ductile field, then it becomes apparent that strength decreases with decreasing strain rate. The presence of water, as will be seen in the sequel, also has an important effect on strength.

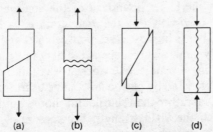

Figure 5.4 Modes of brittle failure: (a) extensional shear fracture, (b) extension fracture, (c) compressional shear fracture, and (d) axial splitting fracture (from Paterson 1978).

91

Table 5.1 Differential stress at fracture in triaxial tests (from compilations by Handin 1966 and Paterson 1978).

Rock	p(MPa)	T (°C)	$\sigma_1 - \sigma_3$ (MPa)
amphibolite (Hudson Highland	101	150	659
Complex, New York, USA)	505	500	1350
anhydrite (Blaine, Oklahoma, USA)	14	24	188
	47	24	275
basalt (Knippa, Texas, USA)	69	24	462
	103	24	551
dolomite (Blair, W. Virginia, USA)	101	24	570
	202	24	694
	275	24	990
gneiss, biotite (St Lawrence County,	101	150	703
New York, USA)	505	500	1075
granite (Westerly, Rhode Island, USA)	507	24	2080
	507	300	1660
limestone (Solenhofen, Bavaria,	30	24	353
Germany)	77	24	474
	81	150	420
	113	150	380
peridotite (Dun Mt, New Zealand)	507	300	1545
	507	500	1080
	507	800	850
quartzite (Eureka, Nevada, USA)	507	500	1690
sandstone (Oil Creek, Texas, USA)	101	150	736
	101	300	691
	202	150	990
	202	300	912
schist, hornblende (New York, USA)	100	24	640
shale (Muddy, Colorado, USA)	100	150	201
	100	300	147
slate (Mettawee, Vermont, USA)	101	150	240
	101	300	275
	202	150	437
	202	300	432

The pressure at the lower boundary of the continental crust (about 35 km of depth) is $p \simeq 1$ GPa; the temperature varies according to tectonic province, but is in most cases between 600 and 1200 K. A question of great geodynamic significance is the depth of the *brittle– ductile transition,* separating a zone above where rocks fail by fracture from a zone below where they fail by plastic flow. (The brittle–ductile transition is, of course, gradual rather than sharp, and should not be confused with the depth at which the yield strength becomes zero, which separates 'elastic' – more accurately, elastic–brittle overlying elastic–plastic – lithosphere above from 'viscous' substratum below.)

The brittle–ductile transition for various rocks can be studied experimentally by seeing how the *fracture stress* (i.e. the stress at failure in the brittle regime) and the *yield stress* (the strength in the ductile regime) vary with pressure and temperature. The picture that emerges is depicted schematically in Figure 5.5 (see Paterson 1978, for a more detailed discussion of the probable mechanisms involved). The effects of pressure and temperature are illustrated separately in Figures 5.5a and b (curves are only indicative and are not to be taken as expressing functional relationships.) The effect of pressure on the fracture stress is much stronger than on the yield stress, while the effect of temperature is much stronger on the yield stress. The mode of failure can be predicted on the basis of the combined (p, T)-effects on fracture stress and yield stress. If fracture stress is less than yield stress (as at low p and T), then it will be the first to be reached as $\sigma_1 - \sigma_3$ increases, and failure will consequently be brittle. If yield stress is less than fracture stress (as at high p and T), then failure will be by ductile flow. The effects of pressure and temperature are combined in Figure 5.5c, which is a plot of the mode of failure in the (p, T)-field. By extrapolating experimental results it is possible to estimate the trend of fracture stress and yield stress with depth, and consequently the probable depth of the brittle–ductile transition, which has an important bearing on the rheology of the upper lithosphere and on the depth distribution of earthquakes (see Section 12.1).

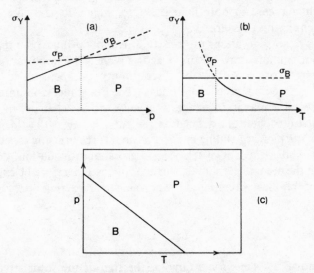

Figure 5.5 Variations of fracture stress (σ_B) and ductile yield stress (σ_P) with (a) pressure and (b) temperature. The combined effects of temperature and pressure are shown in (c): P denotes ductile failure, B brittle fracture.

Another result that emerges from experimental work is the importance of *water* in rock deformation. We refer here to the presence of water occupying the pore space of a rock – the possible effects of lattice or impurity water on the rheology of silicate minerals will be discussed in Section 10.5. The presence of an interconnected fluid phase (liquid or gaseous) decreases the fracture stress of a rock. The decrease is, to a first approximation, linearly proportional to the *pore fluid pressure*, i.e. the hydrostatic pressure exerted by the fluid on the walls of the pores. As discussed in Section 5.4, this observation can be incorporated in phenomenological theories of failure by assuming that pore fluid pressure acts to reduce normal stress.

A related effect is the so-called *high-temperature embrittlement*, that is, a transition from ductile to brittle behaviour at high temperature, accompanied by a fall in strength. This transition is in the sense opposite to that observed under normal conditions, and occurs in rocks containing large amounts of hydrated minerals. A typical example is that of serpentinite (see, for example, Paterson 1978). At confining pressure of a few hundred megapascals serpentinites are ductile up to a temperature $T \simeq 800$ K, above which there is a fall in strength by a factor of about three (from more than 600 MPa to 200 MPa or less), accompanied by a switch to a brittle mode of failure. The embrittlement temperature coincides with the attainment of the critical temperature for the dehydration reaction of serpentine, producing forsterite, talc, and water. The transition is therefore attributed to the release of water vapour and the consequent increase in pore fluid pressure, which acts to decrease the effective confining pressure.

The above considerations on rock failure apply primarily to materials that do not contain a pre-existing fracture plane. The important subject of *friction* across a surface of discontinuity can be studied experimentally by loading precut specimens. The resistance to sliding at a macroscopically planar interface is usually expressed in terms of the *coefficient of friction* μ_0, defined as the ratio between the shear force parallel to the plane of sliding and the normal force pressing together the two sides. The coefficient of friction, to a first approximation, does not depend on the area of the contact surface, and can therefore be expressed in terms of the shear stress τ and the normal stress σ acting on the surface of contact, as

$$| \tau | = \mu_0 \sigma \tag{5.1}$$

This is known as *Amonton's Law* (as the sign of the shear stress does not affect friction, its absolute value is used; $\sigma > 0$ since it is compressive). An immediate consequence of Amonton's Law, well-verified experimentally, is that the shear stress required for sliding increases with increasing normal stress. It should be pointed out that the shear stress to initiate sliding is not necessarily the same as the shear stress to maintain sliding, and that a distinction ought therefore to be made, in principle, between *static* and *dynamic* friction. However, differences are not very large; the coefficient of friction for most rocks is usually in the range $0.5 < \mu_0 < 0.8$.

When experimental results are plotted on σ, τ-co-ordinates, it is often seen that the straight lines of slope μ_0 representing the frictional law are better expressed by an equation of the type

$$|\tau| = S_0 + \mu_0\sigma \qquad (5.2)$$

Therefore, even in the absence of normal stress, a finite shear stress S_0 is required for sliding. The parameter S_0 is termed the *cohesion* of the contact surface, and for most rocks is of the order of 1–10 MPa.

Friction is a consequence of the presence of asperities (such as interlocking grains) on the surface along which sliding takes place. The coefficient of friction and the cohesion are therefore affected by surface conditions, and by the wear and gouge formation resulting from the sliding process itself (*gouge* is the fine-grained detrital material formed at the interface). In particular, occurrence of hydrated platy minerals (clays) in the gouge decreases friction considerably. The presence of water, and consequently of fluid pressure, along the surface of sliding decreases the effective normal stress across the surface and therefore the shear stress required for sliding. This weakening effect of fluid pressure plays a role in the faulting process (see Section 8.6).

More comprehensive treatments of experimental results can be found in the previously mentioned books by Paterson (1978) and Jaeger & Cook (1979). However, the brief survey in this section is sufficient for a consideration of the phenomenological theories of brittle failure of a homogeneous, isotropic body, which will be treated next.

5.3 Empirical failure criteria for shear fracture

Brittle or semibrittle shear fracture under a triaxial compressive stress system is the most commonly occurring type of failure in the upper lithosphere. An *empirical failure criterion* is a rule that allows prediction of the occurrence of fracture under given conditions, without involving an analysis of the underlying microscopic processes. The situation in the study of rock fracture is different from the one prevailing in the case of solid-state flow (see Chs 9 & 10), where a vast physical and metallurgical

95

literature provides microphysical creep theories applicable to crystalline Earth materials. Although physical theories of brittle fracture exist (e.g. the Griffith Theory, mentioned briefly later), empirical failure criteria are usually sufficient to account for at least those characteristics of fracture which are of geodynamic significance.

The simplest empirical failure condition is provided by *Tresca's criterion* (which was originally proposed for the onset of plastic flow, but is equally applicable to brittle shear fracture), stating that fracture occurs when the maximum shear stress reaches the shear strength σ_Y of the material

$$|\tau_{max}| = \tfrac{1}{2}(\sigma_1 - \sigma_3) = \sigma_Y \tag{5.3}$$

The strength, of course, depends not only on the material but also on extrinsic conditions.

According to Equation 5.3, neither the intermediate principal stress nor the hydrostatic pressure affects strength. Moreover, Tresca's criterion predicts that shear fracture should occur along either (or both) of two planes containing the intermediate principal stress axis (see Section 2.5) and bisecting the directions of maximum and minimum principal stress, whereas experiments show that the angle of the plane of fracture to the σ_1-direction is usually less than $\pi/4$.

A more general condition for shear fracture is *Mohr's criterion,* which states that fracture occurs across the plane (or planes) where the shear stress first reaches a value that depends on material parameters and on the normal stress on that plane

$$|\tau| = f(\sigma) \tag{5.4}$$

The explicit form of Equation 5.4 can be determined experimentally, being the envelope of Mohr circles at failure. It often turns out that the envelope is approximately linear (Fig. 5.6), and consequently Mohr's criterion can be written explicitly in the form

$$|\tau| = S + \mu' \sigma \tag{5.5}$$

where S and μ' are parameters of the material termed *cohesive strength* and *internal friction*, respectively. The condition expressed by Equation 5.5 is usually referred to as the *Coulomb–Navier criterion*. It is represented in stress space by two straight lines intersecting the τ-axis at $\pm S$, with slopes $\pm \mu' = \tan \phi$, where ϕ is the *angle of internal friction* (in most rocks, $\phi \simeq 30°$). The lines are the envelope of all critical Mohr circles. The orientation of the fracture planes can be determined

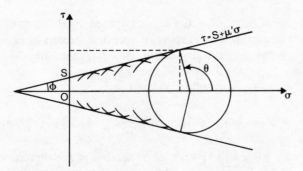

Figure 5.6 Coulomb–Navier criterion for shear fracture.

graphically: on the basis of the considerations developed in Section 2.5, their normals make an angle of $[(\pi/4) + (\phi/2)]$ to the direction of maximum compressive stress.

Although the form of Equation 5.5 is identical to the frictional law (Eqn 5.2), its physical meaning is not the same. The law of friction refers to sliding along a pre-existing surface, while the Coulomb–Navier criterion refers to the formation of a fracture plane. The material parameters S and μ' are not therefore necessarily the same as the frictional parameters S_0 and μ_0. (While the values of μ' and μ_0 are in the same range, $S > S_0$ by one order of magnitude or more.) The cohesive strength may be regarded as the inherent strength of the material, or the shear stress required for fracture when $\sigma = 0$. The coefficient of internal friction is an empirical measure of the increase in fracture strength brought about by the increase in normal stress. We shall see that it can be assigned a physical meaning on the basis of the Griffith Theory.

The Coulomb–Navier criterion, despite its limitations (to mention just one, the internal friction is seen in practice to be a decreasing function of normal stress, and therefore the envelope is slightly curved with concavity toward the σ-axis), gives an empirically satisfactory account of brittle shear failure of rocks in the laboratory and in the field. Equation 5.5 may be formulated in several equivalent ways. One that is particularly useful, because it gives some insight into the physical meaning of internal friction, is in terms of the maximum and minimum principal stresses. The normal and shear stress on a plane with normal n_i forming an angle θ with the σ_1-axis are obtainable immediately from Equations 2.23, i.e.

$$\sigma = \tfrac{1}{2}(\sigma_1 + \sigma_3) + \tfrac{1}{2}(\sigma_1 - \sigma_3) \cos 2\theta$$

$$\tau = -\tfrac{1}{2}(\sigma_1 - \sigma_3) \sin 2\theta$$

where, in compliance with the notation used in this book, we have reverted to the convention of taking compressive stresses as negative, and therefore σ_1 is the algebraically largest principal stress, i.e. the *minimum* compression. Equation 5.5 should then be written as $|\tau| = S - \mu'\sigma$, and the quantity $|\tau| + \mu'\sigma$ can be expressed as

$$|\tau| + \mu'\sigma = \tfrac{1}{2}(\sigma_1 - \sigma_3)(\sin 2\theta + \mu'\cos 2\theta) + \tfrac{1}{2}\mu'(\sigma_1 + \sigma_3) \quad (5.6)$$

The maximum value of $|\tau| + \mu'\sigma$ as a function of θ is therefore attained for $\tan 2\theta = 1/\mu'$. Expressing $\sin 2\theta$ and $\cos 2\theta$ in terms of this maximum value, Equation 5.6 becomes

$$(|\tau| + \mu'\sigma)_{max} = \tfrac{1}{2}(\sigma_1 - \sigma_3)(\mu'^2 + 1)^{1/2} + \tfrac{1}{2}\mu'(\sigma_1 + \sigma_3) \quad (5.7)$$

By Equations 5.5 and 5.7, therefore, fracture will occur when

$$\sigma_1[(\mu'^2 + 1)^{1/2} + \mu'] - \sigma_3[(\mu'^2 + 1)^{1/2} - \mu'] = 2S \quad (5.8)$$

Equation 5.8 gives the Coulomb–Navier criterion in terms of principal stresses, and shows explicitly how the existence of internal friction ($\mu' \neq 0$) modifies the simpler Tresca criterion (Eqn 5.3). An equation of the same form can be derived on the basis of the microscopic theory for brittle fracture known as the *Griffith Crack Theory* (for its complete derivation see, for example, Jaeger & Cook 1979). The Griffith Crack Theory postulates the existence in the material of *microcracks*, considered as ellipsoidal cavities with very large aspect ratios, which generate stress concentrations at their tips, where the *local* stress becomes much larger than the applied stress. (Microcracks of all sizes and shapes are known to exist in brittle polycrystals.) The stress concentrations at the tips of Griffith cracks favourably oriented with respect to the applied stress cause them to propagate when the critical stress is reached locally, and macroscopic shear fracture occurs as a result of this propagation and the coalescence of several cracks. The stress components at the crack tips can be calculated by solving the relevant elastic problem. Assuming that under compressive states of stress the cracks are closed, and taking into account the friction between crack walls, an expression identical to Equation 5.8 is found, where cohesive strength is given by twice the uniaxial tensile strength of the material, and internal friction is replaced by the coefficient of sliding friction between closed crack walls. Thus, the interpretation of the empirical Coulomb–Navier criterion in terms of closed Griffith cracks provides a physical explanation for the coefficient of internal friction.

5.4 The role of pore fluid pressure

Although several other developments in the theory of brittle fracture could be considered, it is more important to turn our attention to the role of pore fluids (water, melt, oil, and gaseous phases) in the shear fracture of materials. Pore fluid pressure, as mentioned in Section 5.2, affects the discontinuous deformation of the upper lithosphere in processes such as overthrusting and earthquakes.

A fluid-saturated porous medium consists of a solid matrix and an interconnected fluid phase. This system can be treated as a continuum if the dimensions of the elemental volume over which the rheological quantities are defined are sufficiently large compared with the average pore dimensions. The theory of the elastic behaviour of such a continuum results in a generalization of Hooke's Law, incorporating the pore fluid pressure p_f (see, for example, Jaeger & Cook 1979). An *effective stress* σ_{ij}^e is defined as

$$\sigma_{ij}^e = \sigma_{ij} - p_f\,\delta_{ij} \tag{5.9}$$

This effective stress controls the rheological behaviour of the porous medium.

The notion of effective stress can be applied to both the initiation of fracture and the sliding along a pre-existing surface. In the first case, for instance, it can be proven that, if Griffith cracks are filled with a fluid with pore pressure p_f, then Equation 5.8 applies provided that the effective stress (Eqn 5.9) is used. Consequently, the Coulomb–Navier criterion (Eqn 5.5) can be written in terms of effective stress as

$$|\tau| = S + \mu'(\sigma - p_f) \tag{5.10}$$

where we have reverted to the customary 'experimental' notation for stress. In the second case the law of sliding friction (Eqn 5.2) may be similarly modified. In both cases the pore pressure acts against the 'solid' normal stress, effectively reducing the resistance to shear fracture and the resistance to sliding.

Thus, the occurrence of pore pressure increases the likelihood of fracture and sliding, because it reduces the shear stress required to produce them. Failure can occur not only because a steadily increasing shear stress reaches a critical value, but (perhaps more commonly in geodynamics) because this critical value decreases as a consequence of an increase in pore fluid pressure.

5.5   Criteria for ductile flow and rheological models for plasticity

If the intrinsic and extrinsic rheological parameters are such that the material behaviour is characterized by ductility rather than brittleness, then yielding is by plastic (ductile) flow. The simplest criterion for plastic yielding is Tresca's (Eqn 5.3): the material flows when the maximum stress difference reaches a critical value depending on the rheological parameters.

A more general criterion has been proposed by von Mises and Hencky (see, for example, Reiner 1960, Jaeger & Cook 1979). Plastic yielding is independent of the co-ordinate system and therefore, in the most general case, is a function of the invariants of the stress tensor (defined in Section 2.4), i.e.

$$f(I_1, I_2, I_3) = 0$$

For ductile materials, yielding is relatively unaffected by hydrostatic pressure, and therefore the invariants of the stress deviator can be used; as I_1' is identically zero, this reduces the number of invariants in the yield criterion to two. Furthermore, if there is no Bauschinger effect (that is, the yield strength is the same for compressive and tensile states of stress – a reasonably accurate assumption for ductile polycrystals), then the yield criterion must be an even function of I_3'. The *Mises–Hencky criterion* assumes that plastic yielding depends only on the second invariant of the stress deviator

$$f(I_2') = 0 \qquad\qquad (5.11)$$

that is, yielding occurs when I_2' reaches a value characteristic of the material. This is equivalent to stating that yielding occurs when the *deviatoric strain energy* (per unit volume) reaches a critical value (see Section 3.3). The body can be thought of as a system that can store deviatoric strain energy only up to a given maximum amount. If the actual energy is less than the critical value, then deformation is recoverable and the body behaves elastically; if the critical value is reached, then the body yields and exhibits permanent plastic deformation. (A good analogy is that of the body as a vessel, and of deviatoric strain energy as a liquid with which the vessel is filled.)

Recalling the expression for I_2' derived in Section 4.4, we can rewrite the Mises–Hencky criterion (Eqn 5.11) as

$$2I_2' = \sigma_{ij}'\sigma_{ij}' = 2k^2 \qquad\qquad (5.12)$$

where k is a parameter of the material (not to be confused with bulk modulus). The quantity $\sigma_{ij}'\sigma_{ij}'$ is related directly to the deviatoric strain energy (Eqn 3.22, Section 3.3). The Mises–Hencky criterion may also be expressed in terms of stress differences, since

$$\sigma_{ij}'\sigma_{ij}' = \sigma_{ij}\sigma_{ij} - \tfrac{1}{3}\sigma_{kk}\sigma_{kk}$$

and therefore Equation 5.12 becomes

$$(\sigma_{11} - \sigma_{22})^2 + (\sigma_{22} - \sigma_{33})^2 + (\sigma_{11} - \sigma_{33})^2 + 6(\sigma_{12}^2 + \sigma_{23}^2 + \sigma_{13}^2) = 6k^2$$

$$(5.13\text{a})$$

or, in terms of principal stresses,

$$(\sigma_1 - \sigma_2)^2 + (\sigma_2 - \sigma_3)^2 + (\sigma_1 - \sigma_3)^2 = 6k^2 \qquad (5.13\text{b})$$

The Mises–Hencky criterion is not too different from Tresca's; however, yielding is also affected by the magnitude of the intermediate principal stress. This seems to fit the available experimental results better.

Yield criteria such as those of Tresca and of Mises–Hencky specify the stress conditions required for plastic flow and, if there is no strain hardening (i.e. no increase in yield stress with deformation), prescribe the relations among stress components during flow. The rheological behaviour of a body without strain hardening is termed *ideal* or *pure plasticity*. More precisely, an *elastic–plastic* body is a material exhibiting linear elasticity for $\sigma < \sigma_Y$, and pure plasticity for $\sigma = \sigma_Y$, where σ_Y is the yield strength, whose relation with k can be derived from the yield criterion. Its rheological equations are

$$\sigma_{ij}' = 2\mu\varepsilon_{ij}' \qquad \sigma < \sigma_Y$$

$$f(\sigma_{ij}') = \sigma_Y \qquad \sigma = \sigma_Y \qquad (5.14)$$

Such a body is sometimes referred to as a *Saint-Venant body*, and its rheological analogue is shown in Figure 5.7a. The shaded box symbolizes the yield stress element σ_Y. If the deviatoric stress is less than the strength, then only recoverable elastic deformation of the elastic element with rigidity μ occurs; at $\sigma = \sigma_Y$, the strength element yields and unrecoverable plastic flow ensues. On a stress–strain rate diagram, this type of behaviour is represented by the curve $O\sigma_Y V$ of Figure 5.7c: as $\dot{\varepsilon} = 0$ for any stress in the elastic range, there is no flow until $\sigma = \sigma_Y$, then indefinite flow occurs at any strain rate. If elastic strain is negligible compared with plastic strain, then the rheological behaviour can be approximated by a *rigid–plastic* model (no deformation for $\sigma < \sigma_Y$; plastic

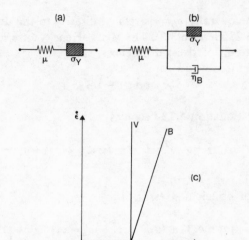

Figure 5.7 Rheological models for (a) elastic–plastic (Saint-Venant) and (b) viscoplastic (Bingham) bodies. Their stress–strain rate curves in (c) are denoted by $0\sigma_Y V$ and $0\sigma_Y B$, respectively.

deformation for $\sigma = \sigma_Y$). The analogue for a rigid–plastic body is simply a strength element (as in Fig. 5.7a, but without the elastic element), and the stress–strain rate diagram is the same as for elastoplasticity.

Another possible type of plastic behaviour is modelled by the *viscoplastic* (or *plasticoviscous*, or *Bingham*) *body*, the rheological properties of which are described by the equations

$$\sigma'_{ij} = 2\mu\varepsilon'_{ij} \qquad \sigma < \sigma_Y$$

$$\sigma'_{ij} = \sigma_Y + 2\eta_B\dot{\varepsilon}'_{ij} \qquad \sigma \geqslant \sigma_Y$$

(5.15)

The Bingham body behaves elastically at stresses lower than the strength, and flows linearly above the strength with a strain rate proportional to $(\sigma'_{ij} - \sigma_Y)$. Its analogue model, consisting of an elastic element μ in series with a viscoplastic element (σ_Y, η_B), is shown in Figure 5.7b: viscoplastic flow occurs only when the strength is exceeded. The stress–strain rate diagram is shown by the curve $O\sigma_Y B$ in Figure 5.7c. Viscoplasticity is a characteristic of several paints and polymers, but does not seem to occur commonly in crystalline Earth materials. The 'plastic viscosity' η_B is a true parameter of the material (i.e. it does not depend on σ or $\dot{\varepsilon}$), and is given by the inverse of the slope of the segment $\sigma_Y B$ in Figure 5.7c. However, the 'apparent viscosity', as determined from the ratio $\sigma/2\dot{\varepsilon}$ at different stresses (or strain rates), varies. The presence of a finite yield stress can therefore, in principle, be detected by the variation of apparent viscosity under different loads.

In metals the rheological behaviour is often elastic below the strength and shows non-linear flow for $\sigma \geqslant \sigma_Y$. This type of plastic flow is known as *Andradean* creep. If the strength is vanishingly small, Andradean creep reduces to non-linear viscosity (cf. Section 4.4), which is a common high-temperature rheology for polycrystalline Earth materials.

5.6 Plane plastic strain

A class of problems in plasticity that has interesting applications to geodynamics (see, for example, Section 8.7) deals with plastic bodies deforming in plane strain (i.e. essentially in two dimensions; see below for a precise definition).

Consider an isotropic, homogeneous, incompressible material, rigid (or elastic) where the stress is below the yield point, and plastic, with the stress state specified by some yield criterion such as those of Tresca or of Mises–Hencky, where the critical value for the stress is reached. The governing equations are the equations of equilibrium

$$\frac{\partial \sigma_{ij}}{\partial x_j} = 0$$

(where the body force is neglected because gravity does not affect the plastic solution), the continuity equation for an incompressible material

$$\frac{\partial v_i}{\partial x_i} = 0$$

and the yield criterion (say, that of Mises–Hencky, Eqn 5.12) which must be satisfied in the plastically deforming zones. The relations between strain increments and deviatoric stresses are expressible by the *Lévy–Mises equations* (see, for example, Hill 1950)

$$d\varepsilon_{ij} = d\lambda \, \sigma'_{ij} \tag{5.16}$$

where $d\lambda$ is a scalar proportionality factor which depends on position and time. Equation 5.16 can be rewritten as

$$\dot{\varepsilon}_{ij} = \tfrac{1}{2} \left(\frac{\partial v_i}{\partial x_j} + \frac{\partial v_j}{\partial x_i} \right) = \dot{\lambda} \sigma'_{ij} \tag{5.17}$$

Equations 5.16 and 5.17 imply that the principal axes of strain increment and stress coincide.

A state of *plane strain* is defined by two conditions: the flow is everywhere parallel to a given plane (e.g. x_1, x_2), and the flow is independent of the third axis (x_3). This situation applies, for instance, to the deformation of geologic structures when most or all flow occurs in the plane perpendicular to the strike of the structure. Therefore, in plane strain $v_3 = 0$, $\sigma_{13} = \sigma_{23} = 0$, and consequently σ_{33} is a principal stress (the other two lie in the x_1, x_2-plane). Furthermore, as $\dot{\varepsilon}_{33} = \partial v_3/\partial x_3 = 0$, it follows by Equation 5.17 that $\sigma'_{33} = 0$ and, recalling the definition of deviator, that the stress component σ_{33} is the intermediate principal stress, i.e.

$$\sigma_{33} = \tfrac{1}{2}(\sigma_{11} + \sigma_{22}) \tag{5.18}$$

In plane strain the equations of equilibrium, continuity, and the Mises–Hencky yield criterion become, respectively,

$$\frac{\partial \sigma_{11}}{\partial x_1} + \frac{\partial \sigma_{12}}{\partial x_2} = 0 \qquad \frac{\partial \sigma_{12}}{\partial x_1} + \frac{\partial \sigma_{22}}{\partial x_2} = 0 \tag{5.19}$$

$$\frac{\partial v_1}{\partial x_1} + \frac{\partial v_2}{\partial x_2} = 0 \tag{5.20}$$

$$\tfrac{1}{4}(\sigma_{11} - \sigma_{22})^2 + \sigma_{12}^2 = k^2 \tag{5.21}$$

The angle θ between the x_1-axis and the algebraically largest principal stress is given by (Eqn 2.24)

$$\tan 2\theta = \frac{2\sigma_{12}}{\sigma_{11} - \sigma_{22}}$$

Similarly, the angle θ' between the x_1-axis and the largest strain increment is

$$\tan 2\theta' = \frac{(\partial v_1/\partial x_2) + (\partial v_2/\partial x_1)}{(\partial v_1/\partial x_1) - (\partial v_2/\partial x_2)}$$

and, as $\theta \equiv \theta'$ by the Lévy–Mises equations,

$$\frac{2\sigma_{12}}{\sigma_{11} - \sigma_{22}} = \frac{(\partial v_1/\partial x_2) + (\partial v_2/\partial x_1)}{(\partial v_1/\partial x_1) - (\partial v_2/\partial x_2)} \tag{5.22}$$

Equations 5.19–5.22 are the basis for the calculation of the stress and velocity fields in regions undergoing plane plastic strain. Boundary conditions may involve only stresses, in which case the state of stress in the plastic region can usually be determined without considering the velocities; or may involve both stresses and velocities, in which case solutions for stresses and velocities have to be obtained together. Since Equations 5.20 and 5.22 are homogeneous in v_i, the stresses are independent of the rate of strain; i.e. time does not enter explicitly the plastic solutions.

Outside the plastic region, where stresses are below the yield point, the yield criterion is replaced by the equation of compatibility (see Section 3.4)

$$\nabla^2(\sigma_{11} + \sigma_{22}) = 0$$

At the rigid–plastic (or elastic–plastic) interface the stress components must, of course, satisfy conditions of equilibrium and continuity.

5.7 Stress and velocity characteristics; slip-line fields

The solution to plastic plane-strain problems is obtained in terms of stress and velocity characteristics. Without going into the full details of the theory of hyperbolic differential equations, we develop here the elements needed to understand the significance and properties of characteristics. For a fuller treatment, refer to Hill (1950).

First we examine the stress equations. From Equation 5.18 and the fact that σ_{kk} is invariant, it follows that

$$\sigma_2 = \tfrac{1}{2}(\sigma_1 + \sigma_3) \tag{5.23}$$

where σ_1 and σ_3, the maximum and minimum principal stresses, lie in the (x_1, x_2)-plane. When the relation between principal stresses is expressed by Equation 5.23, the Mises–Hencky criterion (in this case equivalent to Tresca's) and the mean pressure are, respectively,

$$k = \tfrac{1}{2}(\sigma_1 - \sigma_3) \qquad p = -\tfrac{1}{2}(\sigma_1 + \sigma_3) \tag{5.24}$$

Recalling the transformation equations for the stress tensor given in Section 2.5, and making use of Equations 5.24, we can write for the stress components in the (x_1, x_2)-plane

$$\sigma_{11} = \tfrac{1}{2}(\sigma_1 + \sigma_3) + \tfrac{1}{2}(\sigma_1 - \sigma_3)\cos 2\theta = -p + k \cos 2\theta$$

105

$$\sigma_{22} = \tfrac{1}{2}(\sigma_1 + \sigma_3) - \tfrac{1}{2}(\sigma_1 - \sigma_3)\cos 2\theta = -p - k \cos 2\theta$$

$$\sigma_{12} = \tfrac{1}{2}(\sigma_1 - \sigma_3)\sin 2\theta = k \sin 2\theta$$

where θ is the anticlockwise angle beween the x_1-axis and the direction of the maximum principal stress σ_1 (see Fig. 5.8a). Noting that $\theta = \phi + \pi/4$, where ϕ is the angle between the x_1-axis and the direction of maximum shear stress, we can express the stress components as

$$\sigma_{11} = -p - k \sin 2\phi$$

$$\sigma_{22} = -p + k \sin 2\phi \qquad (5.25)$$

$$\sigma_{12} = k \cos 2\phi$$

Equations 5.25 express the stress components in plane plastic strain as a function of the average compressive stress, the yield criterion, and the orientation of the trajectories of maximum shear stress (which make an angle $\pm \pi/4$ with the σ_1-direction). The problem then becomes that of determining the orientation of these trajectories within the plastic zone. A *stress characteristic* in a plastic plane-strain problem is a curve whose tangent at every point is a direction of maximum shear stress. There are two families of characteristics, which are generally curvilinear and always form an orthogonal network: those making an angle of $-\pi/4$ with the σ_1-direction at every point (α-*lines*), and those making an angle of $+\pi/4$ with the σ_1-direction (β-*lines*)(see Fig. 5.8a).

We now write the equilibrium condition (Eqn 5.19) in terms of Equa-

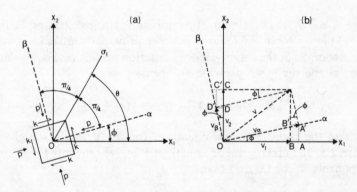

Figure 5.8 (a) Stress and (b) velocity characteristics in plane plastic strain.

106

tions 5.25, and obtain

$$-\frac{\partial p}{\partial x_1} - 2k \cos 2\phi \frac{\partial \phi}{\partial x_1} - 2k \sin 2\phi \frac{\partial \phi}{\partial x_2} = 0$$

$$(5.26)$$

$$-\frac{\partial p}{\partial x_2} - 2k \sin 2\phi \frac{\partial \phi}{\partial x_1} + 2k \cos 2\phi \frac{\partial \phi}{\partial x_2} = 0$$

It is proved in the theory of partial differential equations that Equations 5.26 are hyperbolic; their characteristic equation has roots $\tan \phi$ and $\tan(\phi + \pi/2)$, which give the orientation of α- and β-lines (cf. for example, Hill 1950). If we define a local co-ordinate system x_α, x_β, such that the co-ordinate axes are tangent to the characteristics ($\phi = 0$), then we can see from Equations 5.25 and 5.26 that

$$\frac{\partial}{\partial x_\alpha}(-p - k \sin 2\phi)\Bigg|_{\phi=0} = -\frac{\partial p}{\partial x_\alpha} - 2k \frac{\partial \phi}{\partial x_\alpha} = \frac{\partial}{\partial x_\alpha}(-p - 2k\phi) = 0$$

$$\frac{\partial}{\partial x_\beta}(-p + k \sin 2\phi)\Bigg|_{\phi=0} = -\frac{\partial p}{\partial x_\beta} + 2k \frac{\partial \phi}{\partial x_\beta} = \frac{\partial}{\partial x_\beta}(-p + 2k\phi) = 0$$

and therefore

$$p + 2k\phi = \zeta_\alpha \qquad p - 2k\phi = \zeta_\beta \tag{5.27}$$

where ζ_α and ζ_β are constant on an α- and β-line, respectively, although in general their values change from one α-line (or β-line) to another. Equations 5.27, known as *Hencky's equations,* are of considerable importance in the numerical solution of plastic plane-strain problems.

We now consider the velocities. With reference to Figure 5.8b, if v_i is the velocity vector, with components (v_1, v_2) and (v_α, v_β) in the (x_1, x_2)- and (x_α, x_β)-co-ordinate systems, respectively, then the following relations hold:

$$v_1 = \overline{OB} = \overline{OA} - \overline{BA} \qquad \overline{OA} = v_\alpha \cos \phi \qquad \overline{B'A'} = \overline{BA} = v_\beta \sin \phi$$

$$v_2 = \overline{OC} = \overline{OD} + \overline{DC} \qquad \overline{OD} = v_\beta \cos \phi \qquad \overline{D'C'} = \overline{DC} = v_\alpha \sin \phi$$

and therefore

$$v_1 = v_\alpha \cos \phi - v_\beta \sin \phi \qquad v_2 = v_\alpha \sin \phi + v_\beta \cos \phi \tag{5.28}$$

Equations 5.28 give the velocity components in terms of the orientation of the characteristics.

As the directions of maximum shear stress and maximum shear strain rate coincide, the stress characteristics are also lines of maximum shear strain rate, across which the gradient of the tangential component of velocity can be very large and, under appropriate boundary conditions, discontinuous (in physical terms this may be taken to be the limit of a narrow zone where the tangential shear strain rate is very large). In other words, the characteristics of stress and of velocity coincide.

Combining the incompressibility condition (Eqn 5.20) with Equation 5.22, expressing the coaxiality of stress and strain rate, the following condition is found to hold along a characteristic ($\phi = 0$):

$$\frac{\partial v_1}{\partial x_1} = \frac{\partial v_2}{\partial x_2} = 0$$

This relation shows that the rate of elongation along a characteristic is zero. Moreover, when taken together with Equation 5.28, it implies that in the (x_α, x_β)-system, tangent to the characteristics,

$$\frac{\partial}{\partial x_\alpha} \left(v_\alpha \cos \phi - v_\beta \sin \phi \right) \bigg|_{\phi=0} = \frac{\partial v_\alpha}{\partial x_\alpha} - v_\beta \frac{\partial \phi}{\partial x_\alpha} = 0$$

$$\frac{\partial}{\partial x_\beta} \left(v_\alpha \sin \phi + v_\beta \cos \phi \right) \bigg|_{\phi=0} = v_\alpha \frac{\partial \phi}{\partial x_\beta} + \frac{\partial v_\beta}{\partial x_\beta} = 0$$

that is,

$$dv_\alpha - v_\beta \, d\phi = 0 \qquad dv_\beta + v_\alpha \, d\phi = 0 \tag{5.29}$$

on α- and β-lines, respectively. Equations 5.29, known as *Geiringer's equations,* determine the components of velocity once the characteristics are known.

The two orthogonal families of curves with direction at every point coinciding with the direction of maximum shear stress and shear strain rate (i.e. the characteristics) are known as *slip lines*. In plastic plane strain, the *slip-line field* under given boundary conditions is the fundamental unknown to be determined. Both analytical and numerical methods of solution are available (see Hill 1950, Johnson *et al.* 1970). Two examples of slip-line fields are given in Figure 5.9. Figure 5.9a shows the slip-line field in a plastic medium being indented by a flat rigid die; Figure 5.9b shows the slip-line field in the cross-section of an infinite plastic slab, where the stress difference is at the yield point everywhere

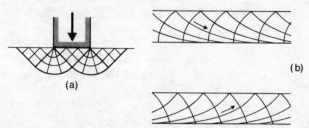

Figure 5.9 Examples of slip-line fields: (a) plastic medium indented by a rigid die (shaded); (b) active (top) and passive (bottom) Rankine states; slip-lines are suggestive of normal and thrust faults, respectively (arrows).

and the horizontal normal stress is either algebraically larger (so-called 'active Rankine state') or smaller ('passive Rankine state') than the vertical normal stress.

The physical occurrence of slip lines can be demonstrated in experiments on easily-flowing plastic model materials. The fact that, under suitable boundary conditions, the tangential component of velocity may be discontinuous across a slip line implies that slip lines can be used to model faults. For instance, the two solutions shown in Figure 5.9 are of great relevance to geodynamics (see Sections 8.6 & 7). The first has been used to model the fault pattern in China and Central Asia where India can be considered to act as an indenter, and also at the boundaries between different tectonic provinces of the Precambrian Canadian Shield. Active and passive Rankine states have obvious connections with regions of crustal stretching and shortening, respectively.

PART II

The continuum approach to Earth rheology

6 The short timescale: seismological Earth models

Information on the rheological properties of the Earth over the short timescale (seconds to days) comes mainly from seismic waves and free oscillations. At high frequencies the Earth behaves elastically, with some imperfections attributable to delayed elastic response (anelasticity). In this chapter we survey the fundamentals of elastic wave propagation, and discuss the ways in which seismological data are used to infer the distribution of elastic properties and density with depth.

On the basis of evidence from body waves, surface waves, free oscillations, and anelastic damping, a spherically symmetric Earth model such as PREM (Preliminary Reference Earth Model) can be obtained. The variations of physical parameters with depth may then be interpreted in terms of the mineralogical and chemical composition of the interior.

6.1 Elastic wave theory

In the next three chapters we give a brief survey of the rheological and thermal behaviour of the Earth, as inferred from geophysical observables, together with the basic theory on which the observations are based. The material covered can be found, developed to different degrees, in books on the physics of the Earth (for example, Jacobs 1974, Stacey 1977a, Garland 1979). Advanced discussions are provided by Jeffreys (1976) and Dziewonski & Boschi (1980). Bullen (1963) gives a unified treatment of the fundamentals of seismology. More-descriptive surveys are given by Brown & Mussett (1981), Bolt (1982), and Bott (1982).

The characteristic times of stress cycles in the Earth vary from the order of seconds (seismic waves), to the order of hundreds of millions of years (mantle flow), i.e. over a range of more than 10^{15} s. It is therefore not surprising that the Earth shows different rheological responses on different timescales. For stresses of short duration (seconds to hours), the best tool for looking at the rheological properties of the Earth is provided by seismology. An earthquake involves a sudden change of stress at the focus, resulting in the generation and propagation of seismic waves. Experience shows that seismic waves are elastic to a first approximation, although anelastic damping does occur. The bulk of the Earth is therefore elastic (with imperfections) in the short time range pertaining to seismology. In this chapter we focus on seismic waves and related phenomena, and examine how observation of these vibrations leads to the determination of the elastic properties of the interior of the Earth and – indirectly – of its probable composition.

The general form of the wave equation was introduced in Section 3.4. The simplest physical instance of elastic waves is that of a uniform, homogeneous, elastic string, fixed at its two ends (separated by a distance L) and made to vibrate in the vertical plane (Fig. 6.1). The configuration of the string at any time t is $u(x, t)$, where the vertical deflection u is assumed to be infinitesimal. Let T_1 and T_2 be the tangential tensions at the ends of a small element of the string. Since there is no displacement in the x-direction, the horizontal component of tension T_h is constant, i.e.

$$T_1 \cos \gamma_1 = T_2 \cos \gamma_2 = T_h$$

In the vertical direction, by applying Newton's Law to the same small element we have

$$T_2 \sin \gamma_2 - T_1 \sin \gamma_1 = \rho \, \Delta x \ddot{u}$$

where ρ is the density of the string. It follows that

$$\tan \gamma_2 - \tan \gamma_1 = \rho \ddot{u}(\Delta x / T_h)$$

Using the definition of derivative and taking the limit for $\Delta x \to 0$, we can

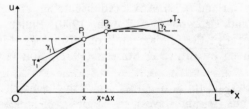

Figure 6.1 Vibrations of an elastic string.

114

write

$$\frac{\partial^2 u}{\partial x^2} = \frac{1}{c^2} \ddot{u} \qquad (6.1)$$

where $c = (T_h/\rho)^{1/2}$ has the dimensions of a velocity, $[lt^{-1}]$. Equation 6.1 is the *one-dimensional wave equation*. (In three dimensions, the left-hand side becomes the Laplacian of the deflection).

The general solution to Equation 6.1 is of the form

$$u = f(x + mt)$$

which, when substituted back, yields the condition $m = \pm c$. Therefore the displacement function

$$u = f_1(x - ct) + f_2(x + ct) \qquad (6.2)$$

known as the *d'Alembert solution,* satisfies the one-dimensional wave equation. Since $f(x - ct) = f(x - c(t + \Delta t))$ if x is increased by the amount $c\,\Delta t$, the first term on the right-hand side of Equation 6.2 represents a *plane wave* in which the disturbance travels in the positive x-direction with speed c; similarly, the second term represents a plane wave proceeding in the negative direction.

The solution can be written in an alternative form as follows. By using the boundary condition $u(0, t) = 0$ in Equation 6.2, we have $-f_1(-ct) = f_2(ct)$; by using the second boundary condition $u(L, t) = 0$, we have $f_2(ct - L) = f_2(ct + L)$, i.e. f_2 is a periodic function with period $2L$ (and similarly for f_1). Then the displacement function can be expressed by an infinite series

$$u(x, t) = \sum_{n=1}^{\infty} u_n(x, t) \qquad (6.3)$$

i.e. by the sum of simple *harmonics*

$$u_n(x, t) = \left(a_n \cos \frac{n\pi ct}{L} + b_n \sin \frac{n\pi ct}{L} \right) \sin \frac{n\pi x}{L} \qquad n = 1, 2, 3, \ldots$$

$$(6.4)$$

which are themselves solutions of Equation 6.1. These harmonics are termed the eigenfunctions of the vibrating string; each represents a harmonic motion with frequency $cn/2L$ cycles per unit time, called the nth normal mode of the string.

115

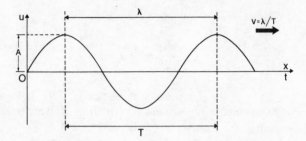

Figure 6.2 Parameters of a simple harmonic wave.

The parameters that describe a simple harmonic wave are defined in Figure 6.2. The *amplitude* (maximum displacement from the equilibrium position) and *wavelength* (distance between peaks) are A and λ, respectively; the time for the wave to travel the distance of one wavelength is the *period* T; and the *frequency* $\omega = T^{-1}$ (measured in hertz, $1\ Hz = 1\ s^{-1}$) is the number of waves per second. The velocity of a simple harmonic is $v = \lambda T^{-1}$. In nature, vibrations consist of the superposition of many simple harmonics of different amplitude, frequency, wavelength, and phase. Wave motion can be analysed in the time domain or in the frequency domain, analysis in the frequency domain allowing separation of the individual harmonics forming the overall wave pattern.

To extend the previous considerations to a continuous body such as the Earth, consider the equations of motion in a homogeneous, isotropic elastic medium, neglecting body forces since they are mostly balanced by the hydrostatic pressure field (Eqn 3.14, Section 3.2):

$$(\lambda + \mu)\,\frac{\partial \theta}{\partial x_i} + \mu \nabla^2 u_i = \rho \ddot{u}_i \tag{6.5}$$

Taking the divergence $\partial/\partial x_i$ of Equation 6.5 we have

$$(\lambda + \mu)\nabla^2\theta + \mu\,\frac{\partial}{\partial x_i}\,(\nabla^2 u_i) = \rho\,\frac{\partial \ddot{u}_i}{\partial x_i}$$

that is, inverting the order of differentiation, and since $\partial u_i/\partial x_i \equiv \theta$,

$$\nabla^2\theta = \frac{\rho}{\lambda + 2\mu}\,\ddot{\theta} \tag{6.6}$$

Consequently, the quantity θ (cubical dilatation) follows the wave equation; i.e. in an isotropic, homogeneous, elastic body a disturbance involving changes in volume propagates with velocity

$$\alpha = \left(\frac{\lambda + 2\mu}{\rho}\right)^{\frac{1}{2}} = \left(\frac{k + \frac{4}{3}\mu}{\rho}\right)^{\frac{1}{2}} \tag{6.7}$$

116

where k, μ, and ρ, are the bulk modulus, shear modulus, and density, respectively.

We now take the curl of the equation of motion. In index notation, the curl of a vector z_k is given by

$$e_{ijk}\frac{\partial z_k}{\partial x_j} = \zeta_i$$

where the (third-order) *permutation tensor* e_{ijk} is defined by the condition that if any two of its indices are equal, then $e_{ijk} = 0$; if all the indices are different and in cyclic order, then $e_{ijk} = 1$; and if they are different but not in cyclic order, then $e_{ijk} = -1$. Inverting the order of differentiation and noting that $e_{ijk}\partial/\partial x_j(\partial\theta/\partial x_k) = 0$, Equation 6.5 becomes

$$\mu\nabla^2\left(e_{ijk}\frac{\partial u_k}{\partial x_j}\right) = \rho\frac{\partial^2}{\partial t^2}\left(e_{ijk}\frac{\partial u_k}{\partial x_j}\right)$$

i.e.

$$\nabla^2\zeta_i = (\rho/\mu)\ddot{\zeta}_i \qquad (6.8)$$

The quantity ζ_i therefore follows the wave equation and propagates in the elastic medium with velocity

$$\beta = (\mu/\rho)^{\frac{1}{2}} \qquad (6.9)$$

(By the definition of curl, the components of the vector ζ_i are

$$\zeta_1 = \frac{\partial u_3}{\partial x_2} - \frac{\partial u_2}{\partial x_3} \qquad \zeta_2 = \frac{\partial u_1}{\partial x_3} - \frac{\partial u_3}{\partial x_1} \qquad \zeta_3 = \frac{\partial u_2}{\partial x_1} - \frac{\partial u_1}{\partial x_2}$$

that is, they represent an equivoluminal, rotational pattern of displacement.)

Two types of elastic *body waves* can therefore occur within an elastic medium: *longitudinal* or *compressional* waves (Eqn 6.6), in which the disturbance is propagated through the medium with velocity α (Eqn 6.7), and *shear* or *rotational* waves (Eqn 6.8), in which the disturbance travels with velocity β (Eqn 6.9). These are illustrated in Figure 6.3. The particles reached by the wavefront vibrate in the direction of propagation of the wave in the case of longitudinal waves, and in the plane normal to the direction of propagation in the case of shear waves. As $\alpha > \beta$, compressional waves arrive first at any seismic station, and therefore

117

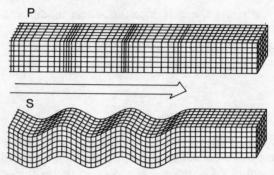

Figure 6.3 Material deformation during the passage of compressional (*P*-) and shear (*S*-) waves (from Bolt 1976).

appear first on seismograms; hence they are termed *P-waves* ('*undae primae*'). Shear waves arrive later and are termed *S-waves* ('*undae secundae*').

The velocities α and β are functions of the elastic parameters and the density of the medium in which the waves travel. The basic importance of seismology to the study of the interior of the Earth follows from this dependence. If the velocity variations of *P*- and *S*-waves with depth can be determined, then the elastic properties and the density can be inferred, under not too limiting assumptions. Using collateral information, it then becomes possible to place limits on the physical and chemical properties of the interior. However, before looking at the velocity distribution with depth we must clarify some aspects of seismic wave propagation.

First, we notice that the simple solution given by Equation 6.2 is valid only for the one-dimensional wave equation (Eqn 6.1). In the more general case (see, for example, Eqns 6.6 & 8), the wave equation for a quantity ψ is of the form

$$\nabla^2\psi = (1/c^2)\ddot{\psi} \tag{6.10}$$

with solution $\psi = f(x_1, x_2, x_3, t) = f(r, t)$, where $r = (x_i x_i)^{1/2}$ is the distance from the origin. For spherical symmetry in spherical co-ordinates, the Laplacian is

$$\nabla^2\psi = \frac{\partial^2\psi}{\partial r^2} + \frac{2}{r}\frac{\partial\psi}{\partial r} = \frac{1}{r}\frac{\partial^2(\psi r)}{\partial r^2}$$

and therefore Equation 6.10 can be written as

$$\frac{\partial^2(\psi r)}{\partial r^2} = \frac{r}{c^2}\ddot{\psi} = \frac{1}{c^2}\frac{\partial^2(\psi r)}{\partial t^2} \tag{6.11}$$

Equation 6.11 is formally identical to Equation 6.1, and consequently the solution for (ψr) is the same as that for u in the one-dimensional case (Eqn 6.2). Therefore ψ is of the form

$$\psi = (1/r)[f_1(r - ct) + f_2(r + ct)] \qquad (6.12)$$

It follows that, for spherical waves, the amplitude of the disturbance is inversely proportional to the distance from the origin, as a consequence of the geometrical spreading of the travelling wavefront.

Secondly, it should be noted that the stresses associated with the transmission of seismic waves occur in a medium that is under considerable initial stress from the weight of the overlying material (of the order of 1 GPa at a depth of about 33 km, increasing to more than 100 GPa in the deep interior). The initial stress involves finite strains, and causes change in density and elastic parameters in otherwise homogeneous regions. However, seismic stresses can be regarded as infinitesimal perturbations from the equilibrium stress, which is approximately of the type $- p\,\delta_{ij}$, where p is the hydrostatic pressure. The finite effects can accordingly be neglected, as changes in density and elastic parameters within a wavelength are usually small.

6.2 Seismic phases and spherically symmetric velocity distribution

When considering the transmission of seismic waves within the Earth, a considerable simplification is introduced by the notion of *seismic ray;* that is, the direction of propagation of a given wave, perpendicular to the wavefront at any point. The laws of geometrical optics can be applied to seismic rays. In particular, they are refracted and reflected at surfaces of discontinuity. However, the situation is more complicated than in geometrical optics, as – except in particular cases – the energy of the incident ray is partitioned into two refracted and two reflected rays. A generalized form of *Snell's Law* holds

$$v/\sin \gamma = f \qquad (6.13)$$

where v is the wave velocity (α or β as the case may be), γ is the angle between the ray (incident, refracted, or reflected) and the normal to the surface of discontinuity, and f is a constant.

Discontinuities in the interior of the Earth generate an abundance of *seismic phases*. A *first-order seismic discontinuity* is one where the change in elastic properties and density is sharp, i.e. it occurs over a

119

length scale (usually a few kilometres) that is small compared with the wavelength, and this results in a discontinuity in seismic wave velocity. A *second-order discontinuity* is a zone where the velocity gradient undergoes an abrupt change.

From the seismological viewpoint the interior of the Earth can be thought of as approximately spherically symmetric, excluding lateral variations in velocities that are especially prominent in the crust and upper mantle. Figure 6.4 shows the three worldwide first-order discontinuities (the Moho discontinuity separating the crust from the mantle, the mantle/core or Gutenberg discontinuity and the outer/inner core boundary), together with some of the most frequently observed seismic phases for body waves.

The terminology for seismic phases is as follows. The symbols P and S denote longitudinal and shear waves propagating in the mantle (and crust); K denotes a P-wave travelling in the outer core (which does not transmit S-waves); I and J denote P- and S-waves, respectively, propagating in the inner core. An upward reflection from the mantle/core boundary is denoted by c; from the inner core boundary, by i. For example, ScS denotes a shear wave that travels downward from the focus, is reflected at the mantle/core boundary, and crosses the mantle again as a shear wave; $PKIKP$ denotes a P-wave that has travelled through mantle, outer and inner core; and SKS a shear wave that has been refracted at the mantle/core boundary and has travelled in the outer core as a compressional wave. As a rule, seismic rays are concave upward, i.e. the velocity usually increases with depth.

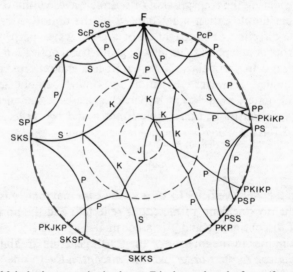

Figure 6.4 Main body-wave seismic phases; F is the earthquake focus (from Bolt 1982).

The determination of spherically symmetric velocity distributions, $\alpha(r)$ and $\beta(r)$, is one of the central problems of seismology. The basic data are the *travel times* of seismic phases, i.e. the time t taken by a given seismic phase to cover the angular distance Δ between the epicentre and any given point on the surface. Years of observation and analysis have resulted in the *Jeffreys–Bullen travel-time curves* (1940) which give the travel times of seismic body phases. An example of such travel-time curves is shown in Figure 6.5. In practice travel times may show *residuals* with respect to 'standard' travel times (i.e. spherically averaged). These residuals are important, since they imply departures of seismic properties from spherical symmetry which may be of regional tectonic interest.

The velocity distribution of a spherically symmetric elastic Earth can be derived from the travel times. Assume the Earth to consist of many thin layers $i = 1, 2, 3, \ldots, n$, each with velocity v_i, $v_i < v_{i+1}$. Applying Snell's Law to the refraction at point P_i we have (Fig. 6.6a)

$$(r_i \sin \gamma_i)/v_i = (r_i \sin \gamma_i')/v_{i+1}$$

Considering the right triangles ODP_i and ODP_{i+1}, we can write

$$r_i \sin \gamma_i' = r_{i+1} \sin \gamma_{i+1} = d$$

that is,

$$(r_i \sin \gamma_i)/v_i = (r_{i+1} \sin \gamma_{i+1})/v_{i+1}$$

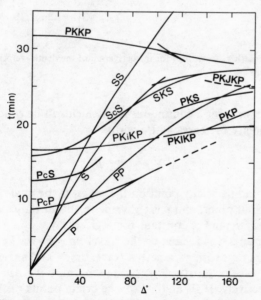

Figure 6.5 Jeffreys–Bullen travel-time curves for some of the main seismic phases.

121

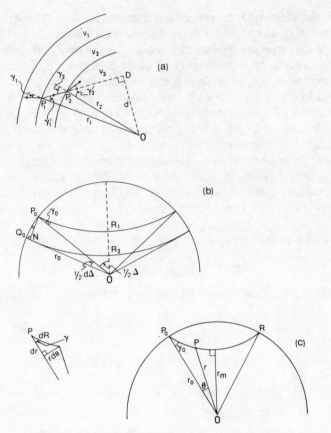

Figure 6.6 Geometric relationships for the derivation of the depth–velocity distribution. See text for details.

for all i. Therefore, for a continuous ray (which will be concave upward as long as velocity increases with depth)

$$(r \sin \gamma)/v = p \qquad (6.14)$$

where r is the radius at any point on the ray, γ is the angle between ray and radius at that point, and v is the velocity at that radius; p is termed the *ray parameter,* and is constant for a given ray.

The ray parameter is related to the travel-time curve. Let R_1 and R_2 be two adjacent rays with parameters, travel times, and angles subtended at the centre of the Earth given by p, t, Δ, and $p + \mathrm{d}p$, $t + dt$, and $\Delta + \mathrm{d}\Delta$, respectively. If P_0 and Q_0 are the origin points of the two rays, and γ_0 is the angle between the radius r_0 and R_1 at P_0 (Fig. 6.6b), it

122

follows from the right triangle P_0NQ_0 that

$$\sin \gamma_0 = \frac{\overline{Q_0N}}{\overline{Q_0P_0}} = \frac{v_0 \; dt}{r_0 \; d\Delta}$$

that is, by Equation 6.14,

$$p = dt/d\Delta \tag{6.15}$$

The ray parameter is therefore the slope of the travel-time curve.

Inversion of travel-time curves consists of determining $v(r)$ from $p(\Delta)$. The element of distance dR along a ray with parameter p is given by

$$dR = r \; d\theta/\sin \gamma$$

where $d\theta$ is the change in angle corresponding to dR (Fig. 6.6c). Therefore, by the definition of ray parameter (Eqn 6.14),

$$p = \frac{r^2}{v} \frac{d\theta}{dR}$$

The above equation, together with the relation

$$(dR)^2 = (dr)^2 + r^2 (d\theta)^2$$

implies that

$$d\theta = \pm \; (p/r)(\eta^2 - p^2)^{-\frac{1}{2}} \; dr \tag{6.16}$$

where $\eta = r/v$. Integrating Equation 6.16 between the surface (r_0) and the deepest point of the ray (r_m) we have

$$\int_{r_m}^{r_0} \frac{p \; dr}{r(\eta^2 - p^2)^{\frac{1}{2}}} = \frac{\Delta}{2} \tag{6.17}$$

which is an integral equation for $\eta(r)$, and hence $v(r)$, usually referred to as the *Wiechert–Herglotz integral*. It can be proven (cf. for example, Bullen 1963) that it is possible to rewrite it as

$$\int_0^{\Delta_m} \cosh^{-1}(p/p_m) \; d\Delta = \pi \ln(r_0/r_m) \tag{6.18}$$

where Δ_m is the value of Δ for the ray with deepest point at r_m, and p_m

123

the corresponding value of the ray parameter. The left-hand side of Equation 6.18 can be evaluated numerically for successively larger values of Δ_m, as $p(\Delta)$ and p_m are obtainable from travel-time curves. This allows determination of r_m, the radius at the deepest point of the ray emerging at Δ_m. The velocity at the deepest point is given by $v_m = r_m/p_m$ (as can be seen from Eqn 6.14 with $\sin \gamma = 1$), and consequently the velocity distribution $v(r)$ for a given kind of ray is determined.

The inversion of travel-time curves is subject to certain limitations which place constraints on $v(r)$. The velocity usually increases with depth $(dv/dr < 0)$; if it decreases with depth $(dv/dr > 0)$, then the decrease must be such that the downward curvature of the ray must be less than the curvature of the level surface (r^{-1}) at that depth; furthermore, zones of very rapid increase may cause complications (see, for example, Bullen 1963 for further discussion).

On the basis of the $v(r)$ distributions, the Earth has been subdivided into several layers. Bullen (1963), in his classical models, named them sequentially A (crust); B, C, and D (upper mantle, transition zone, and lower mantle); and E, F, and G (outer core, core transition zone, and inner core). Two models for $\alpha(r)$ and $\beta(r)$ are shown in Figure 6.7. These are based on body-wave travel times, and are also constrained by data on surface waves and on the free oscillations of the Earth. Noteworthy is the presence of a low-velocity zone (LVZ) in the upper mantle, approximately at depths between 100 and 300 km (revealed also by surface wave studies; cf. Section 6.4). The mantle transition zone, extending from about 400 to 700 km, exhibits at least two steep increases in velocity. The mantle/core boundary is clearly the most important discontinuity within

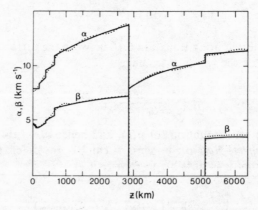

Figure 6.7 Two models of the velocity distribution with depth. Dotted line, 1066A (Gilbert & Dziewonski 1975); solid line, PEM-A (Dziewonski *et al.* 1975) (from Jordan 1980).

the Earth. More details about the properties of the interior are given in Section 6.6, where a more recent seismic model (the 'Preliminary Reference Earth Model', or PREM) will be discussed.

6.3 Density and elastic parameters

The velocities $\alpha(r)$ and $\beta(r)$ are related to density and elastic parameters by means of Equations 6.7 and 6.9, which allow determination of the ratios k/ρ and μ/ρ as functions of radius. In order to determine $\rho(r)$, $k(r)$, and $\mu(r)$ – and consequently, by means of the relationships developed in Section 3.1, the other elastic moduli – it is necessary to have a third relation which relates density and depth.

The parameters ρ, k, and μ are functions of the initial strain, temperature, pressure, and composition. In the interior of the Earth they are considered to be slowly-varying functions of r except at discontinuities. The values of the elastic moduli depend on the thermodynamical conditions during deformation. In seismic wave propagation, since the period of any wave is much less than the characteristic time of heat flow across one wavelength, the conditions are approximately adiabatic; but differences between adiabatic and isothermal parameters are very small. Density is constrained by the known mass of the Earth (5.974×10^{24} kg) which yields the average density $\langle \rho \rangle = 5.515 \times 10^3$ kg m^{-3}, and by the moment of inertia I, which yields a ratio $I/MR^2 = 0.3308$ (M and R are the mass and radius of the Earth, respectively). This is less than the value for a uniform sphere (0.40), and therefore both average density and moment of inertia show that density increases with depth.

The variation of density with depth in a region where the temperature gradient is *adiabatic* (i.e. it is that resulting from self-compression without heat exchanges with the surroundings) can be determined using as a starting point the variation of hydrostatic pressure p:

$$\frac{\mathrm{d}p}{\mathrm{d}r} = -\rho g \qquad (6.19)$$

where $\rho = \rho(r)$ and g is the acceleration of gravity, given by $g(r) = Gm/r^2$, m being the mass enclosed by the sphere of radius r. Therefore, in a homogeneous, adiabatic region

$$\frac{\mathrm{d}\rho}{\mathrm{d}r} = \frac{\mathrm{d}\rho}{\mathrm{d}p}\frac{\mathrm{d}p}{\mathrm{d}r} = -\frac{\rho^2 g}{k} \qquad (6.20)$$

since $\mathrm{d}\rho/\mathrm{d}p = \rho/k$ from the definition of incompressibility.

125

From the equations for seismic wave velocities (Eqns 6.7 & 9) it follows that

$$\alpha^2 - \tfrac{4}{3}\beta^2 \equiv \phi = k/\rho \tag{6.21}$$

which, when substituted into Equation 6.20, yields

$$\frac{d\rho}{dr} = -\frac{\rho g}{\phi} = -\frac{G\rho m}{\phi r^2} \tag{6.22}$$

This equation, known as the *Adams–Williamson equation*, gives the density gradient in terms of the *seismic parameter* ϕ which is a known function of radius. Together with the equation for the mass inside radius r

$$m = \int_0^r 4\pi r^2 \rho(r) \; dr$$

it determines the density distribution in homogeneous, adiabatic regions. Integration proceeds step by step, downward from the Moho, with initial value for ρ of about $3.32 \times 10^3 \; \mathrm{kg\,m^{-3}}$, and for m given as the mass of the Earth stripped of its crust.

The two possible deviations from the Adams–Williamson equation are non-adiabaticity and inhomogeneity. If the temperature gradient is larger than the adiabatic, the excess gradient is

$$\chi = \left(\frac{\partial T}{\partial p}\right)_s \frac{dp}{dr} - \frac{dT}{dr} \tag{6.23}$$

where $(\partial T/\partial p)_S$ denotes the adiabatic increase; note that, as both dp/dr and $dT/dr < 0$, $\chi > 0$ when the temperature increases with depth more steeply than the adiabat. The density gradient is

$$\frac{d\rho}{dr} = \left(\frac{\partial \rho}{\partial p}\right)_s \frac{dp}{dr} + \left(\frac{\partial \rho}{\partial S}\right)_p \frac{dS}{dr} \tag{6.24}$$

where S is entropy and the subscripts S and p denote adiabatic and constant-pressure conditions, respectively. The first term on the right-hand side of Equation 6.24 is the adiabatic density variation. The second term is caused by departures from adiabaticity, and can be expressed in terms of the thermal expansion coefficient α (see, for example, Bullen 1963). The total density variation is

$$\frac{d\rho}{dr} = -\frac{\rho g}{\phi} + \alpha\rho\chi \tag{6.25}$$

126

A superadiabatic temperature gradient therefore decreases the rate of increase of density with depth, but this effect is unlikely to be large. Furthermore, the effect of inhomogeneity on the density gradient works in the opposite sense (assuming that materials are normally stratified in the field of gravity), and therefore the Adams–Williamson equation can be used as a first approximation to determine the density gradient in shells free from discontinuities and transition regions.

Once the density distribution is determined, it is possible to determine pressure and elastic parameters as a function of radius. One such model for ρ, k, and μ is shown in Figure 6.8. Limiting our attention to the

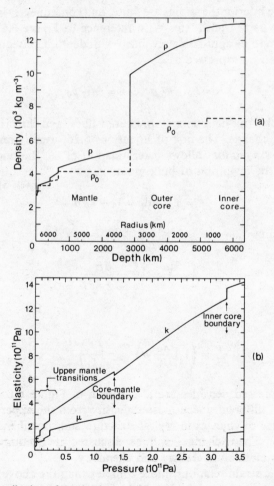

Figure 6.8 Distribution of (a) density and (b) elastic parameters for model PEM-A of Dziewonski *et al.* (1975) (from Stacey 1977a).

127

mantle, we now consider the general problem of inferring the *uncompressed density* of a shell, given the density and the elastic parameters.

One observation should be made at the outset about the depth distribution of any parameter in a spherically symmetric Earth model. The parameter value at a given depth is the average over a finite depth range, and there is a trade-off between depth range and accuracy (i.e. between resolution and precision). This should be borne in mind when interpreting parameter variations with depth.

Inferences on the composition of the interior require the reduction of $\rho(r)$ to ρ_0 (the corresponding density at zero pressure); that is, the establishment of an equation of state $\rho = \rho(p)$. At the high pressures pertaining to the interior the strains are finite, and consequently infinitesimal strain theory is not adequate. With reference to Figure 6.8, $k(p)$ and $\mu(p)$ are seen to be approximately linear with depth. Consequently, their variations can be expressed as

$$k = k_0 + k_0' p \qquad \mu = \mu_0 + \mu_0' p \tag{6.26}$$

where the subscript denotes zero-pressure values, and the prime denotes their rate of change. The first of Equations 6.26, usually referred to as *Murnaghan's equation,* allows determination of ρ_0, the uncompressed density. By the definition of bulk modulus,

$$k = \rho \, dp/d\rho = k_0 + k_0' p$$

and integrating

$$\int_{\rho_0}^{\rho} \frac{d\rho}{\rho} = \int_0^p \frac{dp}{k_0 + k_0' p}$$

we obtain

$$\frac{\rho}{\rho_0} = \left(1 + \frac{k_0'}{k_0} p\right)^{1/k_0'} = \left(\frac{k}{k_0}\right)^{1/k_0'} \tag{6.27}$$

The uncompressed densities are also shown in Figure 6.8. The lower mantle has a different uncompressed density from the upper mantle, as a consequence of changes in crystal structure and, possibly, changes in composition. These changes will be discussed in Section 6.6, when dealing with mantle mineralogy and chemistry.

Other finite strain relations, more complex than the above, have been proposed on the basis of interatomic potential functions and their relation to $k(p)$. However, Equations 6.26 can be used as a first approximation, and are roughly verified by shock-wave experiments which allow compressibility measurements up to pressures of the order of hundreds of gigapascals (for a discussion see Stacey 1977a).

6.4 Surface waves and free oscillations: elastic Earth models

Spherically symmetric Earth models giving elastic properties and density as functions of radius, such as those shown in Figures 6.7 and 6.8, are constrained not only by body-wave travel-time and whole-Earth data, but also by surface waves and free oscillations. The theory of surface waves and free oscillations may be found in classical seismological tests (e.g. Bullen 1963), and an excellent qualitative account is given by Bolt (1982).

The most prominent seismic waves on a seismogram are usually not body waves, but later-arriving *surface waves* which have travelled along the outer shell of the Earth. While body waves have typical periods of, say, 1 s for *P*-waves and 5 s for *S*-waves, the periods of surface waves vary from tens to hundreds of seconds; their wavelengths can correspondingly be up to hundreds of kilometres. As the geometrical spreading is less, the amplitude of surface waves decreases with distance from the focus less rapidly than the amplitude of body waves does.

Two types of surface waves are most commonly originated by earthquakes (see Fig. 6.9): *Rayleigh waves* (*R*), in which the particle displacement is a retrograde ellipse in the vertical plane, with major axis vertical; and *Love waves* (*L*) which are horizontally polarized *S*-waves. Typically, the first onsets of *R*- and *L*-waves have velocities of approximately 3.9 and 4.4 km s^{-1}, respectively, that is, less than β for the uppermost

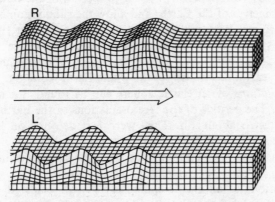

Figure 6.9 Material deformation during the passage of Rayleigh (*R*-) and Love (*L*-) waves (from Bolt 1976).

129

mantle. It can be proven in the theory of surface waves (cf. Bullen 1963) that the velocity of R-waves is $v_R \simeq 0.9\,\beta$ for a uniform half-space with shear-wave velocity β; and the velocity of L-waves is in the range $\beta_1 < v_L < \beta_2$ for a layer over a half-space with shear velocities β_1 and β_2, respectively.

All surface waves, except R-waves propagating along the surface of a homogeneous half-space (a case of no relevance to the Earth), show *dispersion* – i.e. phases with different wavelength travel with different velocity. The velocity normally increases with the wavelength, as longer waves penetrate more deeply and β increases with depth. One must therefore distinguish between *phase velocity* c (the velocity of propagation of a pure harmonic of wavelength λ), and *group velocity* U (the velocity of propagation of the wavetrain consisting of a spectrum of frequencies). The velocity of a single harmonic of period T and frequency ω is $c = \lambda T^{-1} = \lambda \omega$; by superimposing harmonics it can be proven that the group velocity is $U = d\omega/dk$, where $k = 1/\lambda$ is the *wavenumber* of any harmonic. The relation between group and phase velocities is therefore

$$U = c - \lambda\ \partial c/\partial\lambda \qquad (6.28)$$

In the absence of dispersion ($\partial c/\partial\lambda = 0$), peaks of different λ (or period T) travel with the same velocity as the wavetrain as a whole; for normal dispersion ($\partial c/\partial\lambda > 0$), individual peaks travel faster than the wave packet.

Both phase and group velocities of surface waves can be measured, and it is therefore possible to construct *dispersion curves* giving c and U as a function of T (Fig. 6.10a). These curves provide important information on lateral variations of physical properties in the outermost few hundred kilometres of the Earth, and allow the subdivision of the outer shell into different 'tectonic' regions (Fig. 6.10b; see Bott 1982). In the context of spherically symmetric Earth models, surface wave data impose additional constraints on the structure of the outermost mantle.

Very large earthquakes also set the whole Earth vibrating. The theory of *free oscillations* of an elastic sphere can be found, for instance, in Love (1944). The periods of the free oscillations of the Earth vary from a few minutes to about 1 h. They were first detected on long-period seismograms after the large earthquake in Chile in May 1960. Since then their study has become an important part of the analysis of the rheology of the Earth at 'seismic' frequencies, i.e. at $\omega \gtrsim 10^{-4}$ Hz.

The vibrations of an elastic sphere are of two kinds. In *spheroidal* oscillations (*S-modes*) the displacements of the vibrating particles have, in general, both radial and tangential components (a purely radial

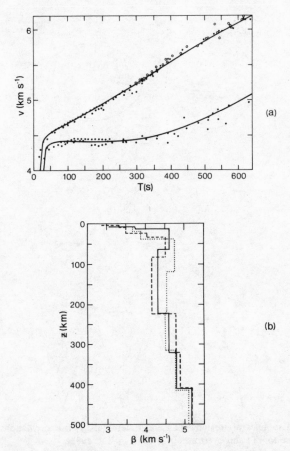

Figure 6.10 (a) Theoretical dispersion curves for Mongolia–Pasadena path (solid lines) compared with observations for Love waves (closed circles) and free oscillations (open circles). Upper and lower curves give c and U respectively (from Toksöz & Anderson 1966). (b) Velocity distribution of shear waves in different tectonic provinces: Canadian Shield (dotted), Pacific Ocean (solid), and Alps (broken) (from Dorman 1969).

oscillation is a particular case of a spheroidal oscillation with zero tangential component), and are therefore analogous to those occurring in Rayleigh waves. In *torsional* oscillations (*T-modes*) particle displacements are tangential, that is, analogous to those occurring in Love waves.

A vibrating sphere may possess nodal surfaces on which there is no displacement. The intersection of nodal surfaces with the surface of the sphere gives rise to nodal lines, i.e. lines on the surface along which there is no displacement. There can also be concentric nodal surfaces inside the sphere. Figure 6.11 gives some examples of free oscillations. Spheroidal

131

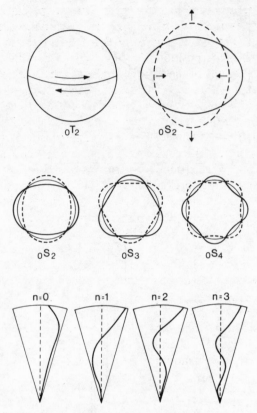

Figure 6.11 Examples of torsional (T) and spheroidal (S) free oscillations, with fundamental mode ($n = 0$) and overtones ($n \geqslant 1$)(from Bolt 1982).

and torsional oscillations are expressed by the symbols $_nS_l$ and $_nT_l$, respectively, where n and l are integers ($0 \leqslant n, l < \infty$, so that in theory there is a double infinity of vibrational modes). The integer l denotes the number of surface nodal lines in spheroidal oscillations, and the number of surface nodal lines plus one in torsional oscillations (we are restricting attention here to zonal harmonics in which nodal lines are circles of latitude). The integer n denotes the number of internal nodal surfaces, if any. Any fundamental mode $_0S_l$ or $_0T_l$, associated with a given pattern of surface displacements, may have overtones if $n > 0$, i.e. if there are internal spherical surfaces on which the displacement is zero.

The importance of free oscillations in the context of the determination of the internal structure of the Earth is that the period of any mode is a function of the internal structure. Furthermore, the lower modes, with periods up to almost 1 h, stress the whole Earth, while the lower-period higher modes increasingly affect the outer parts of the Earth, effectively merging, at periods of a few minutes, into the spectrum of surface waves. (Indeed, free oscillations can be formally regarded as standing waves resulting from the superposition of surface waves travelling in opposite directions.) The periods of the various modes are thus determined by the density and elastic parameters of the interior, with layers in different depth ranges being relatively more important for different modes.

We are now in a position to appreciate how an elastic Earth model is obtained. A spherically symmetric, non-rotating, elastic, isotropic Earth model (SNREI, for short) is constructed by inverting travel times of body waves to obtain a $\{k, \mu, \rho\}$-model which is then refined by matching it to surface-wave and free-oscillation data. This inversion procedure requires that the related direct problem be solvable: given $k(r)$, $\mu(r)$, and $\rho(r)$, the periods of several free-oscillation modes must be calculated and compared with observed periods. The model is then modified to minimize the residuals between calculated and observed values.

The construction of seismological Earth models has led to a fairly detailed knowledge of the density and elasticity of the interior. A review of problems and procedures can be found in Jordan (1980). We note that, besides the assumption of spherical symmetry, SNREI models are subject to two further limitations: the rheology is taken as perfectly elastic, and the elastic properties as isotropic. The remainder of this chapter discusses the relaxation of these assumptions.

6.5 Anelasticity

If a body shows imperfections of elasticity, as is the case for the Earth at seismic frequencies, then part of the elastic strain energy is dissipated as heat. This anelastic absorption results in an amplitude decay of vibrations, i.e. in the *damping* (attenuation) of seismic waves and free oscillations. The term *anelasticity* (or *internal friction*) is used to denote such imperfections of elasticity. Only the intrinsic damping due to anelastic behaviour is of interest here, although other causes of amplitude decay exist (geometrical spreading of spherical waves – as discussed in Section 6.1 – scattering, reflection, *etc.*).

The microphysics of internal friction will be briefly discussed in Section 10.1. Here we limit ourselves to a phenomenological approach. Anelasticity not only affects the seismic signal, but also contains information on the physical conditions and material parameters of the Earth. The construction of spherically symmetric anelastic Earth models is therefore an important step towards the specification of the rheological properties of the interior.

For seismological purposes a unified measure of anelastic damping is given by the *seismic quality factor* Q, defined by the relation

$$2\pi Q^{-1} = -\Delta E/E \qquad (6.29)$$

where ΔE is the absorbed elastic energy per cycle ($\Delta E < 0$) and E is the total energy. A high Q therefore denotes low attenuation, and vice versa. Writing Equation 6.29 in differential form

$$-2\pi Q^{-1} = \frac{T}{E}\frac{dE}{dt}$$

where T is the period of the vibration, and integrating, we obtain

$$E = E_0 \exp(-2\pi t/TQ) \qquad (6.30)$$

i.e. the energy (and therefore the amplitude, as the energy is proportional to the square of the amplitude; Bullen 1963) decays exponentially at a rate determined by Q^{-1}.

A distinction should be made between attenuation of spheroidal modes (or Rayleigh waves) and of P-waves (Q_S or Q_α) on the one hand, and attenuation of torsional modes (or Love waves) and of S-waves (Q_T or Q_β) on the other. The former involve both compressional and shear strains, while the latter involve only shear. In general, it is observed that, at a given period, $Q_S > Q_T$ and $Q_\alpha > Q_\beta$. Moreover, purely compressional vibrations (the fundamental radial mode and its overtones, $_nS_0$) show exceedingly low damping. Therefore, to a first approximation it can be assumed that attenuation in the mantle is due entirely to anelastic absorption of shear strain energy ($Q_\mu = Q_\beta$), while attenuation in pure compression is negligible ($Q_k = \infty$).

A model for attenuation as a function of depth in pure shear and pure compression (Q_μ and Q_k, respectively) is shown in Figure 6.12. Despite the scatter of the data, and problems of uniqueness, features established with confidence include a low-Q_μ upper mantle underlying a high-Q_μ lithosphere, a smooth increase of Q_μ with depth in the mantle, and a decrease in the lowermost mantle.

Models such as the above depict the attenuation properties of the Earth in the period range $10 \leqslant T \leqslant 4000$ s, and are compatible with a frequency-independent Q in this frequency band. (We refer here to intrinsic, not to apparent, frequency-dependence: the latter is related to the deeper penetration of the longer periods.) However, attenuation may be dependent on frequency; as, for instance, in metals, where strong absorption peaks can often be recognized. By contrast, experiments on

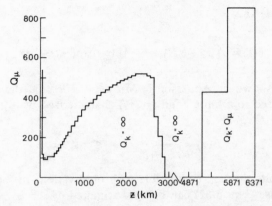

Figure 6.12 Depth distribution of seismic quality factor (from Anderson & Hart 1978).

silicates and seismological studies show that in Earth materials the frequency-dependence of Q is weak over a rather wide frequency band. As the problem of frequency-dependence is related not only to attenuation, but also to transient creep, we give here a brief account of the elementary results of phenomenological anelasticity theory. A fuller treatment may be found in Minster (1980).

The term *anelastic relaxation* can be used to denote a time-dependent adjustment of a continuous medium to a change in an external variable. Viewed in this context, attenuation and transient creep are two aspects of the same phenomenon. Since the strains involved in seismic wave propagation and free oscillations are of the order of 10^{-6} or less, it is permissible to use linear theory. Recalling the results established in Section 4.6, the transient creep response of a linear body may be written in general terms as (compare, for example, with Eqns 4.35 or 4.41)

$$\varepsilon(t) = \sigma\{ J_u + \delta J[1 - \exp(-t/\tau)] \} \tag{6.31}$$

where J_u is the unrelaxed compliance (the reciprocal of the unrelaxed elastic modulus), δJ is the difference between the relaxed (long-time or low-frequency) and the unrelaxed (short-time or high-frequency) compliance and τ is a characteristic time. An analogous expression can be written for the stress. Despite the common usage of the term 'relaxation time', a distinction should be made between the *strain retardation time* at constant stress, and the *stress relaxation time* at constant strain – the quantity appearing in the previous equation is the former.

The standard linear body has a single retardation (or relaxation) time τ. Weak frequency-dependence can be achieved only by a spectrum of characteristic times. Denoting their distribution by $F(\tau)$, one can modify

135

Equation 6.31 thus

$$\varepsilon(t) = \sigma\left(J_u + \delta J \int_0^\infty F(\tau)[1 - \exp(-t/\tau)]\,d\tau\right) \qquad (6.32)$$

It has been shown by Anderson & Minster (1979) that the choice that $F(\tau)$ be limited to a finite band (τ_1, τ_2) characterized by a parameter α

$$F(\tau) = \frac{\alpha \tau^{\alpha - 1}}{\tau_2^\alpha - \tau_1^\alpha} \qquad \tau_1 \leqslant \tau \leqslant \tau_2 \qquad (6.33)$$

leads to results that are compatible with observation. In this case, the creep response (Eqn 6.32) can be approximated by

$$\varepsilon(t) \simeq \sigma[J_u + \delta J\,\Gamma(1 - \alpha)(t/\tau_2)^\alpha] \qquad (6.34)$$

where Γ is the gamma function, defined as

$$\Gamma(\gamma) = \int_0^\infty [\exp(-\xi)]\,\xi^{(\gamma - 1)}\,d\xi$$

Equation 6.34 is analogous to the Modified Lomnitz Law (Eqn 4.30), which could therefore arise from a similar spectrum of characteristic times.

The corresponding Q in the frequency band $\omega\tau_1 \ll 1 \ll \omega\tau_2$, for the spectrum of characteristic times given by Equation 6.33, can be approximated by (Anderson & Minster 1979)

$$Q(\omega) \simeq \frac{2J_u}{\pi\alpha\delta J}\left(\cos\frac{\alpha\pi}{2}\right)(\tau_2\omega)^\alpha \qquad (6.35)$$

However, since characteristic times depend on temperature (see Section 10.1), the relationship $Q \propto \omega^\alpha$ holds only at intermediate temperatures. At low temperature, where characteristic times are long, $Q \propto \omega$; at high temperature, $Q \propto \omega^{-1}$.

Concerning the value of the parameter α, Anderson & Minster (1979) compared attenuation data for normal modes, solid Earth tides, and the Chandler wobble (see Stacey 1977a for a discussion of the last two). These processes cover the frequency range $10^{-8} \leqslant \omega \leqslant 10^{-3}$ Hz. In agreement with the results of Jeffreys (1976) and transient creep data (see Section 4.5), they found that $0.2 \leqslant \alpha \leqslant 0.4$ accounts best for the observations.

6.6 PREM and the structure and composition of the mantle

The information obtained from seismology is presented again in Figure 6.13, which shows the spherically symmetric 'Preliminary Reference Earth Model' (PREM) developed by Dziewonski & Anderson (1981). The parameters α, β, and ρ are shown as functions of depth. PREM is the most recent seismological Earth model, and should serve as a starting point for studies of the physical properties of the interior and of deviations from spherical symmetry.

PREM has been constructed from a large data set, consisting of about 100 normal mode periods, more than 10^6 body-wave travel-time observations, 100 normal mode Q-values, and the mass and moment of inertia of the Earth. The depth of the Moho is placed at 19 km (weighed average of continents and oceans). Discontinuities within the mantle are predetermined and set at 220, 400, and 670 km of depth. A second-order discontinuity some 150 km above the mantle/core boundary is necessary to fit the data. The Gutenberg discontinuity is placed at 2891 km, and the inner/outer core boundary at 5150 km. Parameters are assumed to vary smoothly between the discontinuities.

A noteworthy feature of the model is that it is impossible to satisfy the data without some elastic anisotropy, at least in the uppermost 200 km of the mantle, because of discrepancies between L- and R-wave observations. When *transverse isotropy* (vertical velocity different from horizontal velocity, the latter being the same in all directions) is introduced, the fit to the data becomes satisfactory. The required anisotropy is of the order of 2–4 per cent for both P- and S-waves, with the horizontal

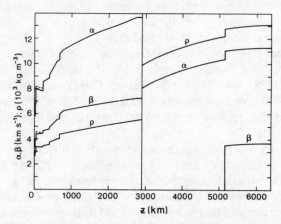

Figure 6.13 PREM – Preliminary Reference Earth Model (from Dziewonski & Anderson 1981).

137

velocities faster than the vertical velocities; no anisotropy need be present below the 220 km discontinuity. This suggests the presence in the upper mantle of anisotropic minerals such as olivine, with a preferred orientation probably related to flow.

Regional studies reveal that the uppermost mantle is, in places, also anisotropic in the horizontal plane. *P*-wave velocities just below the Moho may vary with direction by up to 7 per cent both in continental and in oceanic lithosphere. This anisotropy is probably related to the preferred orientation of olivine crystals. That the pattern of anisotropy is related to mantle flow is confirmed by the observation that, in oceanic lithosphere close to a ridge axis, the horizontal velocity normal to the strike of the axis is larger than the velocity parallel to it (see Crampin 1984 for a review).

The introduction of upper-mantle transverse isotropy renders unnecessary the velocity decrease required by most surface wave models somewhere in the 100–300 km depth range (mantle low-velocity zone, or LVZ). This does not imply that the LVZ is not present under tectonic and oceanic regions, but only that a ubiquitous LVZ is not a necessary feature in a spherically symmetric model.

Of the three discontinuities situated in the upper mantle, those at depths of 400 and 670 km are associated with mineralogical, and perhaps chemical, changes. The nature of the so-called Lehmann discontinuity at 220 km is not altogether clear, but its presence is unequivocally shown by seismic data.

The lower mantle appears to be rather homogeneous, except for its lowermost 150 km or so (region D'' of Bullen 1963), marked by a decrease in the seismic velocity gradient. Because this feature is probably associated with the occurrence of a thermal boundary layer above the mantle/core boundary, it will be discussed in Section 7.7 when dealing with the internal temperature of the Earth.

Limiting our attention to the mantle, we now consider its *composition*. Discussions of this topic can be found in the books by Wyllie (1971), Ringwood (1975), and Brown & Mussett (1981), and in a review by Oxburgh (1980). We mention here only those aspects of the problem likely to be of significance in long-term geodynamic processes such as convection currents.

It is generally accepted that the upper mantle is ultrabasic in composition, the main mineral phase being magnesium-rich olivine (forsterite). Possible mineralogical assemblages vary, resulting in a variety of peridotitic rock types that include dunite (consisting practically of olivine only), harzburgite and lherzolite (olivine and orthopyroxene, and olivine and clinopyroxene, respectively, with or without spinel), and their garnet varieties. Ringwood's 'pyrolite' model, based on the assumption that the upper mantle must produce basalt on melting and leave behind a dunite residuum, belongs to this category.

An upper-mantle model including rock types rich in olivine satisfies seismic and density data and the observed seismic anisotropy. Moreover, slabs of uppermost mantle tectonically emplaced within the crust (e.g. ophiolites) are broadly of peridotitic composition. In this model, the 400 km discontinuity is associated with a phase change from olivine to β-spinel, and the 670 km discontinuity with a phase change from spinel to perovskite + periclase + wüstite, the latter possibly coupled with a chemical change.

An alternative model has been developed by Anderson (1984) in which the Lehmann discontinuity is taken to be a chemical boundary between garnet lherzolite above and eclogite (garnet + pyroxene) or olivine eclogite below. The upper mantle between 220 and 670 km is supposed to be eclogite (the high-pressure equivalent of basalt), and to be the source and sink of the oceanic lithosphere. The proposed layering of lherzolite over eclogite is gravitationally stable, and Anderson argues that the high-pressure equivalent of subducted basalt can sink only to a depth of 670 km, where it accumulates during geological history. However, it should be noted that according to Ringwood (1982) only the subcrustal lithosphere (consisting of harzburgite) becomes relatively buoyant and stays in the upper mantle, while the oceanic crust (consisting of basalt) eventually sinks into the lower mantle.

In either case the chemical composition of the upper mantle relative to the lower mantle still remains much of a problem. Its solution is of fundamental importance to geodynamics, as chemical layering would be an obstacle to mantle-wide convection.

Anderson's model leads to a stably stratified mantle. In the peridotitic model, the question hinges on the nature of the 670 km discontinuity. Can the changes in physical parameters observed at this depth be explained purely in terms of phase transformations, or is a change in chemistry also necessary? The small variations in physical parameters brought about by minor changes in chemistry are not resolvable by geophysical observation; therefore all that can be said from geophysical data is that changes in elasticity and density appear to be compatible with, but do not require, a mantle of uniform composition.

A different approach consists of studying the phase equilibria of mantle minerals at relevant pressures and temperatures. According to Jeanloz & Thompson (1983) no known phase change satisfies the sharpness of the 670 km discontinuity. Thus, either a yet-unobserved phase change must be invoked, or a chemical discontinuity is required. However, this conclusion is subject to many uncertainties.

Geochemical data suggest that MORB (mid-ocean ridge basalts) and

OIB (oceanic island basalts) originate from two different reservoirs, the former being more depleted in K, U, and Th than the latter. Allègre *et al.* (1983a, b) have tried to estimate, from Rb/Sr and Sm/Nd systematics, the volume of mantle that has been involved in the differentiation of the crust. They conclude that the depleted mantle could be anywhere between 30 and 90 per cent of the mantle. However, such a 'three-box model' (crust, depleted mantle, and less-depleted mantle) is not necessarily indicative of stable stratification, as chemical domains could be randomly distributed within the mantle. If, on the other hand, 'depleted' mantle is identified with 'upper' mantle, then the required thickness of this layer varies between 700 and 1700 km.

This is where the matter stands at present. In Section 7.9 we shall return to the problem of mantle stratification, and discuss how it may affect the convection pattern.

The assumption of spherical symmetry is only a convenience for modelling. From the geodynamic viewpoint *departures from spherical symmetry* are of the utmost importance. Many, but by no means all, of these lateral variations are associated with plate boundaries – upwelling mantle material below diverging boundaries, and sinking lithospheric slabs at converging boundaries. The long-wavelength variations are the most useful in the study of the large-scale dynamic evolution of the Earth. For instance, a still-unanswered question is whether the thickness of the *tectosphere* or tectonic lithosphere (that is, the outermost shell that translates coherently during plate tectonics) is determined exclusively by physical (essentially thermal) factors, or whether lateral thickness variations are brought about by chemical factors. In the former case the thickness of the lithosphere in the 'tectonic' sense may be equated to the depth to the LVZ or asthenosphere; this identification is not necessarily valid in the latter case.

There is evidence that lateral variations within the upper mantle may quite possibly extend to the 400 km discontinuity. Jordan (1978), on the basis of travel-time observations of *ScS* waves, concluded that differences between subcontinental and suboceanic mantle extend to hundreds of kilometres of depth. Knopoff (1983), from inversion of regional surface wave dispersion data in the period range $30 \leqslant T \leqslant 250$ s, found the thickness of the *seismological lithosphere* (that is, the thickness of the shell above the LVZ) to be 100 km or less in regions other than Precambrian shields; however, no evidence exists of a LVZ under Precambrian shields down to a depth of 300 km, which is the depth of resolution of the data.

Some geochemical data can be interpreted to favour the existence of a continental 'keel'. Bell *et al.* (1982), for instance, showed that the source region of Precambrian alkaline rocks of deep origin (>100 km) beneath the Superior Province of the Canadian Shield has behaved as a closed geochemical system for the LIL (large-ion lithophile) elements since 2.6 Ga BP. This implies that their source (depleted sub-continental upper mantle, or SCUM) has retained its isotopic identity during this long period. A simple way to achieve this is to assume that SCUM has travelled coherently with the overlying crust.

Deviations from spherical symmetry are not confined to the lithosphere and upper mantle. The recently developed technique of *seismic tomography* is enhancing considerably our knowledge of mantle structure. In essence, seismic tomography combines the information contained in large numbers of criss-crossing seismic rays, to obtain a three-dimensional image of the medium through which the seismic energy has been propagated. As wave velocity is inversely related to temperature, the pattern of velocity anomalies is related to the pattern of mantle flow. Lateral variations in velocity have been detected both in the upper (Woodhouse & Dziewonski 1984) and lower (Dziewonski 1984) mantle. Furthermore, the varying degree of velocity anisotropy of surface waves in different tectonic regions (Regan & Anderson 1984) may be used to infer the direction of flow in the upper mantle. Thus, information from seismology can be correlated with gravity and heat flow to bear directly on the problem of mantle convection (see Sections 7.9 & 8.1). The high-frequency and low-frequency material properties of the mantle are related through the temperature.

7 Temperature and its effects

Temperature has a central role in geodynamics. It strongly controls the rheology of the Earth; it is related to the thickness of the lithosphere, and to viscous flow in the mantle; and, through various kinds of energy flow, it affects all geotectonic processes.

This chapter examines the thermal properties of, and processes in, the Earth. First the thermal parameters of silicates are introduced. Then the conduction equation is derived, and applied to the determination of the temperature profile of both the continental and the oceanic lithosphere. Finally a review of the thermal balance of the Earth, and of the constraints on the internal temperature distribution, leads to a discussion of thermal convection in the mantle.

The capacity of the mantle to flow over the long timescale ($t \geqslant 10^3$–10^4 as an order of magnitude) in response to thermally or mechanically induced stresses, makes convection central to any analysis of the dynamic, chemical, and geologic evolution of the Earth.

7.1 The importance of temperature in rheology

The three extrinsic rheological parameters affecting the mechanical behaviour of materials are *time*, *temperature*, and *pressure* (to which one should add the *chemical environment*, notwithstanding the fact that its effects are not well known, let alone understood). In Chapter 6 we studied the evidence for the conclusion that the Earth in the short time range is elastic with some anelastic attenuation, and examined the distribution with depth of elastic properties, damping, and density. All of these parameters are functions of the material, pressure, and temperature. The pressure at any depth can be obtained once the density distribution is known (see Section 6.3); and in Section 6.6 we discussed – limiting our attention to the lithosphere and the mantle – the materials most likely to form the interior. However, temperature has not yet been considered explicitly, although seismological and other properties depend on it.

Parameters such as t, p, and T (time, pressure, and temperature, respectively) affect rheological behaviour not only because intrinsic material parameters (for instance, in the case of elasticity, rigidity, bulk modulus, etc.) depend on them, but also because the rheological equation obeyed by a given material may be different under different (t, p, T) conditions. Temperature is therefore a fundamental quantity in rheology and geodynamics. The determination of its depth distribution is complicated because $T(z)$ (or $T(r)$) depends on several variables that are only imperfectly known. The depth distribution of heat sources and of the thermal properties of constituent materials, the mode of heat transfer, and the evolution of the system in time all play a part in the problem.

The two most basic observations with regard to the thermal state of the Earth are that there is an outflow of heat from the Earth which must be maintained by suitable processes, and that plates move. The latter observation implies, by the principle of conservation of mass, that there exist large-scale motions in the mantle, and as such can be taken as direct evidence of convective heat transfer in the interior.

The depth distribution of temperature may be inferred in two ways, either through a consideration of heat sources and modes of heat transfer, or by measuring the depth variations of temperature-related properties such as electrical conductivity. Observed surface heat flow furnishes a basic constraint, and any deduced $T(r)$ curve must be compatible with 'fixed points' such as phase transformations and melting.

In this chapter we examine the fundamentals of temperature and heat transfer in the Earth, review briefly the relevant evidence, and discuss some rheological and geodynamic implications. A good qualitative account of these topics can be found in Bott (1982). At a more advanced level, see Stacey (1977a) and O'Connell & Hager (1980). Several geothermal problems are treated in Turcotte & Schubert (1982). Other references will be given as the need arises.

7.2 Thermal properties of silicate polycrystals

Material thermal properties that will be of frequent use in the sequel are defined here. The subscripts p, V, T, and S denote constant pressure, constant volume, isothermal, and constant entropy (i.e. reversible adiabatic) changes, respectively.

A change in temperature in an unconstrained body results in a change in volume and therefore in density. In the general case a single crystal is thermally anisotropic, and its linear expansion coefficient is a tensor of rank two, α_{ij}, whose three principal components are related to the axes of an ellipsoid into which an imagined sphere within the crystal is transformed as a consequence of the expansion. The *volume expansion coefficient* is the first invariant of this tensor, i.e. $\alpha \equiv \alpha_{kk}$. Under the assumption of isotropy, usually permissible in polycrystals, α_{ij} reduces to a scalar, say α^L, and the volume expansion coefficient is $\alpha = 3\alpha^L$.

In terms of the volume (or density) change, the volume expansion coefficient is defined as

$$\alpha = \frac{1}{V} \left(\frac{\partial V}{\partial T}\right)_p = -\frac{1}{\rho} \left(\frac{\partial \rho}{\partial T}\right)_p \tag{7.1}$$

Consequently, as $\alpha > 0$, a temperature increase ($\Delta T > 0$) generates a volume increase (or density decrease). The expansion coefficient has dimensions $[T^{-1}]$ and is of the order of $10^{-5}\,K^{-1}$ for most silicates (K denotes the kelvin).

Recalling the definition of the bulk modulus or incompressibility (Section 3.1), we can define its inverse or *compressibility* as

$$\beta_{T,S} = -\frac{1}{V} \left(\frac{\partial V}{\partial p}\right)_{T,S} = \frac{1}{\rho} \left(\frac{\partial \rho}{\partial p}\right)_{T,S} \tag{7.2}$$

From Equations 7.1 and 7.2 it can be seen that

$$\left(\frac{\partial \alpha}{\partial p}\right)_T = -\left(\frac{\partial \beta_T}{\partial T}\right)_p$$

i.e. increasing pressure reduces α in a way related to the change in β_T with temperature.

Seismic Earth models give the adiabatic incompressibility k_S, as conditions during seismic wave propagation are approximately adiabatic. The distinction between adiabatic and isothermal values is usually neglected, as differences do not exceed a few per cent for most solids. If needed, the isothermal compressibility can be obtained from the thermodynamic relationship between β_T and β_S (see, for example, Birch 1966)

$$\beta_T = \beta_S + \frac{\alpha^2 T}{\rho c_p}$$

where c_p is the *specific heat* (per unit mass) at constant pressure, which is of the order of $1\,kJ\,kg^{-1}\,K^{-1}$ for silicates. Relations between thermodynamic parameters have been reviewed by Stacey (1977b).

Of particular interest in geodynamics is the dimensionless *Grüneisen parameter*, defined as

$$\gamma = \frac{\alpha k_S}{\rho c_p} = \frac{\alpha k_T}{\rho c_V} \tag{7.3}$$

k and c being the incompressibilities and the specific heats under the appropriate conditions. The classical definition of γ expresses it as the ratio of thermal pressure to thermal energy per unit volume. On the basis of considerations on thermal pressure arising from atomic vibrations, it is possible to obtain γ as a function of the incompressibility and its pressure derivative only (see Stacey 1977a). In principle, this makes the Grüneisen parameter derivable from seismological data; its value in the mantle shows very little variation with depth and is close to unity.

The relations between thermal parameters can be expressed simply by means of the Grüneisen parameter. For instance, from the above relation between β_T and β_S, and Equation 7.3, it follows immediately that

$$\frac{\beta_T}{\beta_S} = \frac{k_S}{k_T} = 1 + \gamma\alpha T$$

Also, the Grüneisen parameter finds application in the study of the internal temperature of the Earth (Section 7.7), since it is more nearly independent of p and T than its constituent quantities.

Another important thermal quantity is the *thermal conductivity K*. Fourier's Law of Heat Conduction establishes the linear proportionality between heat flow q_i and temperature gradient in a thermally isotropic medium

$$q_i = - K \, \partial T / \partial x_i \tag{7.4}$$

where the minus sign denotes that heat flows down the temperature gradient. As heat flow is a flux of energy through a unit area per unit time, K has dimensions $[mlt^{-3}T^{-1}]$, and is measured in $W\,m^{-1}\,K^{-1}$ in the SI system. Its value for silicates is in the range $1-5\ W\,m^{-1}\,K^{-1}$. (For anisotropic media the thermal conductivity is a tensor K_{ij}, which is symmetric because heat flow from an isolated point occurs along straight lines. Isotropy may be safely assumed for polycrystals.)

The conductive properties of a material are also conveniently expressed by its *thermal diffusivity x*, defined as

$$x = K/\rho c_V \tag{7.5}$$

The dimensions of x are $[l^2 t^{-1}]$, and its value is of the order of $10^{-6}\ m^2\,s^{-1}$ in silicates.

Thermal properties are themselves functions of temperature and pressure, and therefore their values change with depth. A table of the variations of the thermal quantities of the Earth in various depth ranges is presented in Section 7.7, in the context of a discussion of the internal temperature distribution. As discussed in Sections 7.7–7.9, the predominant mode of heat transfer in the Earth is not conduction but convection. Conduction is the relevant heat transfer mechanism only in the thermal boundary layers of the convective circulation, i.e. in the lithosphere and in non-adiabatic layers at the base of (and possibly within) the mantle.

7.3 Thermal stresses in the upper lithosphere

Any change in temperature tends to generate a change in volume (Eqn 7.1). If the medium is confined and elastic (a condition that applies to rocks in the upper crust down to a depth of 15–20 km), thermal stresses arise from the resistance to the temperature-induced volume change. In an isotropic body subject to a small temperature change ΔT, the strain can be regarded as the sum of a component due to thermal expansion, ε_{ij}^T, and a component due to the resistance of the medium (because changes in volume cannot be fully accommodated) ε_{ij}^E, that is

$$\varepsilon_{ij} = \varepsilon_{ij}^T + \varepsilon_{ij}^E$$

The thermal strain is related to the linear expansion coefficient by

$$\varepsilon_{ij}^T = \alpha^L \, \Delta T \, \delta_{ij}$$

and the elastic strain is related to stress in the usual Hookean manner (Eqn 3.12)

$$\varepsilon_{ij}^E = \frac{1+\nu}{E} \, \sigma_{ij} - \frac{\nu}{E} \, \sigma_{kk} \, \delta_{ij}$$

so that the total strain is

$$\varepsilon_{ij} = \frac{1+\nu}{E} \, \sigma_{ij} - \left(\frac{\nu}{E} \, \sigma_{kk} - \alpha^L \, \Delta T \right) \delta_{ij} \tag{7.6}$$

This equation can be solved for stress to yield

$$\sigma_{ij} = \lambda\theta \, \delta_{ij} + 2\mu\varepsilon_{ij} - \alpha^L (3\lambda + 2\mu) \, \Delta T \, \delta_{ij} \tag{7.7}$$

which is known as the *Duhamel–Neumann Law* (θ is cubical dilatation, and elastic parameters are denoted by the usual symbols; cf. Section 3.1).

146

Substituting Equation 7.7 into the equation of equilibrium (Eqn 2.7) one obtains

$$(\lambda + \mu)\,\frac{\partial \theta}{\partial x_i} + \mu \nabla^2 u_i + \rho X_i - \Lambda\,\frac{\partial T}{\partial x_i} = 0 \tag{7.8}$$

where $\Lambda = (3\lambda + 2\mu)\alpha^L$. Equation 7.8 is formally equivalent to the Cauchy–Navier equation derived in Section 3.2, with the body-force term replaced by $(\rho X_i - \Lambda \partial T/\partial x_i)$, and the tractions in the boundary conditions augmented by $\Lambda\,\Delta T n_i$, equivalent to an hydrostatic pressure. Thus, the thermoelastic problem is reduced to the elastic one, provided that $T(x_1, x_2, x_3)$ is known.

Are thermoelastic stresses of any relevance to geodynamics? Consider an elastic half-space confined in the horizontal directions ($\varepsilon_{11} = \varepsilon_{22} = 0$) and unconstrained in the vertical direction ($\sigma_{33} = 0$). Then, by Equation 7.6, the horizontal normal strains can be written as

$$\varepsilon_{11} = \frac{1+\nu}{E}\,\sigma_{11} - \left(\frac{\nu}{E}\,(\sigma_{11} + \sigma_{22}) - \alpha^L\,\Delta T\right) = 0$$

$$\varepsilon_{22} = \frac{1+\nu}{E}\,\sigma_{22} - \left(\frac{\nu}{E}\,(\sigma_{11} + \sigma_{22}) - \alpha^L\,\Delta T\right) = 0$$

i.e.

$$\sigma_{11} - \nu\sigma_{22} + E\alpha^L\,\Delta T = 0$$

$$\sigma_{22} - \nu\sigma_{11} + E\alpha^L\,\Delta T = 0$$

Subtracting the second of these equations from the first,

$$(1 + \nu)(\sigma_{11} - \sigma_{22}) = 0$$

that is, $\sigma_{11} = \sigma_{22}$. Adding,

$$(\sigma_{11} + \sigma_{22})(1 - \nu) + 2E\alpha^L\,\Delta T = 0$$

that is

$$\sigma_{11} = \sigma_{22} = -\frac{E}{1-\nu}\,\alpha^L\,\Delta T \tag{7.9}$$

From this equation it can be seen that an increase in temperature

147

generates a compression; for $E = 70$ GPa, $\nu = 0.25$, $\alpha^L = 10^{-5}$ K^{-1}, this thermally-induced normal stress is $\sigma_{11} = \sigma_{22} \simeq 10^6\ \Delta T$, i.e. a change $\Delta T \simeq 100$ K generates stresses of the order of 100 MPa (see also Turcotte & Schubert 1982).

These stresses are far from negligible, even in the high-stress environment of the upper crust (where they would be superimposed on stresses of different origin, such as those from plate boundary forces, geometric and physical inhomogeneities, sedimentation, and erosion, etc.). However, they would be relieved − partially or totally − if the medium was not constrained laterally. Also, we have assumed perfect elasticity. If the medium is viscoelastic, with a Maxwellian relaxation time (see Section 4.6) of the order of the age of the thermal event or shorter, then the thermal stresses are substantially relaxed. On a shorter timescale they can be relieved by fracture or transient creep, or both. Consequently, although their magnitude can be large in the upper crust, it is doubtful whether thermal stresses play a significant role in long-term geodynamic processes except in some special cases.

7.4 The conduction equation

Fourier's Law of Heat Conduction, introduced in Section 7.2, relates heat flow to temperature gradient through the thermal conductivity. It can be used to determine the temperature distribution within a body of prescribed geometry, heat-generating properties, and thermal boundary conditions, when thermal conduction is the relevant mechanism of heat transfer (see Carslaw & Jaeger 1959 for a treatise on this subject).

The heat-conduction equation is set up by considering an elemental parallelepiped of volume $dV = dx_1\ dx_2\ dx_3$ and faces normal to the co-ordinate axes (Fig. 7.1). The net heat flow through the faces perpendicular to the x_2-direction is

$$\left(q_2 + \frac{\partial q_2}{\partial x_2}\ dx_2\right)\ dx_1\ dx_3 - q_2\ dx_1\ dx_3 = \frac{\partial q_2}{\partial x_2}\ dV$$

and similarly for each of the two remaining pairs of faces perpendicular to the x_1- and x_3-directions. The net flow of heat from the elemental volume is therefore $(\partial q_i / \partial x_i)\ dV$.

At equilibrium the heat flow must be equal to the heat produced (or absorbed) by heat sources (or sinks) within dV, plus any change in the heat content of dV. The body is assumed not to move, i.e. convective terms are absent (cf. Section 1.6). Denoting the heat-production function (per unit mass and time) by $F(x_1, x_2, x_3, t)$ and any change of the heat

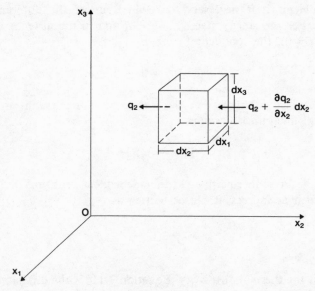

Figure 7.1 Heat conduction in infinitesimal volume dV. Only components in the x_2-direction shown.

content of dV by $\partial Q/\partial t$, we may therefore write

$$\frac{\partial q_i}{\partial x_i}\,\mathrm{d}V = -\frac{\partial Q}{\partial t} + \rho F\,\mathrm{d}V \tag{7.10}$$

where the minus sign signifies that an outflow of heat larger than the heat generated results in a decrease of heat content. Using Fourier's Law (Eqn 7.4) and the fact that the change in heat content is related to the temperature change by $\partial Q/\partial t = c\rho(\partial T/\partial t)\mathrm{d}V$, where c and ρ are the specific heat and density, respectively, Equation 7.10 becomes

$$\frac{\partial}{\partial x_i}\left(-K\frac{\partial T}{\partial x_i}\right)\mathrm{d}V = -c\rho\frac{\partial T}{\partial t}\,\mathrm{d}V + \rho F\,\mathrm{d}V$$

which, dividing by dV and assuming K constant, reduces to

$$K\nabla^2 T = c\rho\frac{\partial T}{\partial t} - \rho F \tag{7.11}$$

This equation is the general form of the *heat conduction equation* (see Section 7.8 for the general equation when convective terms are present). The assumption of constant K may be only an approximation, and in

149

some problems it is necessary to retain it under the differentiation operator. For the steady state ($\partial T/\partial t = 0$) and in the absence of heat sources ($F = 0$), the conduction equation reduces to

$$\nabla^2 T = 0 \qquad (7.11\text{a})$$

i.e. the Laplace equation. For the steady state and a non-zero heat-generating function, it becomes

$$\nabla_2 T = -(\rho/K)F \qquad (7.11\text{b})$$

which is the Poisson equation. In time-dependent problems and in the absence of heat sources, it can be written as

$$\nabla^2 T = \frac{1}{\varkappa} \frac{\partial T}{\partial t} \qquad (7.11\text{c})$$

where $\varkappa$ is the thermal diffusivity. Equation 7.11c is the diffusion equation that has been introduced in Section 3.4.

Heat conduction is of central importance in the determination of the thermal state of the continental and oceanic lithosphere (Section 7.6). Thermal processes in the bulk of the mantle, where convection is the predominant heat transfer process, are considered in Section 7.8. By way of introduction to these problems, we consider next the thermal budget of the Earth.

7.5 Observed heat flow and distribution of heat sources

In the Earth the direction of temperature increase is approximately vertical. If the vertical co-ordinate (positive downwards) is denoted by z, then the surface heat flow is given by

$$q = -K \, dT/dz \qquad (7.12)$$

where the minus sign denotes symbolically that heat flows upwards. The determination of q requires the measurement of both thermal conductivity and increase of temperature with depth in a borehole (see, for example, Bott 1982). The temperature gradient near the surface is very high ($dT/dz \simeq 30 \text{ K km}^{-1}$), but it must obviously decrease with depth, otherwise the Earth would be molten at depths of the order of a few tens of kilometres.

The results of heat flow observations are shown in the first part of

Table 7.1. The ocean floor has been subdivided into regions of the same age using magnetic isochrons; the continents have been classified, more roughly, into tectonic provinces whose age reflects the last tectono-thermal event. Although the scatter of measurements is very large, heat flow generally decreases with increasing age. (We have used mW m^{-2} – milliwatts per square metre – as the unit of heat flow, to conform with the SI system; note, however, that the μcal cm^{-2} s^{-1}, or HFU – heat flow unit – is still in common usage in geophysics: 1 HFU $\simeq$ 42 mW m^{-2}.)

The averages of heat flow observations in continents and oceans are similar, 57 and 66 mW m^{-2}, respectively. However, in the oceans the expected average heat flow (calculated on the basis of thermal models of lithospheric cooling, cf. Section 7.6) is larger: 99 mW m^{-2}. Consequently, approximately one-third of heat loss through the oceans must occur by means other than conduction. It is now accepted that hydrothermal transport, effected by water circulating in the topmost oceanic crust in areas with no or little sedimentary cover (see Sclater *et al.* 1980), accounts for the discrepancy.

The estimated heat loss through oceans, continents, and the whole globe is given in the second part of Table 7.1. The total estimated heat loss for the Earth is 42.0×10^{12} W, of which more than two-thirds escapes through the oceanic lithosphere.

Table 7.2 gives a breakdown of the heat loss in oceans, continents, and the whole globe according to the origin of the heat. The sources contributing to heat loss in continents and oceans are basically different, and accordingly the approximate equality of continental and oceanic heat flow is, to a certain extent, coincidental. In the oceans the greatest source of heat is the process of plate formation and cooling; in the continents it is radioactive heat production, mainly within the upper crust (see Section 7.6). For the Earth as a whole, radioactivity within the lithosphere (concentrated in the upper continental crust) contributes only 19 per cent of the total heat loss. The remaining 81 per cent comes from the mantle, primarily as a result of plate tectonic processes related to mass transfer (lithosphere formation and cooling in the oceans, and in minor part thermal events and orogeneses in continents), and subordinately as conductive flux from below the lithosphere.

Some, but not all, of the terrestrial heat ultimately comes from radioactivity. The main heat-producing isotopes are ^{238}U, ^{235}U, ^{232}Th, and ^{40}K. Table 7.3 lists average heat production in some acid, basic, and ultrabasic rocks. Rocks of acid-to-intermediate composition tend to be concentrated in the upper continental crust, and radioactivity generally decreases rapidly with depth. However, the mantle is not totally devoid of radioactive isotopes, and some ^{40}K may be present in the core. If the

151

Table 7.1 Heat loss from the Earth (from Sclater et al. 1980).

Observed heat flow

oceans age (Ma)	>160	160–125	125–110	110–95	95–80	80–65	65–52	52–35	35–20	20–9	9–4	4–0
q(mW m^{-2})	50±25	49±16	55±21	54±17	54±20	57±28	62±31	57±35	60±38	71±52	117±180	149±126

continents age (Ma)	>1700	1700–800	800–250	250–0
q(mW m^{-2})	46±16	50±10	63±21	76±53

Estimated heat loss

	q(mW m^{-2})	Area (10^6 km^2)	Heat loss (10^{12} W)	Percentage of Earth's loss
oceans	66 (observed)		20.3	48
	+ 33 (hydrothermal)		+ 10.1	+ 24
	= 99 (total)	308.6	= 30.4	= 72
continents	57 (observed)	201.5	11.5	28
Earth	82 (estimated)	510.1	42.0	100

Table 7.2 Heat loss according to sources (from Bott 1982).

Source	Percentage of oceanic (o) or continental (c) heat loss	Percentage of total heat loss
Oceans		
plate formation and cooling	85(o)	62
radioactivity within lithosphere	5(o)	4
heat flux from sublithospheric mantle	10(o)	7
	100(o)	
Continents		
radioactivity within upper crust	40(c)	11
radioactivity from rest of lithosphere	15(c)	4
heat flux from sublithospheric mantle	25(c)	7
cooling lithosphere after thermal events	20(c)	5
	100(c)	100
Earth		
radioactivity within lithosphere		19
heat from the mantle through plate formation, cooling, and thermal events		67
heat conduction across the lithosphere		14
(sum of previous two: heat coming from the mantle)		(81)

Earth as a whole is assumed to have a radioactive isotope concentration equivalent to carbonaceous chondrites, as shown in Table 7.3, then the resulting present heat generation is 31.4×10^{12} W. On the other hand, if the Earth is heavily depleted of potassium, then heat generation could be considerably less. In any event, present heat loss exceeds heat generation, and probably between 25 and 50 per cent of the present heat loss comes from sources other than present radioactivity. Barring yet-undiscovered heat-producing processes, this excess heat is likely to be released through a slow cooling of the Earth (see also Section 7.10).

So far we have been dealing with present values of heat generation and heat loss. The concentration of long-lived radioactive isotopes decreases with time, and consequently the thermal budget is time-dependent: both heat production and loss were higher in the Precambrian. Furthermore, other heat sources (short-lived radioactive isotopes, heat of accretion, core formation – cf. Bott 1982 for a brief discussion and further

153

Table 7.3 Heat generation in various rock types and meteorites, and in the Earth as a whole (from Stacey 1977a).

Material	Heat generation $(10^{-12} \, W \, kg^{-1})$
granites	1050
alkali basalts	180
tholeiitic basalts	27
eclogites	9.2
peridotites	1.5
ordinary chondrites	5.9
carbonaceous chondrites	5.2

Earth				
radioactive isotope	^{238}U	^{235}U	^{232}Th	^{40}K
heat generation ($\times 10^{12}$ W)	11.33	0.49	11.18	8.41
total present heat generation		31.4×10^{12} W		

references) had a role in the early stages of the history of the Earth. However, decay of long-lived radioactive isotopes and secular cooling, in that order, can safely be assumed to be the only important heat sources in the Earth after its early history.

7.6 Temperature in the lithosphere

In a lithosphere plate (refer to the discussion of the term 'lithosphere' in Section 6.6) heat is transferred by conduction, except in anomalous areas where magmatic or hydrothermal processes play a significant role. The temperature distribution can be determined by solving the relevant form of the heat transfer equation (Eqn 7.11). However, conditions are different for the continental and the oceanic lithosphere. In the former the total time derivative of temperature vanishes, and heat production is important; in the latter cooling occurs as the material moves away from the ridge, and heat production is negligible (cf. Bott 1982 and Turcotte & Schubert 1982 for further discussion and references).

Consider first the continental lithosphere. Denoting by z the vertical co-ordinate (positive downwards) and assuming temperature to be a function of depth only, we see that a one-dimensional form of Equation 7.11b applies. Taking into account the decrease in concentration of heat-producing elements with depth, we take the heat-generating function to

154

be of the type

$$F = F_0 \exp(-z/h)$$

where F_0 is the radioactive heat generation at the surface, and h is the depth at which $F = F_0 \exp(-1)$. (Although the choice of F is not determined uniquely by the observed heat flow, an exponential decrease of heat production with depth fits the available data.) The heat-conduction equation then becomes

$$K \, d^2T/dz^2 + \rho F_0 \exp(-z/h) = 0 \qquad (7.13)$$

which, by integration, gives

$$K \, dT/dz - \rho F_0 h \exp(-z/h) = c'$$

Making use of Fourier's Law (Eqn 7.12), and determining the constant of integration c' from the condition $q = -q_m$ for $z \to \infty$, this equation becomes

$$q = -q_m - \rho F_0 h \exp(-z/h)$$

which gives the heat flow at any depth z. The quantity q_m, representing the heat flow at large depth, is termed the reduced heat flow. Since $q = -q_0$ at the surface (q_0 being the observed heat flow), if follows immediately that

$$q_0 = q_m + \rho F_0 h \qquad (7.14)$$

i.e. the observed heat flow in continental areas is a linear function of the radioactive heat generation in the surface layer. Equation 7.14 is found to hold in most continental areas, with parameters clustering around the values $q_m \simeq 20\text{--}30 \text{ mW m}^{-2}$ and $h \simeq 10$ km.

Integrating Equation 7.13 again, we have

$$KT + \rho F_0 h^2 \exp(-z/h) - q_m z = c''$$

where, using the boundary condition $T = T_0$ at $z = 0$, the constant of integration is

$$c'' = KT_0 + \rho F_0 h^2$$

We can write therefore

$$T = T_0 + (\rho F_0 h^2/K)[1 - \exp(-z/h)] + (q_m/K)z \qquad (7.15a)$$

or, since $\rho F_0 h = q_0 - q_m$ (Eqn 7.14),

$$T = T_0 + [(q_0 - q_m)h/K][1 - \exp(-z/h)] + (q_m/K)z \qquad (7.15b)$$

Equations 7.15a and b are the required solution to the heat-conduction equation in the continental lithosphere, and permit the computation of $T(z)$ in terms of parameters which are either directly measurable or can be inferred from observation. Some *continental geotherms*, for different values of the surface heat flow, are shown in Figure 7.2. $T(z)$ is highly variable from one heat-flow province to another. For instance, the temperature at the Moho may vary from as low as 600 K in Precambrian shields to perhaps 1200 K in some Cenozoic orogens. The curves representing T_m and $0.85T_m$, where T_m is the solidus temperature of the upper mantle, are also shown. If the seismic low-velocity zone is defined by the condition $T \geqslant 0.85T_m$ (either because it is due to subsolidus effects as $T/T_m \to 1$, or because the upper mantle solidus is depressed by the presence of volatiles such as H_2O or CO_2), then the thickness of the seismological lithosphere under continents is predicted to vary from less than 50 km in areas of very high heat flow to the total absence of a LVZ beneath Precambrian shields. This inference closely matches the results obtained from surface wave dispersion discussed in Section 6.4. Given the surface heat flow and a petrological model of crust and upper mantle,

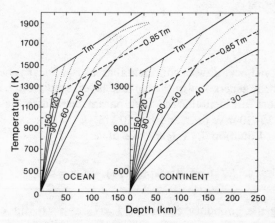

Figure 7.2 Oceanic and continental geotherms for different values of surface heat flow (in $mW\,m^{-2}$). The intersection of geotherms with the curve $T = 0.85T_m$ may be taken as defining the thickness of the lithosphere (from Pollack & Chapman 1977).

temperatures within the lithosphere can probably be considered to be known to within ± 100 K.

Different conditions apply in the oceanic lithosphere, since the material originating at mid-oceanic ridges cools as it moves away by sea-floor spreading, and heat production in the oceanic crust is negligible. If x and z are the horizontal and vertical co-ordinate, respectively, and $v = \mathrm{d}x/\mathrm{d}t$ is the spreading rate, assumed constant and determined in practice from the age t of the crust at distance x from the ridge, then the total time derivative of temperature is (note that the temperature field is steady-state, i.e. $\partial T/\partial t = 0$)

$$\dot{T} = \frac{\partial T}{\partial x}\frac{\mathrm{d}x}{\mathrm{d}t} = v\,\frac{\partial T}{\partial x}$$

Furthermore, since horizontal conduction is negligible compared with horizontal heat transport by plate motion, the Laplacian of temperature can be written as

$$\nabla^2 T = \frac{\partial^2 T}{\partial x^2} + \frac{\partial^2 T}{\partial z^2} \simeq \frac{\partial^2 T}{\partial z^2}$$

so that the heat transfer equation becomes

$$\frac{\partial^2 T}{\partial z^2} = \frac{v}{\varkappa}\frac{\partial T}{\partial x} = \frac{\dot{T}}{\varkappa} \tag{7.16}$$

In other words, the problem of finding $T(x, z)$ in a moving plate becomes, by the transformation $x = vt$, equivalent to that of finding $T(z, t)$ in a stationary plate.

The boundary conditions for the oceanic lithosphere are $T = T_0$ at $z = 0$ (T_0 is the surface temperature of the plate, whose top is assumed to be flat) and $T = T_\mathrm{m}$ (or some fraction thereof) at the lower boundary of the plate (which is not parallel to the upper boundary, but intersects the surface at the ridge axis $x = 0$ and becomes deeper with increasing x). The classical solution to the cooling of a semi-infinite half-space can then be adapted to be a solution of Equation 7.16 (cf., for example, Turcotte & Schubert 1982); it is

$$\frac{T - T_0}{T_\mathrm{m} - T_0} = \mathrm{erf}\left(\frac{z}{2(\varkappa t)^{1/2}}\right) = \mathrm{erf}\left(\frac{z}{2(\varkappa x/v)^{1/2}}\right) \tag{7.17}$$

where $\mathrm{erf}(y)$ is the error function

$$\mathrm{erf}(y) = \frac{2}{\pi^{1/2}}\int_0^y \exp(-\zeta^2)\,\mathrm{d}\zeta$$

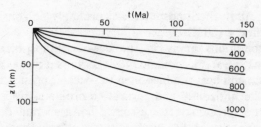

Figure 7.3 Isotherms in oceanic lithosphere; figures on curves give $T - T_0$ in kelvin (from Turcotte & Schubert 1982).

The *isotherms* determined from Equation 7.17 are approximately parabolic in shape, and are shown in Figure 7.3. Results can also be presented in terms of *oceanic geotherms*. Those shown in Figure 7.2 have been calculated including an initial temperature gradient in the lithosphere and a small heat production in the uppermost 10 km. Again, the depth to the LVZ estimated in this way (about 80–100 km beneath old oceanic basins, progressively less for younger ages) agrees well with seismological data. The *thermal lithosphere* can be defined as the material at $T \leqslant T_{crit}$, where $T_{crit} \leqslant T_m$ ($T_m \simeq 1500$ K for dry peridotite at upper mantle pressures). It coincides reasonably well with the seismological lithosphere, and also with the tectonic lithosphere if the seismic LVZ is a zone of decoupling between lithosphere and mantle.

The thermal thickness of the oceanic lithosphere may therefore be defined as the depth to a given critical isotherm. The oceanic lithosphere may be regarded as the moving *thermal boundary layer* of the mantle, where cooling takes place by conduction and the temperature gradient is steeper than in the convecting substratum. The thickness of the boundary layer as a function of its age (or of distance from the ridge) can be obtained directly from Equation 7.17. Taking $T_{crit} = 0.9T_m$, $T_m = 1500$ K, and $T_0 = 300$ K we have

$$\frac{T_{crit} - T_0}{T_m - T_0} = \mathrm{erf}\left(\frac{z_L}{2(xt)^{1/2}}\right) \simeq 0.9$$

where $z_L \equiv h_L$ is the depth to T_{crit}, i.e. the thickness of the thermal lithosphere. Since $\mathrm{erf}(y) \simeq 0.9$ for $y \simeq 1.15$, it follows that

$$h_L/2(xt)^{1/2} \simeq 1.15$$

that is,

$$h_L \simeq 2.3(xt)^{1/2} \quad \text{or} \quad h_L \simeq 2.3(x x/v)^{1/2} \tag{7.18}$$

158

Equation 7.18 gives the thermal thickness of the oceanic lithosphere as a function of its age (or distance from the ridge). The numerical factor depends on the choice for T_0, T_m, and T_{crit}, but it is usually close to the value used here. With our choice of parameters, $h_L \simeq 100$ km for $t \simeq 60$ Ma, an indicative 'average' value for oceanic basins. The thickness of the lithosphere increases according to $t^{1/2}$. This 'square-root law' fits reasonably well the evidence from surface wave dispersion for $t \lesssim 80$ Ma. Older lithosphere tends asymptotically towards a constant thickness, which is probably indicative of some source of heat in the asthenosphere (a temperature gradient, thermal convection, or shear heating).

The relation between h_L and t is only one of several 'square-root laws' holding for the oceanic lithosphere. The predicted surface heat flow can be immediately derived from Fourier's Law of Heat Conduction (Eqn 7.4) and the temperature solution (Eqn 7.17). The surface heat flow $q_0 = -q$ is, according to Fourier's Law,

$$q_0 = K \frac{dT}{dz} \bigg|_{z=0}$$

where the temperature gradient is evaluated at the surface. But $T(z)$ is given by Equation 7.17; therefore, recalling the definition of the error function,

$$\frac{dT}{dz} \bigg|_{z=0} = \frac{T_m - T_0}{(\pi \varkappa t)^{1/2}}$$

Consequently, the surface heat flow is given by

$$q_0 = K(T_m - T_0)(\pi \varkappa t)^{-1/2} = K(T_m - T_0)(\pi \varkappa x / v)^{-1/2} \qquad (7.19)$$

The decrease of heat flow with the square root of age (or distance from the ridge axis) expressed by Equation 7.19 is roughly confirmed by observation. However, for $t \lesssim 40$ Ma, observed heat flow is less than predicted. This is attributable to the convective cooling caused by hydrothermal circulation in the oceanic crust (cf. Sclater et al. 1980, and the discussion in Section 7.5). With increasing age, the increasing thickness of the sedimentary cover (less permeable than fractured basalt) prevents penetration of sea water into the oceanic crust, and accordingly convective cooling no longer operates effectively. For $t \gtrsim 80$ Ma, surface heat flow is down to its asymptotic value, reflecting heat contributed from the asthenosphere, plus the small radioactive heat generation within the oceanic lithosphere.

The conduction equation may also be used to predict ocean floor topography (see Turcotte & Schubert 1982). As the oceanic lithosphere is in approximate isostatic equilibrium, the average depth to the sea floor must increase with age — or distance from the ridge axis — as a consequence of the density increase brought about by cooling. Isostatic equilibrium requires the weight per unit area down to some 'depth of compensation' to be the same for any vertical column. With reference to Figure 7.4, where $h_w \equiv z_w$ is the thickness of the water layer, $h_L = z_L - z_w$ is the thickness of the lithosphere, ρ_m and ρ_w are the reference mantle density and water density, respectively, and ρ is the variable density of the lithosphere, the condition for isostatic equilibrium may be written as

$$\rho_m(h_w + h_L) = \rho_w h_w + \int_0^{h_L} \rho \; dz$$

or, rearranging terms,

$$h_w(\rho_m - \rho_w) = \int_0^{h_L} (\rho - \rho_m) \; dz$$

The lithosphere density is larger than the reference mantle density because of thermal contraction of the lithosphere as it moves away from the ridge.

From the definition of the thermal expansion coefficient (Eqn 7.1), and assuming lithosphere and mantle to have the same composition, we can write

$$\rho - \rho_m = \alpha\rho_m(T_m - T)$$

and the condition of isostatic equilibrium becomes

$$h_w(\rho_m - \rho_w) = \alpha\rho_m \int_0^{h_L} (T_m - T) \; dz$$

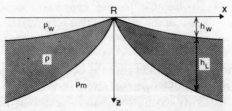

Figure 7.4 Isostatic equilibrium of oceanic lithosphere (shaded); R denotes the position of the ridge (from Turcotte & Schubert 1982).

Substituting Equation 7.17 for T, we obtain

$$h_w(\rho_m - \rho_w) = \alpha\rho_m(T_m - T_0) \int_0^{h_L} \left[1 - \mathrm{erf}\left(\frac{z}{2(xt)^{1/2}}\right) \right] dz$$

Introducing the new variable $\xi = z/[2(xt)^{1/2}]$ under the integral sign, and recalling the definition of the complementary error function

$$\mathrm{erfc}(\xi) = 1 - \mathrm{erf}(\xi)$$

the condition for isostatic equilibrium becomes

$$h_w(\rho_m - \rho_w) = 2\alpha\rho_m(T_m - T_0)(xt)^{1/2} \int_0^\infty \mathrm{erfc}(\xi) \, d\xi$$

where the change in the upper limit of the integral is permissible since $\rho = \rho_m$ at the base of the lithosphere, and consequently the contribution for $z > z_L$ is zero. The value of the definite integral is $\pi^{-1/2}$, and therefore we finally obtain

$$h_w = \frac{2\alpha\rho_m(T_m - T_0)}{\rho_m - \rho_w} \left(\frac{xt}{\pi}\right)^{1/2} = \frac{2\alpha\rho_m(T_m - T_0)}{\rho_m - \rho_w} \left(\frac{xx}{\pi v}\right)^{1/2} \qquad (7.20)$$

Equation 7.20 shows that, if conditions of isostatic equilibrium prevail in the oceanic lithosphere, then the average depth of the ocean floor increases as the square root of the age or distance from the ridge axis. This is confirmed by observation for $t \leqslant 80-100$ Ma (see Bott 1982 for a review of the evidence; cf. also Turcotte & Schubert 1982, for a fuller analytical treatment of the problems treated in this section).

The model of the oceanic lithosphere as a thermal boundary layer that cools as it moves away from the ridge axis therefore leads to predictions of temperature (Eqn 7.17), thickness (Eqn 7.18), heat flow (Eqn 7.19), and depth of the ocean floor (Eqn 7.20) that stand the test of observation. The process of subduction is also related to the temperature in the descending lithosphere. Thermal processes in the oceanic lithosphere affect the whole tectonic cycle. Sometimes the connection may not be obvious at first sight. For instance, changes in sea level of the order of 200–300 m over a timescale of 10–100 Ma are too large and long-lived to be attributable to hypothetical variations of polar ice volume; however, they may be related to periods of faster spreading or more spreading centres, or both, which would result in younger and therefore shallower oceans, with consequent worldwide transgressions (as in the Late Cretaceous, approximately 80 Ma BP).

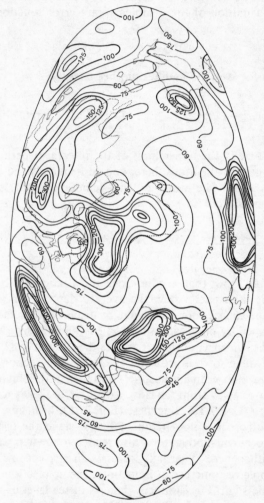

Figure 7.5 Thickness of the thermal lithosphere (contours in km) (from Pollack & Chapman 1977).

Several times during the previous discussion we have referred to the 'thermal lithosphere', defined as the material below a given critical temperature, and we have chosen the latter to be at, or not far below, the estimated solidus temperature. Figure 7.5 shows the thickness of the lithosphere, defined as the depth where $T = 0.85T_m$, calculated from geotherms with surface heat flow as the main independent variable (Pollack & Chapman 1977). Since the thickness is determined from a degree-12 spherical harmonic representation of heat flow, short-wavelength variations are not visible. The thermal lithosphere is thinnest in young oceanic regions, and $h_L \leqslant 100$ km over most of the globe. It thickens considerably in continental areas, and in Precambrian shields the geotherms do not reach the critical temperature at any depth. This matches very well the seismological data discussed in Sections 6.4 and 6.6, and indirectly shows that the seismic LVZ is thermally controlled. As far as the tectonic lithosphere is concerned, it is simplest to assume it to be coincident with the seismological and thermal lithosphere in areas where a seismic LVZ is present. However, under ancient continents there is no obvious layer of decoupling between tectosphere and substratum. This, together with the geochemical data (Section 6.6), suggests the possibility that the shield tectosphere extends to depths of 250–400 km. Such a large difference between continental and oceanic lithosphere, if real, should have geodynamic consequences, detectable for instance in the rate of motion of plates containing ancient continents (cf. the discussion in Section 8.5).

7.7 The thermal state of the interior of the Earth

The geothermal gradient must decrease below the lithosphere, otherwise the bulk of the mantle would be molten. Conduction alone is unable to account for the temperature distribution in the interior. The thermal conductivity is the sum of two terms, $K = K' + K''$, the first related to lattice vibrations, the second to radiation (which becomes important only at high temperature, $T \geqslant 1000$ K). Heat transfer by exciton conductivity, whose contribution to total thermal conductivity is difficult to assess for mantle conditions, may also play a role. As $K' \propto T^{-3/4}$ and $K'' \propto T^3$, thermal conductivity generally increases with depth in the mantle, perhaps by one order of magnitude (representative values of thermal parameters in various depth ranges are given in Table 7.4). However, thermal diffusivity is approximately constant (10^{-6} m^2 s^{-1}) as density increases with depth.

That such a thermal diffusivity is insufficient to remove heat from the deep interior can be seen from the solution to the conduction equation.

163

A half-space, $z \geqslant 0$, initially at temperature T^*, cools according to (see, for instance, Eqn 7.17)

$$T(z, t) = T^* \, \mathrm{erf}\left(\frac{z}{2(\varkappa t)^{\frac{1}{2}}}\right) \tag{7.21}$$

The similarity variable $\xi = z/2(\varkappa t)^{\frac{1}{2}}$ describes the relation between the distance and the time characteristic of diffusion problems. Since $\mathrm{erf}(\xi) \simeq \frac{1}{2}$ for $\xi = \frac{1}{2}$, the temperature at depth z falls to $T = \frac{1}{2}T^*$ for $t = z^2/\varkappa$. The characteristic time may be regarded as the time of propagation of temperature disturbances over a given distance. For $z = 1000$ km, $\varkappa = 10^{-6}$ m^2 s^{-1}, the characteristic time is $t \simeq 30$ Ga, much longer than the age of the Earth. Consequently, heat generated in the interior would not have been removed by conduction in the time available.

It therefore follows that heat must be transported by *convection*, i.e. heat transfer by mass transfer. The dynamics of convection, and its relevance to geodynamics, will be examined in Sections 7.8 and 7.9. Here, we consider the main outlines of the variation of temperature with depth.

In a convecting system the temperature gradient must be close to the *adiabatic gradient* except in thermal boundary layers. The adiabatic gradient is the rate of increase of temperature with depth resulting from increase in pressure. Since the latter does work to reduce the volume of the material, heat is generated, with a resulting increase in temperature, since there is no heat exchange with the surroundings. If the material is incompressible there is no adiabatic temperature increase; however, silicates are sufficiently compressible for significant adiabatic gradients to be formed.

The adiabatic gradient can be derived from the basic thermodynamic relation between entropy (per unit mass) S, temperature, and pressure:

$$dS = \frac{c_p}{T}\,dT - \frac{\alpha}{\rho}\,dp$$

which yields ($dS = 0$ in a reversible adiabatic process)

$$\left(\frac{dT}{dp}\right)_S = \frac{\alpha T}{c_p\rho} = \frac{\gamma T}{k_S} \tag{7.22}$$

In Equation 7.22, use has been made of the definition of the Grüneisen parameter γ (Eqn 7.3). As $dp/dz = \rho g$ in homogeneous regions, where this condition is satisfied the adiabatic increase of temperature with depth can be written as

$$\left(\frac{dT}{dz}\right)_s = \left(\frac{dT}{dp}\right)_s \frac{dp}{dz} = \frac{\alpha g T}{c_p} = \frac{\gamma g T}{\phi} \tag{7.23}$$

where ϕ is the seismic parameter (Eqn 6.21).

The definition of adiabatic gradient in terms of the Grüneisen parameter allows a simple estimate of the adiabatic temperature change across a homogeneous region (Stacey 1977a). Integration of Equation 7.22 gives

$$\ln\left(\frac{T_2}{T_1}\right) \simeq \gamma \int_{p_1}^{p_2} \frac{dp}{k_S}$$

(γ has been taken outside the integral as it is approximately constant). Using the definition of incompressibility, $(\partial\rho/\partial p)_S = \rho/k_S$, and therefore

$$\frac{T_2}{T_1} \simeq \left(\frac{\rho_2}{\rho_1}\right)^\gamma \tag{7.24}$$

Equation 7.24 yields the adiabatic temperature change across a finite depth range in terms of the ratio of densities, given a fixed point in the temperature profile with depth.

The adiabatic gradient is the approximate temperature gradient in a convecting system. A necessary (but not sufficient) condition for convective instability is $(dT/dz) > (dT/dz)_S$, so that material displaced adiabatically upwards (downwards) is hotter (cooler) than the surrounding material, and therefore unstable in the field of gravity. The temperature gradient tends to the adiabat in homogeneous regions; in the lower mantle, for instance, $(dT/dz)_S \simeq 0.25-0.30 \text{ K km}^{-1}$, using the values given in Table 7.4.

An upper limit to mantle temperature is the solidus temperature T_m. Moreover, the melting point of core material brackets the temperature at the Gutenberg discontinuity, since T must be below T_m for mantle material but above T_m for core material. Also, the T and T_m curves must cross at the outer core/inner core boundary. However, the solidus temperature is subject to wide uncertainties, as one must extrapolate $T_m(p)$ to pressure outside the static experimental range (although transient shock-wave experiments can reach core pressures, $p \simeq 10^2 \text{ GPa}$). Several empirical equations for the estimation of T_m at high pressure have been proposed, all connected in some way with *Lindemann's Law of Melting*, according to which melting occurs when the amplitude of atomic vibrations about their equilibrium position becomes larger than

Table 7.4 Spherically symmetric thermal model of the Earth (modified from Stacey 1977c).

z (km)	α (10^{-5} K^{-1})	c_p (kJ kg^{-1} K^{-1})	K (W m^{-1} K^{-1})	$\varkappa$ (10^{-6} m^2 s^{-1})	γ	T_m (K)	T (K)
100–400 (upper mantle)	3.0 1.9	1.2 1.3	3.1 6.9	0.7 1.5	0.85 0.78	1600 2100	1500 1700
400–670 (transition zone)	1.9 1.5	1.3 1.3	6.9 7.3	1.3 1.4	0.77 0.74	2100 2300	1700 2000
670–2891 (lower mantle)	1.9 1.0	1.3 1.3	7.3 9.6	1.2 1.3	1.03 0.91	2300 3200	2000 3000
2891–5150 (outer core)	1.6 0.8	0.7 0.7	27 35	3.9 4.2	1.42 1.27	2900 4200	3000 4200
5150–6371 (inner core)	0.8 0.7	0.6 0.6	36 36	4.6 4.6	1.20 1.19	4200 4300	4200 4300

a given fraction of the interatomic distance. A convenient form of Lindemann's Law is (Stacey 1977b)

$$\frac{1}{T_m}\left(\frac{dT_m}{dp}\right) = \frac{2(\gamma - \frac{1}{3})}{k_m} \tag{7.25}$$

where γ is the Grüneisen parameter and k_m the incompressibility along the melting curve. Estimates of T_m at various depths are given in Table 7.4, but in the lower mantle and core they carry uncertainties of possibly hundreds of kelvins.

Adiabatic gradient (in regions that can reasonably be expected to be near the adiabat) and melting temperature provide constraints on the temperature variation with depth. Other useful information is obtained from electrical conductivity measurements, and from studies of phase equilibria of relevant minerals.

Electrical conductivity is related to temperature by an equation of the type

$$\sigma = \sigma_0 \exp(-E/kT) \tag{7.26}$$

where σ is the conductivity, E is an activation energy of the process, and k is the Boltzmann constant (see Garland 1979; also Gough 1983, for a review). However, uncertainties are very large, being both inherent in the determination procedure and related to factors other than temperature (partial melting, presence of water and volatiles, impurities) affecting the electrical transport properties of silicates. Spherically symmetric profiles of $\sigma(z)$ for the mantle range from $\sigma \simeq 10^{-2}$ S m^{-1} (siemens per metre) in the upper mantle to $\sigma \simeq 1-10$ S m^{-1} in the lower mantle, with the change

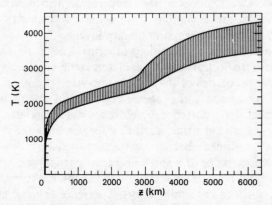

Figure 7.6 Upper and lower bounds of temperature distribution with depth. The area in which most estimates fall is shaded.

concentrated in the transition zone. Determinations of σ in the lithosphere and asthenosphere in different tectonic regions show considerable lateral variations, with relatively high conductivities (10^{-1}–1 S m^{-1}) corresponding to the asthenosphere in oceanic and continental orogenic regions, but absent elsewhere under continents. This may be related to the occurence of partial melting.

The study of mineral phase equilibria is useful in two ways. It is possible to estimate the temperature of equilibration of certain mineral assemblages in upper mantle rocks that have been tectonically transported to the surface (pyroxene geothermometry), which gives an indirect way of estimating temperature down to about 200 km. Furthermore, studies of the (p, T)-relations pertaining to the olivine → spinel and spinel → post-spinel transitions allow 'fixed points' of temperature to be established at $z = 400$ and 670 km ($T = 1700$ and 2000 K, respectively, with uncertainties of the order of ± 200 K).

Although spherically symmetric thermal models cannot achieve the accuracy of seismological models, most estimates of $T(z)$ are within a few hundred kelvins of each other at any depth. Figure 7.6 shows the margins within which most estimates fall. Table 7.4 is a compilation of the variations of thermal quantities in various depth ranges (mostly from Stacey 1977c, but T has been revised downwards to bring it in line with more recent estimates of 'fixed points'). The first value of each parameter in a given depth range refers to the top of the region, the second to the bottom. Values of T_m and T, especially below the transition zone, are subject to uncertainties of hundreds of kelvins.

The main features of the Earth as determined from seismology match well the thermal models. The mantle is solid (but $T/T_m \to 1$ in the upper mantle at $z \simeq 100$–200 km), the outer core is liquid, and the inner core is (just) solid. Two features of the temperature distribution that merit further discussion are deviations from spherical symmetry and occurrence of non-adiabatic thermal boundary layers.

Average continental and oceanic geotherms differ from each other in the lithosphere (see Section 7.6). These lateral temperature variations are reflected in the properties of the seismic low-velocity zone (LVZ) which is present under oceans and orogenic regions, but absent under stable continental areas. Its occurrence therefore seems to be related to the ratio T/T_m reaching a critical value, such as approximately 0.85.

It is usually assumed that the seismic low-velocity, low-Q zone is related to the occurrence of a small degree of partial melting (0.1–1 per cent). That would require a depression of the solidus curve in the LVZ depth range, which can be achieved by the presence of water and volatiles (see Bott 1982 for discussion and references). Partial melting is an attractive explanation also in view of the high conductivity associated with the

LVZ; however, the minimum requirement is for a relatively large T/T_m ratio, as subsolidus effects can adequately explain the observed properties. Tozer (1981), for instance, has argued that free water from amphibole dehydration would be an adequate agent to account for the LVZ, and consequently suggested that temperatures in the upper mantle are lower than are commonly assumed. The problem of geophysical constraints on partial melting has been reviewed in detail by Shankland *et al.* (1981).

Deviations from spherical symmetry are not confined to the lithosphere and seismic LVZ, but affect the whole mantle, as the temperature differences between hot ascending material and cold descending material in a convective system set up horizontal temperature gradients. These will be discussed in Section 7.9, in the context of convection.

Superadiabatic thermal boundary layers occur in the mantle. The lithosphere itself is the upper thermal boundary layer of the convective circulation. There are strong indications of a thermal boundary layer at the base of the mantle (region D''), not only from seismological observations, but also on the basis of thermal considerations. Taking $T = 2000$ K at $z = 670$ km, and a lower mantle temperature gradient $dt/dz \simeq 0.30$ K km^{-1}, the temperature at the mantle/core boundary would be below the melting point of core material. However, if there is heat flowing from the core into the mantle, then a thermal boundary layer occurs at the base of the lower mantle, with temperature probably increasing by 300–800 K over a depth range of 100–200 km. This makes $T > T_m$ in the uppermost core, as shown in Table 7.4 (which takes the occurrence of this boundary layer into account). The estimate $T_m \simeq 2800$–3000 K for the outermost core is based on the assumption that sulphur is the light element present in the core; if, as is entirely possible, the light element is oxygen, then $T_m \simeq 3200$–3500 K, and a thermal boundary layer in the lowermost mantle may no longer suffice (cf. Bott 1982 for a brief discussion of core composition).

If a thermal boundary layer at the base of the mantle is not sufficient to bring the mantle/core boundary temperature to the required level, where can other superadiabatic temperature increases be? Phase changes are not likely to contribute much. The 400-km olivine $\rightarrow$ spinel transition is exothermic, with a latent heat $L \simeq 160$ kJ kg^{-1}, resulting in a temperature increase $\Delta T \simeq 100$–150 K for flow across it. The latent heat for the 670-km phase change is not known; and phase changes in the lower mantle, if present, are likely to be minor. This leaves only two possibilities, as discussed by Jeanloz & Richter (1979). The D'' layer may be chemically distinct from the overlying mantle, with thermal boundary layers both at its top and its bottom, and total $\Delta T > 800$ K; or another superadiabatic layer may exist within the mantle. The most probable

169

position for this hypothetical layer (with thickness and temperature increase comparable with those in the lowermost mantle) is as a double thermal boundary layer between separate upper and lower mantle convective circulations.

The occurrence of a mid-mantle thermal boundary layer would have far-reaching implications for the thermal, chemical, and dynamic evolution of the Earth. However, uncertainties in the estimates of temperature are too large to allow discrimination on the basis of thermal models alone. The matter will be taken up again in Section 7.9, when dealing with convection in the mantle. In Section 12.4 it will be considered from the microphysical viewpoint.

7.8 Thermal convection

The internal temperature distribution of the Earth requires thermal convection to be the predominant mode of heat transfer in the interior. A rheological prerequisite for convection is that the material be a fluid, i.e. able to flow under small differential stresses. The time factor is of paramount importance. In the high frequency range the mantle is approximately elastic; but in the very low frequency range (for periods larger than 1000 a) the mantle below the lithosphere is a fluid. The movements of plates and the thermal profile of the mantle are proof of the validity of this statement. Other evidence comes from geophysical observation and from experiments on high-temperature rock deformation.

Convection is a consequence of mechanical instability related to non-equilibrium density configurations. Density variations of thermal origin generate buoyancy forces that drive convection. In the case of a fluid heated from below (Fig. 7.7), material near the lower boundary is heated

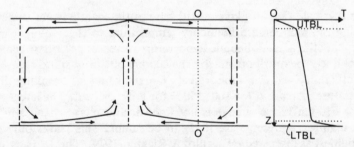

Figure 7.7 Schematic flow pattern (left) and temperature profile along OO′ (right) in a convecting fluid heated from below. Upper and lower thermal boundary layers (UTBL and LTBL, respectively) are delimited by solid lines in the convective circulation and by dotted lines in the temperature profile.

170

and therefore ascends, it gives off heat at the surface, and once cooled it descends to complete the cycle. However, thermal convection can also be established when the fluid is heated from within and cooled from above, leading to a similar configuration of rising hot (lighter) columns and sinking cool (denser) columns.

In this section we focus on fundamental physical aspects of thermal convection, leaving discussion of geodynamic implications for Section 7.9.

The equations governing thermal convection express mass balance (continuity), motion, and energy (thermal) balance. For an incompressible Newtonian fluid in steady-state motion the first two are (see Section 4.3)

$$\frac{\partial v_i}{\partial x_i} = 0 \tag{7.27}$$

$$-\frac{1}{\rho} \frac{\partial p}{\partial x_i} + \nu \nabla^2 v_i + X_i = 0 \tag{7.28}$$

where v_i is the velocity, X_i is the body force per unit mass, p is the fluid pressure, ρ is the density, and $\nu = \eta/\rho$ is the kinematic viscosity (η being the dynamic viscosity). Since convection is driven by buoyancy forces generated by thermally induced density variations, it is necessary to have an equation of state describing the dependence of density on temperature. This is usually expressed as

$$\rho = \rho_0 + \rho' = \rho_0 - \alpha\rho_0(T - T_0) \tag{7.29}$$

where T_0 is the reference temperature, α is the volume coefficient of thermal expansion, and

$$\rho_0 \equiv \rho(T_0) \qquad \rho' \ll \rho_0$$

In the so-called 'Boussinesq approximation' the effects of density variations other than in the body-force term are neglected, and we can use the incompressible continuity condition (Eqn 7.27). Eliminating the hydrostatic pressure related to ρ_0 by means of introducing a new variable (where g is the acceleration of gravity and x_3 is the vertical co-ordinate)

$$P = p - \rho_0 g x_3$$

the equation of motion (Eqn 7.28) becomes

$$-\frac{\partial P}{\partial x_i} + \eta \nabla^2 v_i - \alpha\rho_0 X_i(T - T_0) = 0 \tag{7.30}$$

171

where the only non-zero component of the body force is $X_3 = g$. The buoyancy force per unit mass is given by $\alpha g(T - T_0)$.

Since the buoyancy force depends on T, the velocity field v_i cannot be determined without solving simultaneously for the temperature field; hence the need for an energy balance equation. This is derived by reasoning analogous to that used for the heat-conduction equation in Section 7.4. Heat balance with respect to an elemental volume requires that the heat exchange across its bounding surface be equal to the change in heat content within it. The latter is $(-\partial(\rho cT)/\partial t)$; the former is the sum of a conductivity term $(-K\nabla^2 T)$ and a convective term $(\partial(\rho cTv_i)/\partial x_i)$ (which was absent in the case of conduction). Neglecting heat generation within the elemental volume, and assuming density and specific heat to be constant, the heat balance equation is therefore

$$-K\nabla^2 T + \rho c v_i \frac{\partial T}{\partial x_i} + \rho c T \frac{\partial v_i}{\partial x_i} = -\rho c \frac{\partial T}{\partial t}$$

Using the incompressibility condition (Eqn 7.27) and recalling the definition of thermal diffusivity (Eqn 7.5), this becomes

$$\varkappa\nabla^2 T = \frac{\partial T}{\partial t} + v_i \partial T/\partial x_i \equiv \dot{T} \tag{7.31}$$

which is the heat-transfer equation in a moving incompressible medium.

Solution of a convection problem requires simultaneous solution of Equations 7.30 and 7.31, subject to the continuity condition and the equation of state, for specified boundary conditions. Even for the simplest cases (for instance, an incompressible Newtonian fluid with constant viscosity), one usually needs to use numerical methods. Additional insights are obtained from laboratory experiments of simple 'fish-tank' type (two-dimensional constant-viscosity convection). Reviews of thermal convection in the context of geodynamics may be found in Schubert (1979) and Peltier (1980).

The problem of the onset of convection may be studied in terms of the *Rayleigh Number*, a dimensionless combination of parameters that occurs in linear stability analysis. For a fluid heated from below, the Rayleigh Number is

$$\mathrm{Ra} = \alpha g\Delta T\, h^3/\varkappa\nu \tag{7.32a}$$

and for a fluid heated from within

$$\mathrm{Ra} = \alpha g\rho F h^5/\varkappa\nu K \tag{7.32b}$$

In Equations 7.32a and b the thermal parameters α, K, and $\varkappa$ have the usual meaning; g is the acceleration of gravity; ρ and ν are density and kinematic viscosity; and F, h, and ΔT are heat production per unit mass and time, thickness of the convecting layer, and temperature difference across it, respectively. Convection occurs when the Rayleigh Number is larger than a critical value Ra^*, which depends on boundary conditions and wavelength of the disturbance. In most cases, Ra^* is of the order of 10^3.

The Rayleigh Number for a fluid heated from below is sometimes expressed in terms of the temperature gradient, in which case the layer thickness enters Equation 7.32a as the fourth power. Also, ΔT is to be taken as the total top-to-bottom temperature difference only in the case of incompressible fluids, in which the adiabatic gradient is zero; when the compressibility effect generates a non-negligible adiabatic gradient, ΔT is the superadiabatic temperature difference (and, correspondingly, the gradient is the superadiabatic gradient). As discussed in Section 7.7, a necessary condition for convective instability is superadiabaticity, so that material transported vertically is gravitationally unstable.

The definition of the Rayleigh Number shows that large values of thermal expansion, temperature drop (or heat production), and layer thickness favour the occurrence of convection, whereas high thermal diffusivity and viscosity hinder it. This follows from the fact that the buoyancy force increases in magnitude with increasing thermal expansion and superadiabatic temperature drop, whereas high thermal diffusivity and high viscosity enhance the relative importance of conduction and the mechanical resistance to fluid flow, respectively.

At larger Rayleigh Numbers use may be made of boundary-layer theory, where the assumption is made that most mass transfer occurs near the boundaries of convection cells (see Peltier 1980 for further discussion). A cell is characterized by a near-isothermal core, with upper and lower thermal boundary layers, and ascending and descending columns, which are the agents of vertical heat and mass transfer (refer again to Fig. 7.7).

Besides the Rayleigh Number, there are other dimensionless parameters that arise in the non-dimensionalization of the thermal convection problem. The *Prandtl Number* is the ratio of kinematic viscosity to thermal diffusivity

$$Pr = \nu/\varkappa \tag{7.33}$$

and therefore measures the relative importance of heat and momentum diffusion. The *Reynolds Number* is

$$Re = \nu l/\nu \tag{7.34}$$

173

where v and l are a characteristic velocity and length, respectively. At high Reynolds Numbers the flow is no longer laminar (turbulent flow).

The *Nusselt Number*, a measure of the relative importance of convective heat transport, is defined as

$$\text{Nu} = q/q_c \quad \text{or} \quad \text{Nu} = \Delta T_c/\Delta T \tag{7.35}$$

for a fluid layer heated from below or from within, respectively. In these relations q_c and ΔT_c represent heat flux and temperature drop across the layer in the absence of convection, while their unsubscripted equivalents represent total heat flow and temperature drop in the presence of convection ($q > q_c$, and consequently $\Delta T_c > \Delta T$, since convection transports heat more efficiently than conduction). The Nusselt Number is related to the aspect ratio (ratio of horizontal to vertical dimensions) of convection cells, which in simple cases is between 1 and 3.

The relative importance of convection and conduction is also expressed by the *Péclet Number*

$$\text{Pe} = vl/\varkappa \tag{7.36}$$

where v is a characteristic velocity, and l is a characteristic length of the convecting system. In the following section we shall give order-of-magnitude estimates of these numbers for the mantle of the Earth.

Insight into mantle convection is gained by a consideration of its

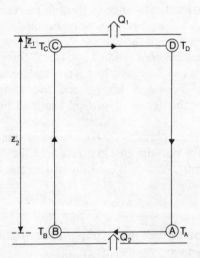

Figure 7.8 Thermodynamic cycle of convection. See text for discussion (from Stacey 1977a).

thermodynamics. The approach formulated by Stacey (1977a, b) is followed here. Convection can be regarded as a heat engine: heat is absorbed at the source, transported vertically, and given off at the sink. Some of this heat is transformed into mechanical work. The *thermodynamic efficiency* of convection is the ratio of potentially usable mechanical (tectonic) work to total heat transported.

To evaluate the efficiency of the convective heat engine, consider a particle of mass m going through the convective cycle ABCDA (Fig. 7.8). Along the lower isobaric limb AB, situated at depth z_2, the particle absorbs the amount of heat Q_2 and its temperature accordingly increases from T_A to T_B. It then rises adiabatically along the vertical limb BC, along which its temperature decreases to T_C (the result of adiabatic decompression from T_B). The amount of heat Q_1 is released along the upper isobaric limb CD at depth z_1, and the particle further cools to T_D. Finally, it descends adiabatically along the vertical limb DA, and in so doing it is heated to T_A. (The assumptions of heating from below and adiabatic vertical transport simplify the argument, but do not alter the basic point.)

If $T(z)$ is an arbitrary temperature profile of the medium, the temperature T_m^* of the particle at any given depth is higher than the ambient temperature in the rising column, and lower in the descending column; that is,

$$T_m^* = \begin{cases} T + T' & \text{along BC} \\ T - T'' & \text{along DA} \end{cases}$$

where T' and T'' are small compared with T. The positive buoyancy of the upward column and the negative buoyancy of the downward column are, respectively,

$$F' = m\alpha g T' \qquad F'' = m\alpha g T''$$

where α is the mean expansion coefficient at a given level. The work done by the buoyancy forces over a complete cycle is therefore

$$W = m \int_{z_1}^{z_2} \alpha g (T' + T'') \, dz \qquad (7.37)$$

Since the vertical limbs are adiabats, the entropy difference between two points on opposite limbs is constant; i.e. considering an arbitrary depth z and the source depth z_2,

$$mc(T' + T'')/T^* = mc_2(T_2' + T_2'')/T_2^* \qquad (7.38)$$

175

where c and c_2 are specific heats at the relevant pressures, and T^* and T_2^* are the mean temperatures at each level along the adiabats. Solving Equation 7.38 for $(T' + T'')$, and substituting in Equation 7.37, we have

$$W = \frac{mc_2(T_2' + T_2'')}{T_2^*} \int_{z_1}^{z_2} \frac{\alpha T^* g}{c} \, dz \qquad (7.39)$$

But $mc_2(T_2' + T_2'') = Q_2$ is the heat input at z_2, and $(\alpha T^* g / c)$ is the adiabatic temperature gradient (Eqn 7.23). Consequently, Equation 7.39 becomes

$$W = \frac{Q_2}{T_2^*} (T_2^* - T_1^*) = Q_2 \frac{T_B - T_C}{T_B} \qquad (7.40)$$

since T_1^* is the temperature at z_1 of the material adiabatically expanded from T_2^* at z_2. The available work is therefore proportional to the adiabatic temperature drop in the hot ascending plume, and the efficiency is

$$\eta_{th} = \frac{W}{Q_2} = 1 - \frac{T_C}{T_B} \simeq 1 - \left(\frac{\rho_C}{\rho_B}\right)^\gamma \qquad (7.41)$$

where the approximate equality is based on Equation 7.24. Values for the efficiency of mantle convection are discussed in Section 7.9.

An important aspect of fluid flow not included in the previous considerations is heat generation by *viscous dissipation*. Consider a slab of material of thickness d, surface area A and volume $V = Ad$, flowing under a shear stress σ, with the upper surface moving at speed v relative to the lower surface. The strain rate is therefore $\dot\varepsilon = v/d$. Accordingly, the power dissipated over the volume V is

$$\dot W = \sigma A v = \sigma A d \dot\varepsilon = \sigma \dot\varepsilon V \qquad (7.42)$$

The viscous heating per unit volume is thus proportional to the strain rate (see also Section 4.3). Since viscosity is a strong function of temperature, viscous dissipation has important consequences for the pattern of flow where shear is highly inhomogeneous, as in the lithosphere. Then, local shear bands can lead to further concentration of shear, because of the associated decrease in viscosity, and considerable softening may occur. However, on a planetary scale, as Tozer (1967) pointed out, the overall effect of viscous heating is to make flow a *self-regulating* process, since an increase in temperature and the associated decrease in viscosity lead to more-efficient heat removal by convection, with a consequent decrease in temperature (and vice versa for an initial decrease in temperature). The system will therefore tend to stabilize at just the viscosity profile that ensures efficient convective heat transport.

7.9 Convection in the mantle

There are several questions about convection in the mantle that have no unequivocal answer. Perhaps the only certain statement that can be made is that convection occurs, powered by some combination of radiogenic heating and secular cooling. This follows not only from what is known of the internal temperature distribution, but also from the simple observation that plates move, and therefore that conservation of mass requires some form of return flow in the interior.

The Rayleigh Number for the mantle is abundantly supercritical. In the case of heating from below, for instance, a kinematic viscosity of the order of 10^{18} m^2 s^{-1} and the usual values of the thermal quantities give Ra $\simeq 10^6$–10^7. The Rayleigh Numbers for separate upper and lower mantle convection are lower, but still supercritical. The orders of magnitude of the mantle Prandtl, Reynolds, Nusselt, and Péclet Numbers are Pr $\simeq 10^{24}$, Re $\simeq 10^{-21}$, Nu $\simeq 10$, and Pe $\simeq 10^3$, respectively, for $\kappa \simeq 10^{-6}$ m^2 s^{-1}, v (speed of plate) $\simeq 10^{-9}$ m s^{-1}, and l (characteristic length) $\simeq 10^6$ m. Therefore, convection is much more efficient than conduction as an agent of heat transfer in the mantle, and inertia forces are negligible. The thermodynamic efficiency varies from about 20 per cent for mantle-wide convection to less than 10 per cent for upper-mantle convection.

Estimates of the order of magnitude of strain rate and viscosity associated with whole-mantle convection can be obtained from a knowledge of the efficiency and the power available (Stacey 1977b). The latter is of the order of 10^{13} W (see Table 7.3; although heat from cooling is available in addition to radiogenic heat, the heat production in the crust does not contribute to convection). Consequently, taking the efficiency to be about 10^{-1}, the convective power used in deforming mantle material is (Eqn 7.42)

$$\dot{W} = \int_V \sigma\dot{\varepsilon} \, dV \simeq 10^{12} \text{ W}$$

where the integral is taken over the volume of the mantle ($V \simeq 10^{21}$ m^3). The average dissipation per unit volume is therefore

$$\langle \sigma\dot{\varepsilon} \rangle = \dot{W}/V \simeq 10^{-9} \text{ W m}^{-3}$$

Assuming a convective speed $v \simeq 10^{-9}$ m s^{-1} (3 cm a^{-1}), and a cell radius $d \simeq 10^6$ m, the average strain rate and stress associated with mantle convection are

$$\langle \dot{\varepsilon} \rangle = v/d \simeq 10^{-15} \text{ s}^{-1} \qquad \langle \sigma \rangle = \dot{W}/V\langle \dot{\varepsilon} \rangle \simeq 1 \text{ MPa}$$

and the average viscosity

$$\langle \eta \rangle = \langle \sigma \rangle / \langle \dot{\varepsilon} \rangle \simeq 10^{21} \text{ Pa s}$$

While the above arguments yield order-of-magnitude estimates of the quantities entering the problems of mantle convection, other questions are still subject to considerable debate. What is the rheology of the mantle, and how does viscosity vary with depth? Is convection mantle-wide, or do upper and lower mantle convect separately? Is there more than one temporal and spatial scale of convection? How are convection currents coupled with lithospheric plates? Is convection approximately steady-state, or episodic? Here we discuss some work bearing on these aspects of the problem.

Any convection model must explain observables such as long-wavelength gravity anomalies, heat flow, size and speed of plates, thickness of the lithosphere boundary layer, and so on. Both geophysical and geochemical evidence must be accommodated within the model. Unfortunately, some of the evidence is not diagnostic, and can thus be interpreted in different ways.

Geophysical evidence, mainly from postglacial rebound (cf. the review by Peltier 1980), shows that viscosity increases with depth in the mantle below the lithosphere, but only by a relatively small amount, so that the average viscosity of the lower mantle is likely to be no more than one to two orders of magnitude higher than that of the upper mantle. The rheology appears to be Newtonian, although interpretations of rebound in terms of a non-Newtonian power-law creep are also possible (see Sections 8.2 & 3). In view of these uncertainties, it is pertinent to ask whether the pattern of convection in a non-Newtonian fluid is significantly different from the linearly viscous case. The possible importance of non-Newtonian convection was originally stressed by Orowan (1965). Recent numerical models by Cserepes (1982) and Christensen (1984) show that, contrary to previous expectations, the differences may indeed be significant. The flow pattern and heat transfer properties in a convecting fluid with power-law rheology are different from those expected in a Newtonian fluid. A non-linear rheology decreases the effect of temperature and pressure on viscosity variations, concentrates deformation, and leads to larger aspect ratios. All of these effects are compatible with plate tectonic evidence.

The most important unresolved question is whether convection is mantle-wide, or whether upper and lower mantle convect separately. The more classical models of convection (cf., for example, Runcorn 1962) assume it to be mantle-wide. The same opinion is held by Peltier (1980), who points out that practically all observations can be most simply explained in terms of mantle-wide convection. However, the evidence can also be accounted for in terms of separate convective circulations in the upper and lower mantle. For instance, Richter & McKenzie (1981) carried out numerical and laboratory experiments on convection in discrete layers, and concluded that circulation is confined to superimposed layers if the density contrast due to changes in chemical composition is larger than that caused by the largest temperature difference within the system. Separate convection would result in a double thermal boundary layer between the upper and the lower mantle.

The factors that would inhibit mantle-wide convection are a large viscosity increase with depth, a phase change with sufficiently negative Clapeyron slope, and lack of mixing between chemically different layers. The geophysical evidence appears to rule out a large increase in viscosity (cf. Section 8.2) and this is confirmed by microphysical considerations (Section 12.4). There is consequently no *rheological* objection to mantle-wide convection. As regards the 400-km discontinuity, the slope of the Clapeyron curve (the curve in (p, T)-space along which two phases of the same material are in equilibrium) for the olivine $\rightarrow$ spinel transition is positive $(\mathrm{d}p/\mathrm{d}T \simeq 4 \ \mathrm{MPa} \ \mathrm{K}^{-1})$, and therefore the phase boundary is displaced upwards in downwelling currents, and downwards in ascending currents. These density perturbations would help the circulation. It is

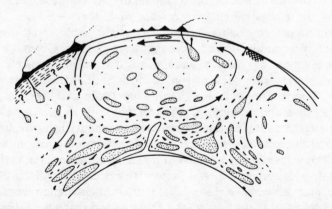

Figure 7.9 Sketch of convective flow in a mantle containing distinct geochemical reservoirs (dotted). Surface volcanism may sample different reservoirs (from Davies 1984).

therefore clear that the phase changes associated with the 400-km discontinuity are not a barrier to convection. However, not much is known about the 670-km discontinuity.

The feasibility of mantle-wide convection must thus be assessed on *geochemical* grounds. We reviewed in Section 6.6 the arguments relevant to possible chemical differences between the upper and the lower mantle. The isotopic evidence suggests at least two (and possibly more) geochemical mantle sources, which have maintained their identities for periods of the order of 1 Ga or more. However, this does not necessarily imply that the mantle is stably stratified. It is equally possible that geochemical heterogeneities occur randomly at all depths, without any compositional layering (Davies 1984). Several studies (Olson 1984, Olson *et al.* 1984, Sleep 1984, Loper 1985) focus on the degree of mixing of compositional anomalies within the mantle. The time scale for mixing appears to be long – of the order of the age of the Earth. It is therefore possible to reconcile mantle-wide convection and the presence of long-lived distinct geochemical reservoirs within the mantle (see Fig. 7.9). On the other hand, Hoffman & McKenzie (1985) conclude that mixing is more efficient, and therefore support separate convective circulations.

The seismic properties of sinking lithospheric slabs may be used to clarify the problem of the extent of convection. No earthquakes in subduction zones occur deeper than about 700 km, but this does not exclude the possibility of aseismic penetration of the lower mantle. Focal mechanisms and energy release of the deepest earthquakes are consistent with the hypothesis that the subducting slab does not penetrate the lower mantle (Richter 1979). However, travel-time residuals seem to indicate that at least some slabs retain their identity to minimum depths of 900–1000 km (Creager & Jordan 1984). Numerical models (Christensen & Yuen 1984) show that a compositional density contrast larger than 5 per cent, or a negative Clapeyron slope of at least $-6\,\mathrm{MPa\,K^{-1}}$, are required to prevent slab penetration, but the same result could be achieved by a combination of more-moderate changes. Given the uncertainties involved, a variety of situations is dynamically possible.

The problem of the scale of convection is not confined to its vertical extent. Temperature-induced viscosity contrasts stabilize the upper boundary layer (lithosphere), which is mechanically strong because of its low temperature, and destabilize the lower boundary layer (region D''), where the viscosity decreases because of the steep increase in temperature. This gives rise to two scales of mantle circulation, even in the case of mantle-wide convection. A slow pervasive circulation affects the bulk of the mantle, while a faster discontinuous circulation consists of isolated 'plumes' that rise from the D''-layer and reach the lithosphere, where they are revealed as 'hot spots' (Morgan 1971, Peltier 1981).

Similar instabilities could arise from the mantle transition zone in the case of layered convection.

There is also uncertainty regarding the degree of coupling between plate movements and mantle convection. In the simplest hypothesis, lithospheric plates (at least the oceanic ones) are merely the thermal boundary layer of mantle flow. However, their mechanical strength makes the situation more complex. A stiff lithosphere may not be related in any simple way to the planform of convection. Correlations between rising and sinking currents, long-wavelength perturbations of topography, and the shape of the geoid (see Section 8.1) make it possible to 'see through' oceanic plates and visualize the planform of convection below. For the Pacific plate, McKenzie *et al.* (1980) have shown that spacing between rising and sinking plumes is approximately 1500–2000 km. The plume signature in the horizontal plane is elliptical, with major axis in the direction of motion of the Pacific plate. Clearly, the motion of the mechanically strong lithospheric plate affects the pattern of flow in the uppper mantle.

The effects of a stiff lithosphere on the convection pattern have been incorporated in some numerical models (see, for example, Hager & O'Connell 1981, Jacoby & Schmeling 1982). Body forces associated with density variations *within* the lithosphere can drive the plates. Moreover, the drag at the base of the plates (of the order of 1 MPa) may actually resist the movement, rather than being one of the driving factors (see also Section 8.5).

Finally, doubts exist about the assumption of steady-state flow. Far from being approximately steady, convection currents could be intrinsically episodic. Bott (1964) and Anderson (1982) point out that the *thermal blanketing* effect of continental plates may result in heat accumulation in the mantle beneath them, which could be relieved by episodic convection. Therefore, if a continental plate is stationary for a sufficiently long time, then conditions will be generated in the mantle favourable to continental break-up. This interesting hypothesis is possibly more consistent with the geological record than steady-state convection.

As mentioned in Section 6.6, it is now becoming possible to 'see' the pattern of convection using seismic tomography. Major shields and ridges persist to depths of the order of 250 km. The direction of flow in the upper mantle, inferred from seismic anisotropy, is horizontal in most of the LVZ, except in the region of the East Pacific Rise, where it is vertical. Lateral variations in seismic velocity occur in the transition zone and in the lower mantle. These velocity anomalies reflect density contrasts associated with flow, and are correlated with geoid anomalies (see Section 8.1). The interdependence of the seismic, gravitational, and thermal fields is providing new insights into the detailed pattern of convection. However, as shown by the previous discussion, many outstanding problems remain.

7.10 Thermal history of the Earth

We conclude this chapter with some considerations on the thermal evolution of the Earth. Present heat loss exceeds heat generation, so that the Earth must be slowly cooling (see Section 7.5). Analyses of the long-term thermal evolution, taking convection into account, are available (cf. among others, McKenzie & Weiss 1975, Cook & Turcotte 1981, Peltier & Jarvis 1982). We follow the treatment of Turcotte & Schubert (1982).

A relation between average mantle temperature and time is developed as follows. For a mantle heated from within with heat generation function F, the Rayleigh Number is (see Section 7.8)

$$Ra = \alpha \rho g F h^5 / x K \nu \qquad (7.43)$$

If T_0, T and T_c are surface temperature, average temperature of the nearly isothermal core of the convection cell, and temperature at an equivalent depth in the absence of convection, respectively, then a dimensionless temperature may be defined as

$$\theta = \frac{T - T_0}{T_c - T_0} \qquad (7.44)$$

i.e. as the ratio between convective and conductive temperature drop. For aspect ratio unity, this is related to the Rayleigh Number as

$$\theta = 2.98 Ra^{-1/4} \qquad (7.45)$$

The above relation, obtained in boundary layer theory, expresses the fact that the more vigorous the convection is, the more efficient the removal of heat, with consequent decrease in the non-dimensional temperature drop. The conductive temperature difference is shown in heat conduction theory to be

$$T_c - T_0 = \rho F h^2 / 2K \qquad (7.46)$$

Combining Equations 7.43–7.46, the following relation is obtained between mantle temperature and heat production:

$$\frac{2K}{\rho Fh^2}(T - T_0) = 2.98\left(\frac{\varkappa \nu K}{\alpha \rho g Fh^5}\right)^{1/4} \qquad (7.47)$$

Time-dependence of heat production is introduced by assuming

$$F = F_p \exp[-\lambda(t - t_p)] \qquad (7.48)$$

where t_p is the present time, F_p is the present heat production rate, and λ is a weighed 'decay constant' for the mantle. But viscosity is a function of temperature of the type

$$\nu = \nu_0 \exp(E/RT) \qquad (7.49)$$

where E is an activation energy and R is the universal gas constant. Viscosity therefore varies with time if temperature does. Substituting Equations 7.48 and 7.49 into 7.47, the following relation between mean mantle temperature and time is obtained:

$$\frac{2K}{\rho F_p h^2}(T - T_0) = 2.98\left(\frac{\varkappa K \nu_0}{\alpha \rho g F_p h^5}\right)^{1/4} \exp\left(\frac{E}{4RT}\right)\exp\left(-\frac{3\lambda}{4}(t - t_p)\right)$$

$$(7.50)$$

As $T = T_p$ at present, the above equation can be simplified to

$$\frac{2K}{\rho F_p h^2}(T_p - T_0) = 2.98\left(\frac{\varkappa K \nu_0}{\alpha \rho g F_p h^5}\right)^{1/4} \exp\left(\frac{E}{4RT_p}\right) \qquad (7.51)$$

Combining Equations 7.50 and 7.51,

$$\frac{T - T_0}{T_p - T_0} = \exp\left[\frac{E}{4R}\left(\frac{1}{T} - \frac{1}{T_p}\right)\right]\exp\left(-\frac{3\lambda}{4}(t - t_p)\right) \qquad (7.52)$$

The rate of cooling of the mantle is the time derivative of Equation 7.52, that is

$$\dot{T} = -\frac{3\lambda}{4}(T - T_0)\left(1 + \frac{E}{4RT^2}(T - T_0)\right)^{-1} \qquad (7.53)$$

As E is of the order of 10^5 J mol^{-1} in ultrabasic silicates (see Section 10.4), the term containing it is much larger than unity; consequently, the

183

following approximate relation holds:

$$\dot{T} \simeq -3\lambda RT^2/E \qquad (7.54)$$

Taking $T_p = 2300$ K, $\lambda = 2.77 \times 10^{-10}$ a^{-1}, and $E = 5 \times 10^5$ J mol^{-1}, the present cooling rate according to Equation 7.54 is $\dot{T}_p \simeq -70$ K Ga^{-1}.

The contribution of cooling to the present heat flow can be estimated by noting the correspondence between rate of heat loss and decrease in thermal energy. Let r_E, $\langle \rho \rangle$ and $\langle c \rangle$ be the radius, mean density, and mean specific heat of the Earth, respectively. If q' is the heat flux generated by cooling, then

$$4\pi r_E^2 q' = -\tfrac{4}{3}\pi r_E^3 \langle \rho \rangle \langle c \rangle \dot{T}_p$$

i.e.

$$q' = -\tfrac{1}{3}(r_E \langle \rho \rangle \langle c \rangle \dot{T}_p)$$

For $r_E = 6371$ km, $\langle \rho \rangle = 5.5 \times 10^3$ kg m^{-3}, $\langle c \rangle = 1$ kJ kg^{-1} K^{-1}, and $\dot{T}_p = -70$ K Ga^{-1}, one has $q' \simeq 26$ mW m^{-2}. Cooling of the Earth contributes approximately 30 per cent of the total heat loss. While other estimates vary slightly (from 25 to 50 per cent), there is convergence on the general conclusion presented here.

Secular cooling of the Earth has important tectonic implications. Since the mean mantle viscosity increases with time, convection was more vigorous in the Archaean. Moreover, higher Archaean mantle temperature would have resulted in a thinner lithosphere. There is no *a priori* assurance that geodynamic processes in the Archaean were the same as in the Phanerozoic.

8 The long timescale: geodynamics and plate tectonics

The rheology of the Earth over the long timescale ($t \geqslant 10^3$ a) is dominated by high-temperature steady-state creep except in the upper lithosphere, where the temperature is low and an elastic–brittle or elastic–plastic rheology applies.

The first part of this chapter deals with the determination of the rheology of the lithosphere and mantle from the observation of the response of the Earth to surface loading and unloading, using models of viscous flow in a fluid half-space and of elastic deflection of a plate under vertical loads. Comparison of results with relative sea-level, gravity, and polar drift data leads to the conclusion that an elastic (or high-viscosity viscoelastic) lithosphere of variable thickness (50–200 km) overlies a viscous mantle, whose rheological equation may be approximated by that of a Burgers body with average viscosity of 10^{21}–10^{22} Pa s.

The second part of the chapter discusses the state of stress in lithospheric plates, and gives examples of the application of fracture mechanics and plasticity theory to the analysis of tectonic processes.

8.1 The geoid and gravity anomalies

The rheological properties of the Earth over the short timescale, discussed in Chapter 6, cannot be extended to the long timescale (say, $t > 10^3$ a), since the thermal state of the Earth implies mantle flow (see Ch. 7). As this conclusion also applies to the core (the geomagnetic field arises out of convection in the liquid outer core), the bulk of the Earth is fluid over geological timescales, with the possible exception of the upper lithosphere, where the temperature is too low to allow significant viscous relaxation.

The long-term rheology of the Earth governs geodynamic processes such as plate tectonics and orogenesis. Incidentally, the term 'geo-dynamics' originally referred to processes affecting the Earth over the whole frequency spectrum; but current practice restricts it to denote long-term processes. Books on geodynamics are available (Scheidegger 1982, Turcotte & Schubert 1982, Artyushkov 1983), and the reader is referred to these for extensive coverage of the various topics. In this chapter we examine the rheology of the mantle and of the lithosphere (as inferred from geophysical observation), the state of stress in the lithosphere, and – as an example – the dynamics of some tectonic features.

Gravity is an important parameter in geodynamics, since it is related to the mass distribution in the interior. While not going into a detailed analysis of the gravity field (see, for example, Garland 1979), we discuss in this section some of the concepts pertaining to the topics that follow.

The gravitational force is *conservative*. The work done by a conservative force F_i in the displacement of a particle along a path L (given by the line integral $\int_L F_i \, dx_i$) depends only on the end-points of the path. A necessary and sufficient condition for $\int_L F_i \, dx_i$ to be independent of the path is that the integrand be an exact differential (Kreyszig 1983):

$$F_i = \partial U / \partial x_i \tag{8.1}$$

The scalar function of position $U(x_1, x_2, x_3)$ is termed the *potential* of F_i. At each point the force is the gradient of the potential, and its direction is therefore perpendicular to surfaces of constant potential (*equipotential surfaces*).

Both the curl and the divergence of a conservative force are subject to certain conditions. Using the permutation tensor introduced in Section 6.1, the curl of F_i is seen to vanish identically

$$e_{ijk} \frac{\partial F_k}{\partial x_j} = e_{ijk} \frac{\partial}{\partial x_j} \left(\frac{\partial U}{\partial x_k} \right) = 0$$

The divergence of F_i is zero in regions where the net force flux is zero. This can be seen by recalling the Divergence Theorem of Gauss:

$$\int_S F_i n_i \, dS = \int_V \frac{\partial F_i}{\partial x_i} \, dV$$

where S is any closed surface with outer normal n_i enclosing a volume V. If the force flux across S is zero, then the above integrals vanish, no matter how small the enclosed volume; this condition requires that

$$\frac{\partial F_i}{\partial x_i} = 0$$

186

Using the above equation in connection with Equation 8.1 we have

$$\frac{\partial F_i}{\partial x_i} = \frac{\partial}{\partial x_i} \left(\frac{\partial U}{\partial x_i} \right) \equiv \nabla^2 U = 0 \qquad (8.2)$$

i.e. the potential satisfies the Laplace equation (Section 3.4) in regions without sources or sinks of lines of force. In the case of gravity, $F_i \equiv (0, 0, g)$, where g is the gravitational acceleration, and the gravity potential satisfies the Laplace equation at points outside attracting masses.

In the case of the Earth, it is more convenient to use spherical co-ordinates r, θ, λ (radius, colatitude, and longitude, respectively). The Laplace equation then takes the form

$$\nabla^2 U = \frac{1}{r^2} \frac{\partial}{\partial r} \left(r^2 \frac{\partial U}{\partial r} \right) + \frac{1}{r^2 \sin \theta} \frac{\partial}{\partial \theta} \left(\sin \theta \frac{\partial U}{\partial \theta} \right) + \frac{1}{r^2 \sin^2 \theta} \frac{\partial^2 U}{\partial \lambda^2} = 0$$

$$(8.3)$$

The solution to Equation 8.3 can be given in terms of a spherical harmonic expansion. The first and third term of the expansion (the second is identically zero) express the potential of a sphere with mass concentrated at its centre, and the contribution due to an axially symmetric equatorial bulge, respectively. They are given by

$$U = - \frac{GM}{r} + \frac{GMa^2}{2r^3} J_2 (3 \sin^2 \phi - 1) \qquad (8.4)$$

where r and ϕ are radius and latitude, respectively ($\phi = \pi/2 - \theta$); $G = 6.67 \times 10^{-11}\,\mathrm{m^3\,kg^{-1}\,s^{-2}}$ is the gravitational constant; $M = 5.97 \times 10^{24}\,\mathrm{kg}$ the mass of the Earth; and $a = 6.378 \times 10^6\,\mathrm{m}$ the equatorial radius. The coefficient J_2 is related to the principal moments of inertia of the Earth (C about the polar axis, and A about the equatorial axis), and can be determined from its effect on the orbits of artificial satellites

$$J_2 = \frac{C - A}{Ma^2} = 1.082\,64 \times 10^{-3}$$

A further contribution comes from the rotation of the Earth, which generates a centrifugal acceleration oriented away from the rotational axis with potential

$$U_r = - \tfrac{1}{2} \omega^2 r^2 \cos^2 \phi \qquad (8.5)$$

187

where ω is the angular velocity of the Earth. The total *geopotential* up to the second harmonic for an axially symmetric rotating Earth is therefore

$$U = - \frac{GM}{r} + \frac{GMa^2}{2r^3} \, J_2(3 \, \sin^2 \phi - 1) - \tfrac{1}{2}\omega^2 r^2 \, \cos^2 \phi \qquad (8.6)$$

Equation 8.6 is important for two reasons. As gravity is everywhere normal to equipotential surfaces, one can define a *reference ellipsoid*, or *spheroid*, that gives the best fit to the actual shape of the Earth as defined by mean sea level. The surface of the reference ellipsoid is an ideal equipotential surface, $U = U_0$. In order to obtain an explicit expression for it, we note that at the equator ($r = a$, $\phi = 0$) Equation 8.6 gives

$$U_0 = - \frac{GM}{a} - \frac{GM}{2a} \, J_2 - \tfrac{1}{2} a^2 \omega^2$$

and at the poles ($r = c$, $\phi = \pi/2$, where c is the polar radius)

$$U_0 = - \frac{GM}{c} + \frac{GMa^2}{c^3} \, J_2$$

Equating the last two equations, and solving for $(a - c)/a = f$ (where f is the *flattening* of the Earth), we obtain, neglecting higher-order terms,

$$f = \tfrac{3}{2} J_2 + \frac{1}{2} \frac{a^3 \omega^2}{GM} \qquad (8.7)$$

The equation for the surface of the ellipsoid is

$$r = a(1 - f \, \sin^2 \phi) \qquad (8.8)$$

The observed value of the flattening is $f = 1/298.256$ ($a = 6378.14$ km, $c = 6356.75$ km). This figure is nearly what is expected for a self-gravitating, rotating fluid in hydrostatic equilibrium.

The second important consequence of Equation 8.6 is the establishment of a *reference value* for the acceleration of gravity on the surface of the ellipsoid as function of latitude. The magnitude of gravity is

$$g = \left[\left(\frac{\partial U}{\partial r}\right)^2 + \left(\frac{1}{r} \frac{\partial U}{\partial \phi}\right)^2 \right]^{1/2}$$

and, since the angular difference between the radial direction and the normal to the ellipsoid is everywhere small,

$$g \simeq \frac{\partial U}{\partial r} = \frac{GM}{r^2} - \frac{3}{2}\frac{GMa^2}{r^4} J_2(3\sin^2\phi - 1) - \omega^2 r(1 - \sin^2\phi)$$

so that on the surface of the ellipsoid, where r is given by Equation 8.8,

$$g_\phi \simeq \frac{GM}{a^2}(1 + 2f\sin^2\phi) - \frac{3}{2}\frac{GM}{a^2} J_2(3\sin^2\phi - 1) - \omega^2 a(1 - \sin^2\phi)$$

$$(8.9)$$

neglecting higher-order terms. In practice the standard value of gravity at sea level as a function of latitude is given by the Geodetic Reference System gravity formula (GRS 67)

$$g_\phi = 9.780\ 318\ 46(1 + 0.005\ 278\ 895\sin^2\phi + 0.000\ 023\ 462\sin^4\phi)$$

$$(8.10)$$

where g_ϕ is measured in m s^{-2}. Other common units in gravity measurements are the *gal* (1 gal = 1 cm s^{-2}) and the milligal (1 mgal = 10^{-3} gal); the latter is the unit in which gravity anomalies are usually given.

Equations 8.8 and 8.10 define the reference ellipsoid and the expected variation of gravity with latitude, respectively. They serve as a basis for comparison with observations. The actual equipotential (mean sea-level) surface of the Earth differs from the reference ellipsoid, and measured values of gravity differ from those predicted from the GRS 67 formula. These differences reflect the occurrence of *lateral density variations* within the Earth, whereas the self-gravitating, rotating model (Eqns 8.8 & 10) lacks such variations.

The actual mean sea-level equipotential surface of the Earth is called the *geoid*. A *geoid anomaly* is the elevation difference between the geoid and the reference ellipsoid. The long-wavelength ($\geqslant 2000$ km) anomalies are determined from orbital perturbations of artificial satellites; their maximum amplitude is of the order of 100 m, and they are shown in Figure 8.1. Much shorter-wavelength geoid anomalies ($\geqslant 100$ km) are measurable over oceanic areas by satellite altimetry. Study of these anomalies has led to the detection of a second scale of convection under the Pacific plate, mentioned in Section 7.9.

Any mass anomaly has both a geoidal and a gravity signature. Some long-wavelength geoid anomalies are correlated with identifiable geo-dynamic features – for instance, there is a strong association between subducted slabs and geoid highs – but others have no obvious connection

Figure 8.1 Geoid heights (in m) with respect to ellipsoid of flattening 1/298.255 (from Wagner *et al.* 1977).

with present plate geometry. The interpretation is complicated by the fact that a geoid anomaly represents the integrated effect of interior lateral density variations and dynamically maintained boundary deformations. For instance, a hot upwelling current is less dense than the surrounding mantle material, but is associated with uplifts of both the upper (free surface) and lower boundaries of the convecting fluid. The amplitude of boundary deformations depends on the distribution of viscosity with depth (cf. Section 8.3) and on the presence or absence of chemical stratification. As pointed out in Sections 6.6 and 7.9, both seismic velocity variations and density anomalies are related to temperature anomalies, and consequently to the pattern of mantle convection. It is therefore possible, at least in principle, to use geoid anomalies to constrain the extent and shape of mantle convection currents. Studies of this kind are beginning to unravel the actual kinematics of mantle flow (for a review, see Hager *et al.* 1985).

At shorter wavelengths, when the causative mass anomalies have typical lateral dimensions of the order of hundreds of kilometres or less, the most convenient geophysical observables are gravity anomalies. In order to compute gravity anomalies, corrections to the measured gravity must be applied (see, for example, Garland 1979). Since the reference value of gravity g_ϕ (Eqn 8.10) is given on the surface of the ellipsoid, it is necessary to apply a correction for elevation – the *free-air correction* – to the observed gravity g_0 before comparing it with the reference value. As the effect of ellipticity on the vertical gradient is negligible, this correction is given approximately by

$$\frac{\partial g}{\partial r} \simeq -\frac{2GM}{r^3} = -\frac{2g}{r} = -0.3086 \text{ mgal m}^{-1}$$

where use has been made of the approximate equality $g \simeq GM/r^2$. The free-air correction must be added to the observed gravity g_0 if the measurement, as is usually the case on land, has been taken at an altitude h above sea level. The *free-air anomaly* is the difference between observed gravity, corrected for elevation, and reference gravity, i.e.

$$\Delta g_{FA} = (g_0 + 0.3086h) - g_\phi \tag{8.11}$$

where g_0 and g_ϕ are in milligals and h is in metres.

Computation of the free-air anomaly does not take into account the additional gravitational attraction of the material above sea level. Considering this to be a laterally infinite slab of density ρ and thickness h, its contribution to gravity is

$$g_B = 2\pi G\rho h = 0.1119h$$

191

where g_B is in milligals, h is in metres, and a 'standard' value of 2.67×10^3 kg m^{-3} has been taken for the density (this is customary in gravity work; note, however, that it is not the average crustal density – which is close to 2.9×10^3 kg m^{-3} – but a representative value of the density of upper crustal rocks). On land, this *Bouguer correction* must be subtracted from the observed gravity in order to reduce it to a sea-level value. The *Bouguer anomaly* is then defined to be

$$\Delta g_B = (g_0 + 0.3086h - 0.1119h) - g_\phi \qquad (8.12)$$

In regions of rugged topography a *terrain correction* is also necessary, to take care of the uneven mass distribution around the point at which gravity is measured.

Gravity anomalies are related to topography. Bouguer anomalies generally show a negative correlation with altitude, and free-air anomalies over broad positive topographic features are often near zero or only slightly positive. This indicates that topographic mass anomalies of sufficiently long wavelength are compensated at depth by corresponding mass anomalies of opposite sign. As state of *isostatic equilibrium* is defined as that in which the net sum of mass anomalies in any vertical column, from the surface down to some 'depth of compensation', is zero.

Although deviations from isostasy occur, most of the lithosphere is in approximate isostatic equilibrium. Combined gravity and seismic studies show that isostatic compensation of passive continental margins and mountain ranges is usually accomplished by variations in crustal thickness (*Airy* mechanism), but lateral variations of crust and upper mantle density (*Pratt* mechanism) play a role in the compensation of mid-oceanic ridges and plateau uplifts.

Isostasy is an equilibrium state, and deviations from it tend to relax in time. It is therefore common to assume that isostatic equilibrium holds (cf. Section 7.6, where this assumption was used to predict ocean-floor topography). The response of the Earth to surface loading and unloading is governed by the tendency toward isostatic equilibrium, as for instance is postglacial rebound (Section 8.2) and lithospheric flexure (Section 8.4). The isostatic compensation mechanism will be further discussed with reference to these processes.

8.2 Rheology of the mantle from geophysical evidence

Although we established in the previous chapter that the mantle is a fluid in the long time range, we have left open the questions of its constitutive

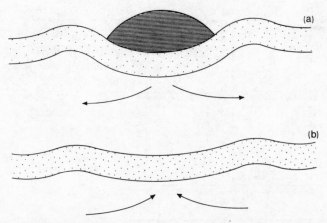

Figure 8.2 Postglacial isostatic rebound: (a) loading and (b) unloading of the lithosphere.

equation and of the variation of viscosity with depth. In this section we examine the answers to these questions that can be obtained from a study of the geophysical evidence. A different approach, using solid-state physics and materials science, will be developed in Chapter 12.

In principle, aspects of the rheology of the mantle can be inferred from any geodynamic process yielding observables that are a function of the rheology. In practice, the most direct quantitative evidence comes from *postglacial isostatic rebound*. The process of postglacial rebound is shown schematically in Figure 8.2. The loading of some regions of the Earth by continental ice sheets during the Pleistocene glaciation resulted in a depression of the lithosphere, accompanied by a peripheral bulge some distance away from the ice margin, the vertical movements being accommodated by flow in the mantle. With the melting of the ice sheets and consequent unloading, the gradual restoration of isostatic equilibrium resulted in an inversion of these movements. Areas of former continental glaciation – the Canadian and Fennoscandian shields – should therefore show a central region of uplift, surrounded by a peripheral region of submergence. If rebound is not yet complete, a large-scale negative free-air anomaly should occupy the centre of the depression. Figure 8.3 shows, for the Pleistocene Laurentide ice sheet over the Canadian shield, how well patterns of vertical movement and gravity anomalies conform to this picture.

In the sequel we give a simplified treatment of postglacial rebound, followed by a brief survey of results. The physics of the process is adequately illustrated by the simple model. The original analysis was done by Haskell (1935). A similar discussion can be found in Turcotte & Schubert (1982).

193

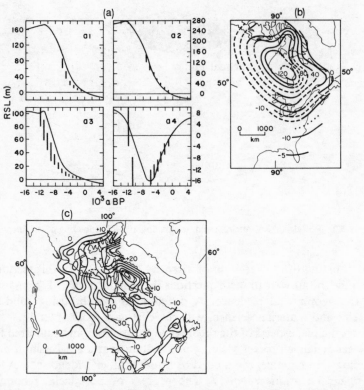

Figure 8.3 (a) Relative sea-level (RSL) curves at selected sites in Canada (solid lines and vertical bars represent predictions and observations, respectively): a1, Ipik Bay, 69°N 74°W; a2, Churchill, 58°N 96°W; a3, Newfoundland, 51°N 56°W; a4, Bay of Fundy, 45°N 65°W (from Peltier 1980). (b) Pattern of postglacial rebound (contours in m) in the Laurentide region in the last 6000 years (from Walcott 1973). (c) Free-air gravity anomalies (in mgal) in the Laurentide region (from Walcott 1972).

Consider the mantle to be a Newtonian, incompressible, constant-viscosity half-space, with upper boundary coinciding with the free surface of the Earth. For slow motion – as applies to postglacial rebound – the velocity field in the half-space satisfies the Navier–Stokes equations in the form (see Section 4.3)

$$-(\partial p/\partial x_i) + \eta \nabla^2 v_i + \rho X_i = 0 \tag{8.13}$$

together with the continuity equation

$$\partial v_i/\partial x_i = 0 \tag{8.14}$$

In Equations 8.13 and 8.14, v_i is the velocity vector, p is the pressure,

η and ρ are viscosity and density, respectively, and $X_i \equiv (0, 0, g)$ is the force of gravity. We limit our attention to the plane strain case, where variables are a function of (x_1, x_3) only, x_1 being taken horizontal on the free surface, and x_3 vertical and positive downwards. Introducing the variable $P = p - \rho g x_3$ to denote the pressure generated by fluid flow, Equation 8.13 becomes

$$- (\partial P / \partial x_i) + \eta \nabla^2 v_i = 0 \qquad (8.15)$$

The incompressible continuity condition (Eqn 8.14) is identically satisfied for

$$v_1 = - \partial \psi / \partial x_3 \qquad v_3 = \partial \psi / \partial x_1 \qquad (8.16)$$

where $\psi(x_1, x_3)$ is the *stream function*, which plays a role somewhat analogous to the potential of conservative force fields, and is of wide use in hydrodynamics. Substituting Equation 8.16 into Equation 8.15, we obtain

$$- \frac{\partial P}{\partial x_1} - \eta \left(\frac{\partial^3 \psi}{\partial x_1^2 \partial x_3} + \frac{\partial^3 \psi}{\partial x_3^3} \right) = 0$$

$$- \frac{\partial P}{\partial x_3} + \eta \left(\frac{\partial^3 \psi}{\partial x_1^3} + \frac{\partial^3 \psi}{\partial x_1 \partial x_3^2} \right) = 0$$

Taking $\partial / \partial x_3$ of the first equation above, $\partial / \partial x_1$ of the second, and subtracting, we have

$$\frac{\partial^4 \psi}{\partial x_1^4} + 2 \frac{\partial^4 \psi}{\partial x_1^2 \partial x_3^2} + \frac{\partial^4 \psi}{\partial x_3^4} \equiv \nabla^4 \psi = 0 \qquad (8.17)$$

i.e. the stream function is biharmonic (see Section 3.5).

Suppose now that the viscous half-space is subject to an initial vertical surface displacement (corresponding to glacial loading) of the form

$$\zeta' = \zeta_0' \cos(2\pi x_1 / \lambda) \qquad (8.18)$$

where the prime denotes initial values, and λ is the characteristic wavelength of the displacement, $\lambda \gg \zeta'$. The return to an undeformed free surface (corresponding to glacial unloading) is governed by Equation 8.17. Subject to the conditions that the stream function be finite for $x_3 \to \infty$, and that $v_1 = 0$ at the upper boundary of the half-space, the

195

solution is

$$\psi = A\left(1 + \frac{2\pi x_3}{\lambda}\right)\exp\left(-\frac{2\pi x_3}{\lambda}\right)\sin\left(\frac{2\pi x_1}{\lambda}\right) \qquad (8.19)$$

from which the velocity components can be obtained immediately by differentiation, according to Equations 8.16,

$$v_1 = A x_3 \left(\frac{2\pi}{\lambda}\right)^2 \exp\left(-\frac{2\pi x_3}{\lambda}\right)\sin\left(\frac{2\pi x_1}{\lambda}\right) \qquad (8.20a)$$

$$v_3 = A \frac{2\pi}{\lambda}\left(1 + \frac{2\pi x_3}{\lambda}\right)\exp\left(-\frac{2\pi x_3}{\lambda}\right)\cos\left(\frac{2\pi x_1}{\lambda}\right) \qquad (8.20b)$$

At the surface (strictly, for $x_3 = \zeta$, but, since ζ is small compared with λ, approximately for $x_3 = 0$), the time rate of change of the vertical displacement equals the vertical velocity, that is (Eqn 8.20b)

$$\dot{\zeta} = v_3 = A \frac{2\pi}{\lambda}\cos\left(\frac{2\pi x_1}{\lambda}\right) \qquad (8.21)$$

The constant of integration A is obtained by recalling the constitutive equation for linear fluids (Eqn 4.6). For vertical normal components, it reads

$$\sigma_{33} = -p + 2\eta\dot{\varepsilon}_{33}$$

At the upper surface, $\dot{\varepsilon}_{33} = (\partial v_3/\partial x_3)$ vanishes, as can be seen from Equation 8.20b. The pressure p is obtained by substituting v_1 (Eqn 8.20a) into the x_1-component of the equation of motion (Eqn 8.13), and integrating:

$$p = 2\eta A \left(\frac{2\pi}{\lambda}\right)^2 \cos\left(\frac{2\pi x_1}{\lambda}\right)$$

But, since the vertical normal stress is simply $\rho g\zeta$, as induced by the topography, $p = -\rho g\zeta$ for $x_3 = 0$, which yields

$$A = -\frac{\rho g\zeta}{2\eta}\left(\frac{2\pi}{\lambda}\right)^{-2}\left[\cos\left(\frac{2\pi x_1}{\lambda}\right)\right]^{-1}$$

Using the above value for A, the equation for the rate of change of the vertical displacement at the surface (Eqn 8.21) becomes

$$\dot{\zeta} = -\rho g\lambda\zeta/4\pi\eta$$

which can be immediately integrated, with the initial condition $\zeta = \zeta'$ at $t = 0$ (time of unloading; see Eqn 8.18)

$$\zeta = \zeta' \, \exp(-\rho g \lambda t / 4 \pi \eta) \qquad (8.22)$$

Equation 8.22 shows that the surface deflection decreases exponentially with time after unloading; it can therefore be rewritten in terms of a *relaxation time* τ:

$$\zeta = \zeta' \, \exp(-t/\tau) \qquad (8.23)$$

where

$$\tau = 4\pi\eta/\rho g \lambda \qquad (8.24)$$

The relaxation time is a function of the viscosity of the fluid (assumed Newtonian and constant) and of the wavelength of the load – the larger the horizontal scale of the disturbance, the shorter the relaxation time. Since the shape of the glacial load is approximately known from geological evidence, determination of the relaxation time allows estimation of viscosity. From Equation 8.24, the latter is given by

$$\eta = \rho g \lambda \tau / 4 \pi \qquad (8.25)$$

The relaxation time can be determined from relative sea-level (RSL) data, which record the relative movement of land and sea, and consist mainly of radiocarbon-dated beach terraces. Tilted strandlines can also be used to determine the pattern of rebound. A general discussion can be found in Cathles (1975), Peltier (1980), and Walcott (1980). Typical uplift curves and the general shape of postglacial uplift for the Laurentide ice sheet are shown in Figure 8.3. Rebound is a function of position, being an uplift for sites formerly beneath the load and its immediate neighbourhood, and a subsidence for sites on the peripheral bulge. Uplift rates near the centre of rebound are typically around 1–2 cm a^{-1}.

In order to evaluate the relaxation time from uplift curves, two additional quantities must be known: the time of unloading (assumed to be instantaneous) and the remaining uplift. The former can be determined from the glacial history; the latter can be estimated from gravity evidence. Large-scale negative free-air anomalies are present in both Fennoscandia and Canada. Moreover, their general shape parallels that of the former ice sheet (Fig. 8.3). The negative anomaly caused by the depression of the lithosphere into the mantle is

$$\Delta g = 2\pi G \rho_m h$$

where ρ_m is the density of the displaced fluid (mantle), and h is the amount of the depression. (The above expression for Δg is simply the Bouguer formula for an infinite slab; see, for example, Garland 1979). It is therefore possible to estimate the remaining uplift from the gravity anomaly in the central parts of the formerly glaciated areas. This procedure, of course, involves the assumption that the gravity anomaly is related to glacial unloading; this is probably correct in view of the aforementioned correspondence between position and shape of the anomaly and of the former ice sheet.

Estimated in this way, the remaining uplift in both Canada and Fennoscandia is of the order of 100–200 m. Relaxation times of the order of 5000 a can then be inferred from uplift curves near the centre of unloading. Consequently, with $\lambda \simeq 3000$ km and $\rho \simeq 3.5 \times 10^3$ kg m^{-3}, Equation 8.25 gives $\eta \simeq 10^{21}$ Pa s as an order of magnitude for the viscosity of the mantle. This value is the same as that inferred from consideration of the power involved in mantle convection (cf. Section 7.9).

The above results have been obtained on the assumption that the mantle is a 'half-space', i.e. that the depth of flow induced by glacial unloading is comparable with the horizontal dimensions of the ice sheet. The situation would be different if flow were restricted to a thin channel below the lithosphere – in particular, the relaxation time of short-wavelength disturbances would be shorter than that of long-wavelength disturbances. The 'thin-channel' model is discussed in detail by Artyushkov (1983). Within its framework the viscosity of a 100–200 km thick asthenosphere (overlying a more viscous substratum) is found to be of the order of 10^{19}–10^{20} Pa s.

Both the constant-viscosity half-space model and the thin-channel model impose artificial restrictions on the viscosity stratification of the mantle. A full theory, taking into account the depth-dependence of viscosity, the presence of a highly viscous lithosphere, the detailed deglaciation history, and the geoidal perturbations caused by melting of the ice sheets, has been developed by Peltier (see, for example, Peltier 1980 for a review). In this self-gravitating, viscoelastic (Maxwell) model the decay of each deformation wavelength is governed by a spectrum of relaxation times. For a given deglaciation history and mantle viscosity profile the model allows prediction of worldwide RSL patterns, which can then be compared with observations at selected sites. Free-air gravity anomalies can also be included in the model (Wu & Peltier 1983). Mantle viscosity profiles able to fit simultaneously RSL and gravity data have a lower mantle viscosity of at most 10^{22} Pa s as an order of magnitude, with a viscosity increase across the transition zone of a factor of ten or less. This result does not prove that the mantle has Newtonian viscosity or, for that matter, that the flow induced by postglacial rebound is steady-state; however, it does show that *observations pertaining to postglacial rebound can be accounted for without recourse to non-Newtonian or transient effects*. Despite years of refinement, postglacial rebound studies have not changed the venerable value of 10^{21} Pa s for the average viscosity of the mantle.

Another set of data is provided by the changes in the rotation of the Earth that are induced by glacial loading. Mathematical descriptions of such rotational response – ranging in period from the Chandler wobble to long-term polar wander – allow theoretical predictions that can be compared with observed polar drift (cf. Sabadini *et al.* 1984 for a discussion of the technique). An analysis based on a four-layer Earth model (elastic lithosphere, viscous upper and lower mantle, and inviscid core) leads to conclusions comparable with those obtained from RSL and gravity data.

In summary, analysis of relative sea-level, gravity, and rotational responses to glacial loading in terms of steady-state Newtonian viscosity yields a consistent picture of the mantle, with average viscosity $\eta \simeq 10^{21}$ Pa s, and an increase from upper mantle to lower mantle viscosity of at most one order of magnitude. However, this conclusion is strongly qualified by uncertainties on the depth-dependence of resolution, on the role of transient creep, and on the assumption of linear viscosity.

8.3 The choice of a rheological constitutive equation for the mantle

Taking the results from glacial loading studies at face value, the mantle has linear (Newtonian) viscosity for characteristic times longer than about 10^3 a, and transient effects, if any, must be limited to shorter timescales. On the other hand, at seismic and free oscillation frequencies the mantle is anelastic, and transient rheology extends to periods at least as long as that of the Chandler wobble (Section 6.5). The timescales of different kinds of mantle deformation are illustrated in Figure 8.4. On this interpretation, the crossover between transient and steady-state rheology must be placed somewhere between the characteristic time of the wobble and that of postglacial processes, although we shall see later in this section that this conclusion is by no means unchallenged.

If transient phenomena are neglected, then the mantle may be taken to show short-term elastic response (as exemplified by seismic waves) followed by long-term linearly viscous response (as exemplified by postglacial rebound and convection currents). With reference to the linear rheological models introduced in Section 4.6, the mantle is therefore a

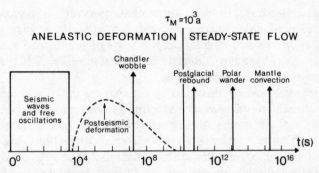

Figure 8.4 Characteristic timescales of mantle deformation processes. The Maxwell time is denoted by τ_M (from Yuen *et al.*, 1982).

viscoelastic (Maxwell) body. Its stress relaxation behaviour can be described in terms of a *Maxwell time* τ (Eqn 4.39). Recalling the basic relations between elastic stress and strain, on the one hand, and between viscous stress and strain rate, on the other (Eqns 4.32 & 33), the Maxwell time can be written as

$$\tau = \eta/\mu = \varepsilon/\dot{\varepsilon} \tag{8.26}$$

where η is viscosity, μ is rigidity, ε is elastic shear strain, and $\dot{\varepsilon}$ is viscous shear strain rate. The Maxwell time can therefore be interpreted as the time necessary for creep deformation at constant strain rate to equal the elastic deformation under the same load. For $t < \tau$ the elastic response predominates; for $t > \tau$ the viscous response does. Consequently τ can be taken as the time above which the rheology is essentially viscous. For the mantle, where (on the average) $\eta \simeq 10^{21}$ Pa s, $\mu \simeq 10^{11}$ Pa, the Maxwell time is of the order of 10^3 a (1 a $= 3.16 \times 10^7$ s), which fits nicely the inferences from postglacial rebound.

A transient response can be introduced by adding a Kelvin element in series with a Maxwell element: this results in a *Burgers body* (Fig. 4.7c). If the mantle is indeed linear, then the Burgers body is the most complete (in the sense that it accounts for elastic, transient, and steady-state deformation) and simplest rheological model for it; as such, it has been proposed by Peltier *et al.* (1981) as an adequate description of mantle rheology. The tensor form of the constitutive equation for the Burgers body is

$$\ddot{\sigma}_{ij} + \left(\frac{\mu_1 + \mu_2}{\eta_1} + \frac{\mu_2}{\eta_2}\right)(\dot{\sigma}_{ij} - \tfrac{1}{3}\dot{\sigma}_{kk}\,\delta_{ij}) + \frac{\mu_1\mu_2}{\eta_1\eta_2}(\sigma_{ij} - \tfrac{1}{3}\sigma_{kk}\,\delta_{ij})$$

$$= 2\mu_2\ddot{\varepsilon}_{ij} + \lambda_2\ddot{\varepsilon}_{kk}\,\delta_{ij} + \frac{2\mu_1\mu_2}{\eta_1}(\dot{\varepsilon}_{ij} - \tfrac{1}{3}\dot{\varepsilon}_{kk}\,\delta_{ij}) \tag{8.27}$$

where material parameters with subscripts 1 and 2 refer to the Kelvin and Maxwell elements, respectively. Equation 8.27 reduces to Equation 4.40 for the uni-dimensional shear case.

The Burgers body is characterized by two viscosities, one (η_1) controlling short-timescale processes, the other (η_2) controlling long-timescale steady flow. For $\eta_1 \to \infty$ the transient stage is negligible, and integration of Equation 8.27 with respect to time gives the Maxwell rheological equation. If both η_1 and $\eta_2 \to \infty$, then one recovers the equation for elasticity. The Burgers model is purely phenomenological, in the sense that it does not provide a microphysical model – that is, an account for the rheology on the basis of the atomic properties of matter.

It is obvious that the Burgers model for the mantle, even within its own context, is only a first approximation. For instance, it leads to a sharp frequency-dependence of the seismic attenuation factor Q (a consequence of the presence of a single relaxation time τ). We saw in Section 6.5 that weak frequency-dependence can be achieved through a continuous spectrum of relaxation times; in the model, this can be represented by several Kelvin elements with different relaxation times.

Although the modelling of mantle rheology in terms of a Burgers body yields a consistent picture, there are important uncertainties associated with it. These ultimately concern the *resolving power* of the geophysical evidence. To illustrate this point, we discuss briefly the possibilities of fine rheological stratification, transient creep, and non-linear rheology.

The resolution of the *fine viscosity structure* of the mantle has not been achieved from relative sea level, gravity, and rotational data. The lower-mantle D'' zone, for instance, is well established on seismological and thermal grounds (Section 7.7) and presumably has low viscosity; however, it has no detectable signature in postglacial processes. The presence or absence of an asthenosphere – a 100–200 km thick low-viscosity zone in the sublithospheric upper mantle – is not clearly resolved (see, for example, Yuen *et al.* 1982). The latter problem is particularly important, as the viscosity of the asthenosphere affects the degree of coupling between lithospheric plates and the substratum. It also bears upon our understanding of the relationship between short-term (low-velocity and high-attenuation) and long-term (low-viscosity?) rheological properties, which is by no means clear.

The role of *transient creep* in postglacial rebound is also a potential source of uncertainty. Since total strains during rebound are of the order of 10^{-3} or less if flow affects most of the mantle, it may well be that the rheology is still in the transient range (see the discussion by Weertman 1978). Analyses so far have not been conclusive. Peltier *et al.* (1981) found no evidence of variation of the relaxation time with time, and therefore of transient rheology, for a single harmonic of the deformation. On the other hand, Sabadini *et al.* (1985) compared theoretical predictions of transient Burgers rheology in the lower mantle with sea-level and rotational data, and concluded that both a steady-state lower mantle rheology (with a thick lithosphere) and a transient rheology (but with a thin lithosphere) can account for the observations. The possibility of transient relaxation in the lower mantle is accepted also by Peltier (1985).

The introduction of a transient rheology raises the long-term (steady-state) lower mantle viscosity necessary to fit the data to 10^{22}–10^{23} Pa s, i.e. to values higher than those estimated neglecting the possible influence of transient effects. Estimates based on geoid anomalies (using the relationship between boundary deformations and radial viscosity distribution mentioned in Section 8.1), yield a lower mantle viscosity at least a factor of 30 larger than the upper mantle viscosity (Hager 1984). Mantle-wide convection with lower mantle viscosity two orders of magnitude larger than upper mantle viscosity, together with seismically-inferred density anomalies, predict satisfactorily the longest-wavelength geoid anomalies (Hager *et al.* 1985). These results indicate that an increase of viscosity with depth by a factor of 10^2 is likely to occur.

The resolving power between Newtonian and non-Newtonian rheologies is another source of uncertainty. Interpretations of data in terms of non-linear creep are possible (cf. Weertman 1978 for a review and discussion). As was emphasized in Section 8.2, Newtonian rheology is an assumption, not a result, of current models of postglacial rebound. Whether the evidence would be equally compatible with a non-Newtonian rheology is an important question, in view of the fact that power-law creep is a common flow mechanism in silicate polycrystals (see Section 10.4). Here we consider only some phenomenological aspects of the problem.

Apparent linearity is possible in power-law bodies. When discussing the power-law rheological equation (Section 4.4)

$$\dot{\varepsilon}_{ij} = A\sigma_E^{(n-1)}\sigma_{ij}' \tag{8.28}$$

we pointed out that under some stress conditions the relation between corresponding components of stress and strain rate can be approximately linear. To illustrate the point in the context of postglacial rebound, assume that, in a two-dimensional frame of reference (x_1-axis horizontal, x_3 vertical), the only non-zero stress components are σ_{13} and σ_{33} (the former associated with convection, the latter with rebound), and that $|\sigma_{33}| \ll |\sigma_{13}|$. Then, by the definition of deviator, the effective stress is

$$\sigma_E' = (\tfrac{1}{2}\sigma_{ij}'\sigma_{ij}')^{1/2} = [\tfrac{1}{2}(\sigma_{33}'^2 + 2\sigma_{13}'^2)]^{1/2} \simeq \sigma_{13}$$

and the corresponding strain rate, by Equation 8.28, is

$$\dot{\varepsilon}_{33} \simeq \tfrac{1}{2} A \sigma_{13}^{(n-1)} \sigma_{33}$$

that is, $\dot{\varepsilon}_{33} \propto \sigma_{33}$, an 'apparently Newtonian' relation (the viscosity, however, is still determined by the dominant component of stress). That this type of situation may be relevant to the mantle was suggested by Weertman (1970). However, rebound stresses and convective stresses are of the same order of magnitude (~ 1 MPa), and consequently apparent linearity is not a likely explanation of the observations.

If the behaviour of the mantle is elastic at short timescales, and non-linearly viscous at long timescales, then its rheological model (neglecting transient effects) is obtained by combining in series a Hooke and a non-linear viscous element. The resulting equation is (Melosh 1980)

$$\dot{\varepsilon}_{ij} = (1/E)\,[\,(1 + \nu)\dot{\sigma}_{ij} - \nu\dot{\sigma}_{kk}\,\delta_{ij}\,] + A\sigma_{\text{E}}^{(n-1)}\sigma_{ij}' \qquad (8.29)$$

where the first term on the right-hand side is related to elasticity (E and ν are the Young modulus and the Poisson ratio, respectively), and the second term is the power-law steady-state strain rate. From the observational viewpoint the major difference between linear and non-linear bodies is the stress-dependence of the viscosity implicit in Equation 8.29. Since the effective viscosity of a non-Newtonian material decreases with increasing stress, it should theoretically be possible to detect this stress-dependence by studying the response of the Earth to different loads. Unfortunately, postglacial rebound does not offer the necessary resolution.

The study of geophysical processes connected with glacial loading appears to be in a paradoxical state. It provides a very important tool in the analysis of the rheology of the mantle. It has firmly established that the mantle is a fluid even at deviatoric stresses as low as 1 MPa, and that the evidence for characteristic times longer than about 10^3 a is compatible with a Newtonian, steady-state rheology. However, it has been less effective in excluding plausible alternatives.

8.4 Flexural thickness and rheology of the lithosphere

Information on the rheology of the lithosphere as a whole is obtained from its flexural response to surface loads. Flexure occurs for loads with horizontal dimension $\lambda \gg H$, where H is the thickness of the lithosphere. While the viscosity of the substratum is the controlling factor for the

recovery of equilibrium, as discussed in Section 8.2, the equilibrium configuration under a constant load depends on the thickness and rheology
of the lithosphere.

In order to analyse the behaviour of the lithosphere in flexure, consider
the equilibrium configuration of a thin *elastic* plate subject to a vertical
load (Fig. 8.5 – a full treatment of the problem can be found in Love
1944; see also Turcotte & Schubert 1982 for a discussion in a geophysical
context). The co-ordinate system is chosen such that x_1 and x_2 are in the
mid-plane of the plate, and x_3 is vertical and positive downwards. In the
two-dimensional case the load per unit area and the deflection are $P(x_1)$
and $\zeta(x_1)$, respectively.

The elongation of an infinitesimal line element l in the x_1-direction is

$$\Delta l = x_3 \; \mathrm{d}\phi = x_3 l / r \tag{8.30}$$

where r is the radius of curvature and $\mathrm{d}\phi$ is the angle subtended by l. The
longitudinal strain is then

$$\varepsilon_{11} = \Delta l / l = x_3 / r$$

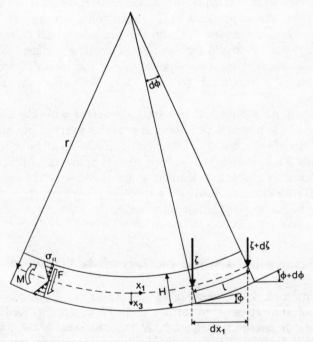

Figure 8.5 Flexure of elastic plate subject to vertical load. See text for details.

The angle ϕ from the horizontal (positive when anticlockwise) is

$$\phi = -\frac{d\zeta}{dx_1}$$

and its rate of change is therefore

$$\frac{d\phi}{dx_1} = -\frac{d^2\zeta}{dx_1^2} = \frac{1}{r}$$

where the second equality is established using Equation 8.30 and the condition $dx_1 \simeq l$ (the deflection gradient is small). Consequently, the longitudinal strain becomes

$$\varepsilon_{11} = -x_3 \frac{d^2\zeta}{dx_1^2}$$

The corresponding stress can be found by using Equation 3.12, with the conditions $\varepsilon_{22} = 0$ and $\sigma_{33} = 0$ (the latter is small compared with the horizontal components). This gives

$$\sigma_{11} = -\frac{E}{1-\nu^2} x_3 \frac{d^2\zeta}{dx_1^2} \tag{8.31}$$

where E and ν are the Young modulus and the Poisson ratio, respectively, for the plate. The horizontal stress is zero on the mid-plane of the plate, compressional above, and tensional below; its magnitude increases with distance from the mid-plane.

Equilibrium of the plate requires that the resultant forces and torques be zero. The vertical shear force F and the bending moment M acting on the plate (per unit width in the x_2-direction) result from the action of σ_{13} and σ_{11}, respectively, across the plate thickness H, and are given by

$$F = \int_{-H/2}^{H/2} \sigma_{13} \, dx_3 \qquad M = \int_{-H/2}^{H/2} \sigma_{11}x_3 \, dx_3$$

If dF and dM are their changes over dx_1, then the balance of vertical forces requires that

$$P \, dx_1 + dF = 0 \quad \text{i.e.} \quad dF/dx_1 = -P \tag{8.32a}$$

and the balance of torques (the minus sign takes account of their opposite sense)

$$dM - F \, dx_1 = 0 \quad \text{i.e.} \quad dM/dx_1 = F \tag{8.32b}$$

Using Equations 8.32a and b, the equilibrium condition becomes

$$d^2 M / dx_1^2 = -P \qquad (8.33)$$

By Equation 8.31 the bending moment is given by

$$M = -\frac{EH^3}{12(1 - \nu^2)} \frac{d^2 \zeta}{dx_1^2} = -D \frac{d^2 \zeta}{dx_1^2} \qquad (8.34)$$

where the parameter

$$D \equiv \frac{EH^3}{12(1 - \nu^2)} \qquad (8.35)$$

is the *flexural rigidity* of the plate, with dimensions $[ml^2 t^{-2}]$; it is measured in Newton metres (N m) in the SI system. Using Equations 8.33–8.35, the equilibrium condition can be expressed in terms of the vertical deflection

$$\frac{d^4 \zeta}{dx_1^4} = \frac{P}{D} \qquad (8.36)$$

In the more general case in which the deflection is two-dimensional, Equation 8.36 becomes

$$\frac{\partial^4 \zeta}{\partial x_1^4} + 2 \frac{\partial^4 \zeta}{\partial x_1^2 \partial x_2^2} + \frac{\partial^4 \zeta}{\partial x_2^4} \equiv \nabla^4 \zeta = \frac{P}{D} \qquad (8.37)$$

which is the differential equation for the surface deflection $\zeta(x_1, x_2)$ of a thin elastic plate of flexural rigidity D subject to a surface load $P(x_1, x_2)$.

If the plate overlies a fluid substratum, as is the case in the Earth, then the displacement of the substratum consequent upon flexure gives rise to a vertical buoyancy force

$$B = -(\rho_m - \rho) g \zeta$$

where ρ_m is the density of the displaced fluid (sub-lithospheric mantle) and ρ is the density of the material on top of the lithospheric flexure (water or rock, as the case may be). The buoyancy force must be included in the vertical force balance (Eqn 8.32a), and the right-hand term of Equations 8.36 and 8.37 is modified accordingly. For instance, for one-dimensional bending, the differential equation for the deflection of a thin

206

plate overlying a fluid substratum is

$$\frac{d^4\zeta}{dx_1^4} = \frac{1}{D} \left[P - (\rho_m - \rho)g\zeta \right] \tag{8.38}$$

The previous analysis applies to an elastic plate. The corresponding case of a *viscoelastic* (Maxwell) plate is fully treated by Nadai (1963). If τ is the relaxation time of the material, then the one-dimensional equation for the deflection is

$$\frac{d}{dt}\left(\frac{d^4\zeta}{dx_1^4}\right) = \frac{1}{D}\left[\left(\frac{d}{dt} + \frac{1}{\tau}\right)P - (\rho_m - \rho)g\left(\frac{d}{dt} + \frac{1}{\tau}\right)\zeta\right] \tag{8.39}$$

which reduces to the elastic case for $\tau \to \infty$. For a viscoelastic plate the shape of the flexure changes with time – there is an increase in amplitude and a decrease in wavelength with increasing duration of loading. The *apparent* flexural rigidity of the plate therefore decreases with time from an initial 'elastic' value immediately upon loading.

Solutions to the problem are relatively simple in the case of harmonic loading (cf. Walcott 1970). For $P = 0$ (i.e. beyond the edge of the load). the uni-dimensional elastic solution has the form of a damped harmonic wave with wavelength $2\pi\alpha$, where α (the *flexural parameter*) is defined as

$$\alpha = \left[\frac{4D}{(\rho_m - \rho)g}\right]^{\frac{1}{4}} \tag{8.40}$$

and has dimensions $[l]$. It determines the wavelength of flexure and affects the amplitude.

The thickness of the lithosphere in flexure is determined by comparing the actual shape of the deflection under a load with solutions to the deflection equation for different values of the flexural parameter. The shape of the deflection is inferred from the resulting gravity anomalies, since the lithosphere is depressed and the Moho discontinuity is assumed to be rigidly embedded in the lithosphere. From the best-fitting α, the flexural rigidity D and the thickness H can be obtained. Surface loads such as ice sheets, postglacial lakes, sedimentary basins, thrust belts, and volcanic chains may be used, but also intralithospheric loads (e.g. thickening and intrusions).

The nature of the best-fitting rheology is still subject to debate. Walcott (1970) observed that D decreases with increasing *age of the load* (from about 10^{25} N m to 5×10^{22} N m for loading ages going from 10^{-3} to 500 Ma), and therefore inferred a Maxwell rheology with relaxation time $\tau \simeq 10^{-1}$ Ma and thickness varying from 110 km (stable continental areas) to 20 km (recently active tectonic areas). However, Watts *et al.* (1982), concluded that an elastic plate with flexural rigidity increasing with the *age of the plate* at the time of loading gives the best fit to the data. This implies that lithospheric thickness is thermally controlled. In the oceanic lithosphere, where D is usually in the range 10^{21}–10^{23} N m, H correlates well with the position of the 720 ± 100 K isotherm, as predicted by thermal studies (see Section 7.6). The flexure of continental lithosphere appears to follow a similar dependence: thickness is correlated with the age of the last tectonothermal event. The large flexural rigidities in Precambrian regions ($D \geqslant 10^{25}$ N m) indicate thicknesses in excess of 100 km. A large continental lithosphere thickness ($H \geqslant 200$ km) is also inferred from relative sea-level data near the margin of former ice sheets, and from the secular drift of the rotational pole (Peltier 1984).

On the other hand, a detailed study of the Alberta foreland basin in western Canada (Beaumont 1981) has shown that the evidence cannot be matched by an elastic plate. Good agreement with observation is obtained when viscoelastic stress relaxation is allowed, with $D \simeq 10^{25}$ N m and $\tau \simeq 20$–35 Ma. The lithosphere responds by flexure to the load of the migrating foreland belt (see also Section 8.7); the occurrence of a foredeep, in which the clastic wedge from the orogen accumulates, is partly a consequence of this flexure. Thus the tectonic pulses in the foreland belt (in this case, from the Late Jurassic to the Eocene) and the sedimentary record in the foreland basin are fully coupled.

A further step in this type of analysis removes the assumption of uniform lithosphere, and assumes a Maxwell rheology with temperature-dependent viscosity. The Palaeozoic Appalachian Basin of the eastern interior of North America, considered as a multistage foreland basin caused by the down-warping of the lithosphere under the load of successive thrust sheets from the Appalachian orogen, has been studied this way (Quinlan & Beaumont 1984). For the chosen geotherms, the viscosity reaches upper mantle values ($\eta \simeq 10^{21}$–10^{22} Pa s) at depths of the order of 100–150 km. The lithosphere above consists of a lower part with relaxation times in the range $1 \leqslant \tau \leqslant 200$ Ma, and an upper part which is too viscous to flow on timescales comparable with the age of the Earth ($\tau \rightarrow \infty$).

The answer to the dichotomy in the interpretation of lithospheric rheology (elastic or viscoelastic) probably lies in the aforementioned variation of rheological properties with depth (see also Section 12.1). Since a material is essentially elastic for loading times less than the relaxation time, the thickness of the 'elastic' lithosphere – defined as the depth at which $t = \tau$ – is a function of load duration t. At one extreme the whole mantle is elastic for $t \lesssim 1$ a; at the other, the uppermost lithosphere maintains stresses for times of the order of the age of the Earth. Therefore thickness is a function of load duration, $H(t)$. The situation is further complicated if the rheology is non-linear. In this case, a non-linear 'Maxwell time' can be defined in a manner analogous to the linear case (Eqn 8.26); however, it will be a function of stress (cf. Eqn 8.28). Consequently, as $\tau(\sigma)$, the thickness will also be a function of the load, $H(\sigma)$. What we call 'lithosphere' therefore depends on the nature of the load, its duration, and its magnitude. In the elastic model the flexural thickness of old continental lithosphere ($H \simeq 100-200$ km) is much larger than the thickness of either oceanic or 'orogenic' lithosphere ($H \leqslant 50$ km). In the viscoelastic model, relaxation times for the lithosphere as a whole vary from 10^{-1} to 10^2 Ma as orders of magnitude, according to the tectonic characteristics of the region, but their value is debatable as viscosity varies with depth.

The degree of isostatic compensation in lithospheric flexure is a function of the wavelength of the topography (cf. Turcotte & Schubert 1982 for a detailed analysis). If the surface topography h and the displacement ζ are both harmonic

$$h = h_0 \sin(2\pi x_1/\lambda) \qquad \zeta = \zeta_0 \sin(2\pi x_1/\lambda)$$

the surface load P is

$$P = \rho_c g h_0 \sin(2\pi x_1/\lambda)$$

where ρ_c is the density of the upper crust. Substitution into Equation 8.38 yields

$$\zeta_0 = \frac{h_0}{(D/\rho_c g)(2\pi/\lambda)^4 + (\rho_m - \rho_c)/\rho_c} \tag{8.41}$$

For short-wavelength topography, $\lambda \ll 2\pi(D/\rho_c g)^{1/4}$, one has $\zeta_0 \ll h_0$, i.e. there is practically no deflection of the lithosphere. The topographic excess mass is uncompensated and the Bouguer anomaly is zero. For long-wavelength topography, $\lambda \gg 2\pi(D/\rho_c g)^{1/4}$, Equation 8.41 reduces to

$$\zeta_0^* = \frac{\rho_c}{\rho_m - \rho_c} h_0 \tag{8.42}$$

i.e. the topography is fully compensated, and the free-air anomaly is zero. (Eqn 8.42 represents hydrostatic equilibrium, i.e. isostasy in the Airy sense – see Section 8.1.) The *degree of compensation C* of the topography may be defined as the ratio of the actual deflection to the maximum (hydrostatic) deflection, i.e. (from Eqns 8.41 and 8.42)

$$C = \frac{\zeta_0}{\zeta_0^*} = \frac{\rho_m - \rho_c}{(D/g)(2\pi/\lambda)^4 + \rho_m - \rho_c} \qquad (8.43)$$

Equation 8.43 expresses the degree of isostatic compensation in terms of the flexural properties of the lithosphere. Taking a typical value $D \simeq 10^{23}$ N m and $\rho_m - \rho_c \simeq 500$ kg m^{-3}, results are $C \simeq 0$ (no isostatic compensation) for $\lambda \simeq 100$ km; $C \simeq \frac{2}{3}$ (about 65 per cent isostatic compensation) for $\lambda \simeq 500$ km; and $C \simeq 1$ (complete isostatic compensation) for $\lambda \simeq 1000$ km. Loads with wavelength less than, or comparable with, the thickness of the lithosphere are supported by the strength of the lithosphere, and are therefore isostatically uncompensated.

8.5 Tectonics I: state of stress in the lithosphere

Plate tectonics provides a unifying framework in which to study geodynamic processes. The lithosphere in the tectonic sense ('tectosphere') consists of a number of approximately rigid plates in relative horizontal motion, with most tectonic, volcanic, and seismic activity concentrated near their margins. Geophysical discussions of plate tectonics can be found in McKenzie (1972), Le Pichon et al. (1973), Bott (1982), and Turcotte & Schubert (1982). The literature is vast, and collections of the most significant papers have been published (Cox 1973, Bird 1980). Useful reviews are given in Dewey (1982) and Turcotte (1983).

In the remainder of this chapter attention is focused on selected aspects of plate tectonics, chosen as examples of the application of continuum mechanics and rheology to geodynamics. Knowledge of plate kinematics and of the main geophysical and geological corollaries of plate tectonics is assumed. We consider first the state of stress in lithospheric plates, which is related to the driving mechanisms of plate motion.

Figure 8.6 shows the main lithospheric plates. The forces on plates resulting from processes at or near their boundaries and from the interaction with the underlying mantle are illustrated in Figure 8.7. They may be classified as follows:

(a) Forces associated with *divergent* plate boundaries, resulting from the gravitational effects of mid-oceanic ridges (ridge push F_{RP}).
(b) Forces associated with frictional resistance to motion at *transform* boundaries (transform fault resistance F_{TF}).

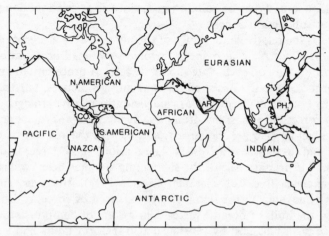

Figure 8.6 Principal lithospheric plates (from Forsyth & Uyeda 1975).

(c) Forces arising out of the complex interactions at *convergent* plate boundaries: negative buoyancy of the cool subducting slab (slab pull F_{SP}); viscous resistance of the mantle to slab sinking (slab resistance F_{SR}); resistive forces originating from plate convergence (colliding resistance F_{CR}); and tensional forces in the overriding plate resulting from the presence of the oceanic trench (trench suction F_{TS}).

(d) Forces deriving from the interaction of a plate and the *underlying mantle* (mantle drag F_{MD}). Given the thickness variations of the plates, it is possible that the drag under continental plates ($F_{MD} + F_{CD}$) is larger than under oceanic plates.

Terms such as 'ridge push' and 'slab pull', although entrenched in common usage, are somewhat imprecise. The forces thus denoted are body forces originating from pressure and density gradients, not surface forces. Examples of the latter are mantle drag, transform fault resist-

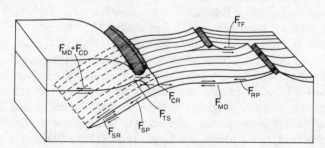

Figure 8.7 Forces acting on lithospheric plates (from Forsyth & Uyeda 1975).

211

ance, and some types of colliding resistance (e.g. friction along thrust faults at subduction zones).

What is the relative importance of the above forces as agents of plate tectonics? Since plate movements involve very small accelerations, the net torque acting on each plate should be zero. A procedure to minimize this torque, given the observed motions and geometries, has been proposed by Forsyth & Uyeda (1975), and yields the relative strength of plate driving forces. The most important conclusions are: (a) forces acting on the subducting slab are the largest, and control the velocity of oceanic plates; and (b) mantle drag resists plate motion, and is larger under continents. The primary role of the subducting slab has been confirmed by other analyses (see, for example, Richter 1977). However, since the negative buoyancy of the slab is nearly balanced by forces resisting slab sinking, the bulk of oceanic plates is in a state of compression, caused by ridge push balanced by resistance at transform faults and trenches, and by a weak drag on its base.

Thermal convection is ultimately responsible, directly or indirectly, for the forces acting on plates. However, if mantle drag resists plate motion, then plates are not simply passive rafts carried by the upper limbs of large-scale mantle convection cells. Plates are an integral part of the circulation, and move under the action of buoyancy forces associated with their diverging and converging margins. There is in all likelihood no one-to-one correspondence between movements of plates and the planform of mantle convection (see also Section 7.9), and the degree of coupling is unknown. The larger mantle drag beneath stable continental areas confirms that the thickness of the continental tectosphere is larger than that of the oceanic tectosphere.

The system of forces described above generates a complex state of stress in the lithosphere. Other stresses originate from processes that are not associated with plate boundaries, such as lithosphere loading by topographic and density variations, flexure, bending, non-sphericity of the Earth (membrane effects), erosion and sedimentation, and thermal events. Stresses in the lithosphere can be conveniently subdivided into two categories (Bott & Kusznir 1984): *renewable stresses*, which persist despite stress relaxation, and *non-renewable stresses*, which are dissipated by brittle or ductile strain relief. To the former category belong those that result from the continued action of the causative force, such as plate boundary processes, mantle drag, and lithosphere loading; to the latter belong bending, membrane, and thermal stresses.

Horizontal stresses originating from lateral thickness and density inhomogeneities within plates, referred to as 'lithosphere loading' above, have been analysed in detail by Artyushkov (1973, 1983). They result from horizontal gradients in overburden pressure, and are perhaps as important as 'plate-boundary' stresses as an agent for orogenesis. We discuss them following the simple analysis of Dalmayrac & Molnar (1981). With reference to Figure 8.8a, consider the crust, of density $\rho(x_1, x_3)$ and 'standard' thickness H, having the thickness inhomogeneity shown (corresponding, for instance, to an isostatically compensated mountain belt). With the co-ordinate axes as in the figure, the equations of equilibrium in two dimensions are

$$\frac{\partial \sigma_{11}}{\partial x_1} + \frac{\partial \sigma_{13}}{\partial x_3} = 0 \qquad \frac{\partial \sigma_{13}}{\partial x_1} + \frac{\partial \sigma_{33}}{\partial x_3} + \rho g = 0$$

To a first approximation, the gradients of σ_{13} are negligible, and therefore the above equations yield the results that σ_{11} is independent of x_1, and that

$$\sigma_{33}(x_1, x_3) = -g \int_S^{x_3} \rho(x_1, x_3) \, dx_3$$

where the limit of integration S denotes the surface. The average horizontal and vertical stresses, from the surface to the depth of compensation L, are

$$\langle \sigma_{11} \rangle = \frac{1}{L} \int_S^L \sigma_{11} \, dx_3 \qquad \langle \sigma_{33} \rangle = \frac{1}{L} \int_S^L \sigma_{33} \, dx_3$$

respectively. Since $\langle \sigma_{11} \rangle$ does not depend on x_1, and $\langle \sigma_{33} \rangle$ does, it follows that the average stress difference $\langle \sigma_{11} \rangle - \langle \sigma_{33} \rangle$ on any cross-section depends on x_1 in the same way as $\langle \sigma_{33} \rangle$. This is illustrated in Figure 8.8b, which shows the variation of σ_{33} with depth in the case of constant ρ, the gradient of $\sigma_{33}(x_3)$ being proportional to density. At any depth down to the level of compensation, the vertical (lithostatic) stress

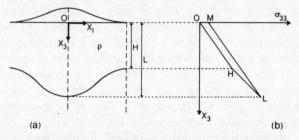

(a)   (b)

Figure 8.8 Stresses originating from lateral inhomogeneities: (a) thickened crust, (b) vertical stress under mountains (ML) and lowlands (OHL).

under mountains (ML) is higher than under lowlands (OHL). For instance, if the horizontal stress is approximately equal to the vertical stress under the lowlands, then the mountains will be in a state of relative (deviatoric) horizontal tension. This *spreading force* (per unit length perpendicular to the plane of the figure) is still present, as a horizontal pressure gradient, when $\sigma_{11} = \sigma_{33}$, and is given by the area OHLM in the stress–depth diagram. These considerations are independent of rheology.

Isostatic equilibrium is therefore not incompatible with horizontally directed forces resulting from variations of $\langle \sigma_{33} \rangle$ with position. Lithospheric mass inhomogeneities may also be described in terms of a *density moment*, defined as (Fleitout & Froidevaux 1982)

$$M = \int_{S}^{L} \Delta\rho x_3 \, dx_3 \tag{8.44}$$

where L is, as before, the depth to which the density anomalies $\Delta\rho$ extend. The stress difference averaged over L varies in proportion to M. These considerations are particularly important when dealing with continental collision, where the whole lithosphere is thickened. Deformation is by flow, except in the top brittle layer. The flow is in response to boundary forces and forces generated by thickness inhomogeneities. England & McKenzie (1982, 1983) introduced the concept of *Argand number* Ar as a measure of the relative importance of crustal thickness inhomogeneities in orogenesis. The Argand number is the ratio between the stress $\sigma_{\Delta H}$ induced by these inhomogeneities and the stress σ_T required to deform the material at tectonic strain rates. If σ_T is large (i.e. if the effective viscosity of the lithosphere is high), Ar $\rightarrow 0$ and flow is not affected by crustal thickening. As Ar increases, so thickening exerts more influence. In the case of a very weak medium, Ar becomes very large and thickness inhomogeneities cannot be sustained.

The magnitudes of the various stresses have been compared by Bott & Kusznir (1984). We saw in Section 7.3 that thermal stresses in the upper lithosphere are of the order of 100 MPa. Similar, or higher, stress differences are associated with bending of the lithosphere at trenches, and with changes in the radius of curvature of plates resulting from latitude changes (membrane effects). All of these stresses are non-renewable. Renewable stresses associated with plate boundary processes, mantle drag, and lithosphere loading are typically one order of magnitude lower (10–40 MPa). However, they can be subjected to *stress amplification*, resulting from creep in the ductile lower lithosphere, which concentrates the stress (by a factor of up to ten) in the upper elastic–brittle lithosphere (Kusznir 1982, Bott & Kusznir 1984, Hasegawa *et al.* 1985). Stress differences of the order of hundreds of megapascals are therefore to be expected in the upper lithosphere.

The stress system can be inferred from observations using geological evidence (neotectonic faults and joints), seismological data (earthquake focal mechanisms), and *in situ* measurements of various kinds (overcoring, hydrofracturing, oil-well breakouts) – cf. Zoback 1983 for a review and extensive bibliography. Despite the increasing importance of *in situ* and neotectonic measurements, it is *earthquake focal mechanisms* that provide most of the information on interplate and intraplate stresses (Fig. 8.9). If the earthquake focus is regarded as a point source, then the distribution of *P*-wave first motions follows a quadrantal pattern with

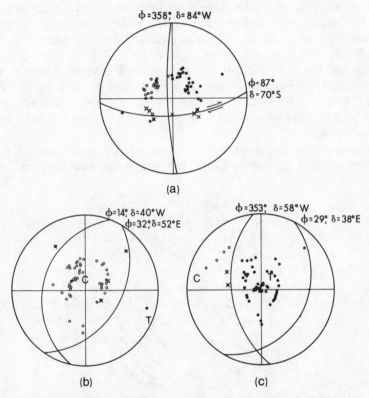

Figure 8.9 Examples of earthquake focal mechanisms. (a) Strike-slip faulting (shock of 15 November 1965, Romanche Fracture Zone, 0.17°S 18.70°W); (b) normal faulting (shock of 20 March 1966, East African Rift, 0.81°N 29.93°E); (c) thrust faulting (shock of 12 September 1964, Macquarie Ridge, 49.05°S, 164.26°E). Lower-hemisphere projection; closed circles denote compressions, open circles dilatations, crosses stations near nodal planes; ϕ, δ, T, and C are strike and dip of nodal planes, and axes of maximum tension and compression, respectively (from Sykes 1967).

215

respect to the focus. This pattern can be displayed on an equal-area lower-hemisphere projection of the *focal sphere* (an imaginary sphere drawn around the focus). The distribution of first arrivals on this projection yields the orientation of two orthogonal planes (the *fault plane* and the *auxiliary plane*) and the type of faulting, from which the stress orientation can be inferred on the basis of shear fracture criteria (Section 5.3). Discrimination between fault plane and auxiliary plane must be achieved on independent evidence. The procedure for seismic focal mechanism studies is discussed in detail by Sykes (1967) and − ever since the classic paper by Isacks *et al.* (1968) − has been applied in countless analyses of plate kinematics and dynamics.

The results of the various methods of *stress orientation* determination are usually consistent over large areas (excluding local fluctuations due to boundary effects and possibly to the presence of remanent stress locked into the rock). For example, continental North America east of the Rocky Mountains Front is in a state of ENE−WSW compression; the Basin and Range Province is in a state of E−W deviatoric tension; and farther to the west, a strike-slip regime predominates (Fig. 8.10). These stress fields can be related to the tectonic environment, and confirm the interrelation between stresses and plate driving forces (Zoback 1983, Gough 1984).

Consistent estimates of *stress magnitude* are also obtained from seismic data. If the seismic fault is modelled as an elastic dislocation of

Figure 8.10 *In situ* stress orientations in North America. NA, region of ENE−WSW compression; BR, region of E−W deviatoric tension; SS, region of strike-slip regime (from Gough 1984).

rigidity μ and area A across which an average slip $\langle u \rangle$ occurs, then a *seismic moment* can be defined as

$$M_0 = \mu \langle u \rangle A \qquad (8.45)$$

The seismic moment can be determined from the amplitude of waves with period longer than the timescale of the faulting process (the latter is of the order of $A^{1/2}/\beta$, where β is the shear-wave velocity). If the fault area can be estimated from independent evidence (for instance, the distribution of aftershocks), then Equation 8.45 yields the average slip $\langle u \rangle$. The average strain drop associated with it is

$$\Delta \varepsilon = C \langle u \rangle / A^{1/2}$$

where C is a dimensionless shape factor of order unity, depending on the geometry of the fault with characteristic length $A^{1/2}$. The *stress drop* associated with the earthquake is then

$$\Delta \sigma = \mu \ \Delta \varepsilon = C\mu \langle u \rangle / A^{1/2} = CM_0 / A^{3/2} \qquad (8.46)$$

Almost all stress drops so determined are in the range 1–10 MPa (cf. the review by Kanamori 1980). This is one to two orders of magnitude less than the ambient stress expected in the upper lithosphere. Consequently, either seismic events release only a small fraction of the tectonic stress, or seismogenic zones – both interplate and intraplate – are weak zones where the stress is substantially lower than elsewhere. Torque-balance calculations of the type previously mentioned indicate a relatively small frictional resistance along transform faults (as does the lack of a large heat flow anomaly), and therefore favour the low-stress model.

In summary, it appears that the tectonic stress in the upper lithosphere is of the order of 100 MPa in plate interiors, but perhaps as much as one order of magnitude less at plate boundaries and in active zones within plates. Although intraplate events occur when stresses in seismogenic zones reach levels of about 10 MPa, the lithospheric plate as a whole can support higher stresses. The orientation of the large-scale stress field reflects the force systems acting at the boundaries of, beneath, and within plates.

8.6 Tectonics II: the dynamics of faulting

The second tectonic process that we examine is faulting. Although the actual process of rupture and sliding takes a short time, the causative stresses are the result of the long-acting tectonic forces discussed in the previous section. Faulting in the upper lithosphere (say, for $T \leqslant 700$ K, corresponding to a maximum depth of 10–30 km according to local geothermal conditions) is a brittle or semibrittle process. At greater depths, the material deforms by flow, and 'faulting' is a ductile shear concentration along relatively narrow bands. We consider here faulting in the upper lithosphere. The relevant rheological model is elastic–brittle; however, in some instances a plastic rheology proves to be a satisfactory assumption.

Experimental evidence on rock failure was reviewed in Section 5.2. As all three principal stresses are generally compressive, faulting is the outcome of shear fracture and sliding under pressure. Shear fracture is adquately described by the Coulomb–Navier criterion (Eqn 5.5), and sliding by Amonton's Law of Friction (Eqn 5.1). Neglecting cohesion and equating, as a first approximation, sliding friction with internal friction, the two criteria can be written as one. Pore water pressure is a ubiquitous factor in the crust, and dry friction is probably very rare. Most faults exhibit *stick–slip* behaviour, alternating quiescent intervals – during which relative motion is either absent or continuous – with discontinuous, seismogenic slippage.

The compressive nature of all three principal stresses in the crust, even in the absence of active lateral compression, can be seen by considering an elemental volume at depth x_3 in a semi-infinite elastic region under gravity. The vertical normal stress is $\sigma_{33} = -\rho g x_3$ where ρ is the average density of the overlying material and g is the acceleration of gravity. If the volume is laterally confined (i.e. $\varepsilon_{11} = \varepsilon_{22} = 0$), then the only non-zero strain is ε_{33}, and the cubical dilatation is $\theta = \varepsilon_{33}$ (see Section 2.7). Therefore Hooke's Law becomes

$$\sigma_{11} = \sigma_{22} = \lambda\theta = \lambda\varepsilon_{33} \qquad \sigma_{33} = \lambda\theta + 2\mu\varepsilon_{33} = (\lambda + 2\mu)\varepsilon_{33}$$

that is, making use of the relations between the elastic parameters λ, μ, and ν (Section 3.1)

$$\sigma_{11} = \sigma_{22} = \frac{\lambda}{\lambda + 2\mu}\,\sigma_{33} = \frac{\nu}{1 - \nu}\,\sigma_{33} \qquad (8.47)$$

For appropriate values of the Poisson ratio ($\nu \simeq \frac{1}{4}$) the horizontal compressive stresses set up within a confined elastic medium are about one-third of the overburden pressure.

The stress system described by Equation 8.47 would originate stress differences of the order of hundreds of megapascals in the crust, greater than the fracture strength of most rocks. Furthermore, it is unrealistic in other ways, since it neglects lateral tectonic forces and stress relaxation and amplification (Section 8.5). It is more convenient to use Anderson's standard state (Section 3.5), where $\sigma_{11} = \sigma_{22} = \sigma_{33}$, as a reference system. Then the stresses associated with the various faulting regimes can be regarded as deviations from the standard state. Also, on a sufficiently large scale, one of the principal stress directions is vertical, and the other two lie in the horizontal plane. This is the basis of *Anderson's Theory of Faulting* (Anderson 1951), which leads to a satisfactory correlation between stress regimes and type and attitude of faults.

Anderson's Theory of Faulting has been expanded by Sibson (1974), whose analysis we follow here. The argument applies both to sliding along pre-existing faults and to the formation of new faults, provided that internal friction is approximately the same as sliding friction, and that cohesive strength is negligible when compared with normal stress.

The frictional law $|\tau| = -\mu'\sigma$ (where μ' is the coefficient of friction) may be expressed in terms of principal stress difference by recalling the expressions for normal and shear stresses on a plane which contains the x_2-axis and whose normal makes an angle θ with the x_1 principal axis, which were derived in Section 5.3. They are

$$\sigma = \tfrac{1}{2}(\sigma_1 + \sigma_3) + \tfrac{1}{2}(\sigma_1 - \sigma_3)\cos 2\theta \qquad \tau = -\tfrac{1}{2}(\sigma_1 - \sigma_3)\sin 2\theta \quad (8.48)$$

(Note that, since $\sigma_1 > \sigma_2 > \sigma_3$ and all negative, σ_1 is the direction of minimum compression.) Since only the absolute value of the shear stress enters the considerations, and the result is symmetric about the x_1-axis, we restrict our attention to angles $0 \leqslant \theta \leqslant \pi/2$. The frictional law can then be written as

$$(\sigma_1 - \sigma_3)\sin 2\theta = -\mu'\left[(\sigma_1 + \sigma_3) + (\sigma_1 - \sigma_3)\cos 2\theta\right] \qquad (8.49)$$

As discussed in Section 5.3, the orientation of the conjugate fault planes is obtained from the condition that $f(\theta) = |\tau| + \mu'\sigma$ be maximum, that is

$$\mathrm{d}f/\mathrm{d}\theta = (\sigma_1 - \sigma_3)(\cos 2\theta - \mu'\sin 2\theta) = 0$$

which gives

$$\tan 2\theta = 1/\mu' \quad \text{i.e.} \quad \theta = \tfrac{1}{2}\tan^{-1}(1/\mu') \qquad (8.50)$$

If we now define a stress ratio $R = \sigma_3/\sigma_1$, then the frictional law in terms of the principal stress difference (Eqn 8.49) may be rewritten as

$$(1 - R)\sin 2\theta = -\mu'\left[(1 + R) + (1 - R)\cos 2\theta\right]$$

219

which yields

$$R = \frac{\sin 2\theta + \mu' (\cos 2\theta + 1)}{\sin 2\theta + \mu' (\cos 2\theta - 1)}$$

or, in terms of μ', by using Equation 8.50,

$$R = [(1 + \mu'^2)^{\frac{1}{2}} - \mu']^{-2} \tag{8.51}$$

For $\mu' \simeq 0.75$ (a fair approximation for most rocks), Equations 8.50 and 8.51 predict that sliding will occur on one (or both) of the planes whose normals make an angle $\theta \simeq 27°$ with the direction of minimum compression, when the stress ratio reaches a value $R \simeq 4$. These results are confirmed by experiment (Section 5.2). The angle θ is also the angle between the fault plane and the direction of maximum compression (see Fig. 8.11).

In dealing with fracture and frictional sliding it is common practice to take compression positive. With this provision, σ_1 and σ_3 can be denoted by σ_{min} and σ_{max}. Then, $R = \sigma_{max}/\sigma_{min}$, and the critical condition for sliding takes the form

$$\sigma_{max} \geqslant R\sigma_{min} \tag{8.52}$$

In terms of maximum stress difference the above inequality can be written as

$$(\sigma_{max} - \sigma_{min}) \geqslant (R - 1)\sigma_{min} \tag{8.53}$$

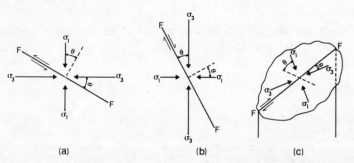

Figure 8.11 Stress distribution and faulting: (a) thrust, (b) normal, and (c) transcurrent faulting. (a) and (b) are diagrams in the vertical plane and FF is the trace of the fault normal to the plane of figure; (c) is an oblique view of the horizontal plane, and the fault is vertical. In all cases, σ_1 and σ_3 are the minimum and maximum compression, respectively, and σ_2 is contained in the fault plane.

The possible occurrence of a fluid phase can be taken into account by the introduction of a *pore fluid factor* λ, defined as the ratio of pore fluid pressure to lithostatic (overburden) pressure

$$\lambda = p_f/\rho g x_3 \tag{8.54}$$

so that all three principal compressions are reduced by an amount proportional to λ.

On the basis of the critical condition given by Equation 8.53, and taking into account the effect of pore fluid pressure, faulting regimes can be classified as follows (Fig. 8.11):

(a) *Thrust faulting.* σ_{min} vertical and σ_{max} horizontal. Faults, accommodating shortening, strike perpendicularly to σ_{max} and dip at a shallow angle from the horizontal. Since $\sigma_{min} = \rho g x_3$, the limiting condition is

$$(\sigma_{max} - \sigma_{min}) \geqslant (R - 1)\rho g x_3 (1 - \lambda) \tag{8.55a}$$

(b) *Normal faulting.* σ_{max} vertical and σ_{min} horizontal. Faults, accommodating extension, strike perpendicularly to σ_{min} and dip at a steep angle from the horizontal. The limiting condition, bearing in mind that $\sigma_{min} = \sigma_{max}/R$, is

$$(\sigma_{max} - \sigma_{min}) \geqslant [(R - 1)/R]\rho g x_3 (1 - \lambda) \tag{8.55b}$$

(c) *Transcurrent faulting.* Both σ_{min} and σ_{max} horizontal. Faults, accommodating strike-slip motion, strike at approximately $30°$ to σ_{max} and are vertical. The limiting condition can be derived by writing the intermediate principal stress (vertical) as

$$\sigma_{int} = \sigma_{min} + \beta(\sigma_{max} - \sigma_{min})$$

where β is a dimensionless factor and $0 < \beta < 1$. It follows that

$$\sigma_{min} = \frac{\sigma_{int}}{1 + \beta(R - 1)}$$

and the limiting condition becomes

$$(\sigma_{max} - \sigma_{min}) \geqslant \frac{R - 1}{1 + \beta(R - 1)} \rho g x_3 (1 - \lambda) \tag{8.55c}$$

221

For the particular case $\beta = \frac{1}{2}$ this reduces to

$$(\sigma_{max} - \sigma_{min}) \geqslant \frac{2(R-1)}{R+1} \rho g x_3 (1 - \lambda)$$

All three limiting conditions (Eqns 8.55a, b & c) can be written in the form

$$(\sigma_{max} - \sigma_{min}) \geqslant \alpha \rho g x_3 (1 - \lambda) \tag{8.56}$$

where α is a factor depending on fault type. For $\mu' = 0.75$ and $\beta = \frac{1}{2}$ the ratio $\alpha(\text{thrust}) : \alpha(\text{transcurrent}) : \alpha(\text{normal})$ is $4.0 : 1.6 : 1.0$. Larger stress differences are therefore associated with thrust faulting than with transcurrent and normal faulting.

Since maximum values of the critical stress difference in crustal fault zones are unlikely to exceed 100 MPa (see Section 8.5), and frictional stick–slip extends to mid-crustal depths, Equation 8.56 shows that pore fluid pressure must play an important role. It is sometimes assumed that the pore fluid pressure is hydrostatic, $p_f = \rho_w g x_3$, where ρ_w is the density of water; in this case, $\lambda = \rho_w / \rho \simeq 0.4$. However, pore fluid pressures of up to 90 per cent of the overburden pressure have been measured in deep wells; such high fluid pressures can be attained if water is confined. Figure 8.12 shows the stresses required for faulting for $\lambda = 0$ and $\lambda = 0.9$. High pore fluid factors appear to be necessary for frictional sliding,

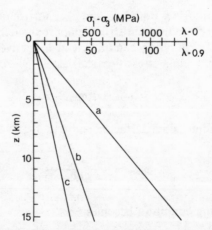

Figure 8.12 Stress differences required for faulting in the limiting cases of no pore pressure ($\lambda = 0$) and high pore pressure ($\lambda = 0.9$). Curves a, b, and c denote thrust, transcurrent, and normal faulting, respectively (from Sibson 1974).

unless the effective friction coefficient in the crust is lower than that measured in the laboratory.

The above analysis of faulting assumes the material to be homogeneous and isotropic, and the fault surfaces to be planar. In practice, anisotropy and pre-existing planes of weakness may change the orientation of faults with respect to the principal stress ellipsoid. However, the main characteristics of different faulting regimes are adequately explained by Anderson's Theory.

Plasticity is sometimes chosen as the rheology in the modelling of faults. In this case, the crust is taken as a rigid–plastic material deforming in plane strain, and the stress system is examined in terms of slip lines, i.e. maximum shear stress trajectories (see Section 5.7). For instance, an analogy has been made between the recent tectonics of Asia

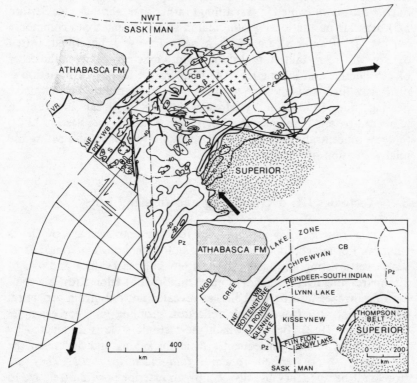

Figure 8.13 Wedge-shaped Superior Province indenter and Churchill Province slip-line field superimposed on tectonic pattern and gravity anomalies, Saskatchewan and Manitoba, Canada. Thick arrows denote the directions of motion of the indenter and the displaced material; thick lines are faults and shear zones; crosses denote batholiths; Bouguer anomalies are in milligals; and letters refer to local geological terms (from Gibb 1983).

223

and the slip-line pattern of a rigidly indented plastic body (the rigid indenter being the Indian craton), and a satisfactory agreement has been found (Tapponnier & Molnar 1976). Similar arguments have been applied to the tectonic pattern near sutures between different provinces of the Precambrian Shield in Canada. Gravity anomalies and strike-slip fault patterns suggest that the suturing of the Superior and Churchill 'plates' may be modelled in terms of indentation tectonics (Gibb 1983, Thomas & Gibb 1983; see Fig. 8.13). Plane-strain plasticity theory finds numerous applications of this type in semi-quantitative geodynamic studies.

The analysis of faulting in terms of Coulomb fracture and frictional sliding is applicable only to the upper crust (in principle, no such limitation applies to plastic analysis). At larger depths a 'fault' becomes a shear zone that deforms in the ductile regime. The depth of the *brittle–ductile transition* (which is gradational rather than abrupt – see Section 5.2) depends on the local geothermal gradient. Shear zones commonly consist of *mylonites*, i.e. fine-grained, intensely deformed, often dynamically recrystallized rocks; their width may vary between the order of centimetres and kilometres. The temperature during deformation is obviously high enough for continuous creep to occur. Furthermore, heating due to viscous dissipation causes a local increase in temperature, and consequently a decrease in viscosity (see Section 12.2), thereby concentrating shear in a relatively narrow band and easing the relative motion of opposing blocks.

8.7 Tectonics III: overthrusting and foreland belts

The term *overthrusting* denotes the tectonic emplacement of sheets ('nappes') of allochthonous rocks over autochthonous material. Large-scale overthrusting, displacing sheets a few kilometres thick and tens of kilometres long over distances of up to hundreds of kilometres, is a common occurrence in orogenic belts, as revealed not only from geological evidence but also from seismic reflection profiling (cf., for example, Oliver *et al.* 1983). Belts in which the tectonic style is dominated by overthrusting – with consequent crustal thickening and flexure of the lithosphere – are termed *foreland fold-and-thrust belts*. They are 'foreland' because they usually occupy an exterior position (relative to the direction of tectonic transport) with respect to the structural and metamorphic culmination (the 'core zone') of the orogen. The Rocky Mountains Foreland Belt of the Canadian Cordillera (Fig. 8.14) may be taken as a typical example of a thin-skinned thrust belt; that is, a belt involving only supracrustal rocks over an unaffected basement (see the

224

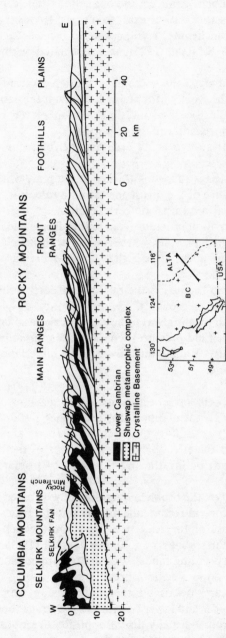

Figure 8.14 Cross section (location shown on inset map) of the Rocky Mountains Foreland Belt, Canadian Cordillera; note numerous west-dipping thrust faults, concave upwards and not affecting the basement (from Price & Mountjoy 1970).

review by Price 1981). Often, however, the deformation also involves the basement (thick-skinned belts). Indeed, the obduction – i.e. detachment and subsequent overthrusting – of crustal 'flakes' is a frequent mode of crustal shortening and thickening, recognizable in most orogenic belts (Oxburgh 1972, Brown & Read 1983) and associated with oceanic closure.

In this section we analyse the tectonic evolution of thin-skinned foreland belts (such as the Rocky Mountains Belt) from the continuum-mechanics viewpoint. We focus on the deformation of the belt as a whole, and not on the mechanics of single thrusts.

The main characteristics of foreland belts are the following.

(a) The belt is usually wedge-shaped – thicker at the rear, which is the direction of provenance of the thrust sheets. As a rule, it is the more oceanward zones that are thrust onto the craton.

(b) The timing of deformation proceeds from rear to front, and a 'piggy-back' mode of transport prevails, where the lowest thrust plane is the most recent, and consequently the highest thrust sheet has travelled the farthest.

(c) The strain field within the belt is the outcome of lateral shortening and vertical thickening, although there are exceptions.

(d) Sometimes, but by no means always, there is a weak basal layer (e.g. evaporites, high pore fluid pressure) separating the thrust belt from the basement; the latter usually dips towards the hinterland whence the thrust sheets came.

Three mechanical models have been proposed to account for over-thrusting: *gravity gliding*, postulating that thrust sheets slide down a weak basal layer; *gravity spreading*, in which thrust sheets are assumed to spread laterally under the effect of uplift at the rear of the belt; and *tectonic compression*, attributing the overthrusting to primary lateral tectonic forces.

Early attempts to analyse the problem quantitatively involved the construction of stress solutions in rectangular blocks under various boundary conditions by the use of the Airy stress function (Section 3.5). A review of these early analyses may be found in Murrell (1981). It soon appeared that the gravity-gliding model is of limited application, as the required forward basal slope is seldom present.

However, a rectangular block is a rather poor analogy for a thrust belt, since in the latter upper and lower boundaries – average topography and basement, respectively – usually dip in opposite directions. Elliott (1976) and Ramberg (1981) pointed out that thrust sheets, similarly to ice sheets, should move in the direction of surface downslope. The 'ice-sheet analogy' leads to the following approximate relationship between basal shear stress and surface slope

$$\tau \simeq \rho g h \alpha \qquad (8.57)$$

where τ is the shear stress at the base of the sheet, ρ is density, g is the acceleration of gravity, and h and α are the thickness and the surface slope, respectively. According to Equation 8.57, thrust sheets spread under the action of gravity.

Gravity spreading is an appealing explanation for the tectonics of foreland belts, but it is difficult to reconcile with the evidence of pervasive shortening. In an attempt to overcome this problem, Chapple (1978) proposed a model combining tectonic compression and lateral spreading, based on the simple consideration that, if a wedge of deformable material is pushed from the rear, then its surface slope increases. This increase contributes to the flow, but the primary agent of tectonic deformation is horizontal compression.

Chapple's model is depicted in Figure 8.15. A wedge with surface slope α and basement slope θ_0 represents the foreland belt (Fig. 8.15a). The rheology is taken as rigid–plastic and the wedge deforms in plane strain. Using polar co-ordinates (r, θ), the equilibrium equations (Eqn 2.7) can be written as

$$\frac{\partial \sigma_r}{\partial r} + \frac{1}{r}\frac{\partial \sigma_{r\theta}}{\partial \theta} + \frac{\sigma_r - \sigma_\theta}{r} + \rho g \sin \theta = 0 \qquad (8.58a)$$

$$\frac{\partial \sigma_{r\theta}}{\partial r} + \frac{1}{r}\frac{\partial \sigma_\theta}{\partial \theta} + \frac{2\sigma_{r\theta}}{r} + \rho g \cos \theta = 0 \qquad (8.58b)$$

The wedge is assumed to be at the yield point. The yield condition is (Section 5.6)

$$(\sigma_r - \sigma_\theta)^2 + 4\sigma_{r\theta}^2 = 4k^2 \qquad (8.59)$$

which is satisfied by a stress distribution of the form

$$\sigma_r - \sigma_\theta = -2k \cos 2\psi \qquad \sigma_{r\theta} = k \sin 2\psi \qquad (8.60)$$

where $\psi(\theta)$ is the angle between the radius and the direction of maximum compressive stress.

The problem is analogous to that of plastic flow through a converging

227

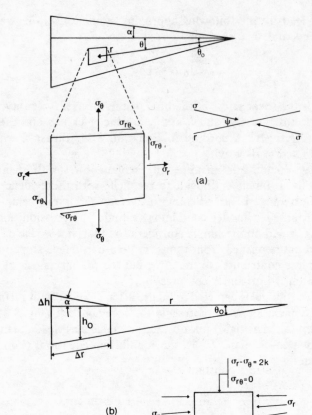

Figure 8.15 Model for foreland deformation: rigid–plastic wedge yielding in compression. See text for discussion (from Chapple 1978).

channel (cf. Hill 1950). The stress solution is

$$\frac{\sigma_r}{2k} = a - c \ln r + \tfrac{1}{2}(c - \cos 2\psi) - \tfrac{1}{2}c \ln(c + \cos 2\psi) - \frac{\rho g r}{2k} \sin \theta$$

$$(8.61a)$$

$$\frac{\sigma_\theta}{2k} = a - c \ln r + \tfrac{1}{2}(c + \cos 2\psi) - \tfrac{1}{2}c \ln(c + \cos 2\psi) - \frac{\rho g r}{2k} \sin \theta$$

$$(8.61b)$$

where a and c are constants of integration. The relation between θ and

228

ψ can be proven to be (see Chapple 1978 for details)

$$\theta = \int_0^\psi \frac{\cos 2\psi}{c + \cos 2\psi} \, d\psi$$

from which it follows that $\psi = 0$ for $\theta = 0$. The boundary conditions are $\sigma_{r\theta} = 0$ for $\theta = 0$, and $\sigma_{r\theta} = \lambda k$ for $\theta = \theta_0$ (Fig. 8.15b). The parameter $\lambda = \sigma_{r\theta}/k$ is the ratio between the yield stress of the basal layer and the yield stress of the wedge, $0 < \lambda \leqslant 1$. Equations 8.60 and 8.61 give a stress distribution that satisfies the equilibrium and yield conditions.

The velocity solutions are constrained by the conditions of coaxiality of principal axes of stress and strain rate (cf. Section 5.6)

$$\frac{2\sigma_{r\theta}}{\sigma_r - \sigma_\theta} = \frac{2\dot\varepsilon_{r\theta}}{\dot\varepsilon_r - \dot\varepsilon_\theta}$$

and must satisfy the incompressibility and boundary conditions ($v_\theta = 0$ for $\theta = \theta_0$). For a constant rate of tectonic shortening an acceptable velocity distribution is

$$v_r = - br(c + \cos 2\psi) \qquad v_\theta = br(\sin 2\psi - \lambda) \qquad (8.62)$$

where b is an arbitrary constant that may be chosen to match a given rate of tectonic shortening.

The physical interpretation of the solution is in terms of compression-induced topographic slope. The vertical normal stress σ_θ does not vanish on the horizontal plane ($\theta = 0$), as can be seen from Equation 8.61b. This is the consequence of the presence of material above the horizontal. The variation of σ_θ can therefore be equated to the effect of topography

$$\left. \frac{\partial \sigma_\theta}{\partial r} \right|_{\theta=0} = - \frac{2kc}{r} = - \rho g \tan \alpha$$

From the stress solution (Eqns 8.61) it also follows that

$$\frac{\partial \sigma_r}{\partial r} = \frac{\partial \sigma_\theta}{\partial r} = - \frac{2kc}{r} - \rho g \sin \theta$$

If Δh is the relief of the element of the wedge of thickness h_0 and width Δr affected by deformation at any stage (Fig. 8.15b), we can write

$$\frac{\Delta \sigma_r}{\Delta r} = \frac{\Delta \sigma_\theta}{\Delta r} = - \frac{2kc}{r} - \rho g \sin \theta = - \rho g \tan \alpha - \rho g \sin \theta \quad (8.63a)$$

229

Also, since the stress difference is $\sigma_r - \sigma_\theta \simeq -2k$ for small ψ and $\theta_0 \simeq \tan \theta_0 = h_0/r$,

$$\frac{\sigma_r - \sigma_\theta}{r} \simeq -\frac{2k\theta_0}{h_0} \tag{8.63b}$$

Finally, as $\sigma_{r\theta} = 0$ for $\theta = 0$ and $\sigma_{r\theta} = \lambda k$ for $\theta = \theta_0$,

$$\frac{\Delta \sigma_{r\theta}}{r \, \Delta \theta} = \frac{\lambda k}{r\theta_0} \simeq \frac{\lambda k}{h_0} \tag{8.63c}$$

Using Equations 8.63 and the approximation $\tan \alpha \simeq \alpha$ (small topographic slope), the radial condition of equilibrium (Eqn 8.58a) takes the form

$$\rho g h_0 \alpha + 2k\theta_0 \simeq \lambda k \tag{8.64}$$

Equation 8.64 shows that the resistance to sliding (λk) is balanced by two driving forces: one related to the surface slope ($\rho g h_0 \alpha$), and the other related to the slope of the basement ($2k\theta_0$). The former is identical to the gravity-spreading stress in the ice-sheet analogy (Eqn 8.57); the latter arises because the horizontal force is greater at the rear of the element, since the radial stress there acts on a larger area. Of course, both average surface slope and basement slope are very small.

According to Chapple's model, therefore, horizontal compressive stresses on a sedimentary wedge cause thickening and the development of a topographic slope; the latter, in turn, causes yielding when it reaches a critical value. As a first approximation the model satisfies the characteristics of foreland belts listed at the beginning of this section (see Chapple 1978 for some geological examples).

Two aspects of the model deserve further comment. First, the lithosphere deforms by flexure under the load of the advancing and thickening thrust belt, reducing the surface slope and increasing the basement slope. Secondly, the yield stress ratio λ affects stress trajectories and velocity distributions. When λ is small (weak basal layer) the critical surface slope is also small. However, the surface slope increases with λ, and this places a limit on the strength of the wedge. For $\lambda \to 1$ (no weak basal layer) the critical slope increases with k, as can be seen from Equation 8.64. As the total relief of the belt is $\Delta h = \alpha R$, where R is the width of the belt, there is a relation between relief and strength. Taking $\theta_0 = 3°$, $\rho = 2.8 \times 10^3$ kg m^{-3}, $h_0 = 6$ km, $R = 100$ km, and no difference in strength between the wedge and its base, the yield strength of the wedge must be low ($k \leqslant 10$ MPa) in order to have realistic values for the relief

($\Delta h \leqslant 5$ km). This result may be interpreted in two ways. Either λ is everywhere considerably less than unity, or the strength of supracrustal sedimentary wedges is low. In the first case weakening of the basal layer could be accomplished in a number of ways (soft material, fluid pressure, dynamic recrystallization, and superplasticity). With respect to the second possibility we note that the model neglects anisotropy due to bedding, and that most thrusts in foreland belts follow bedding planes for considerable distances. It is noteworthy that measurements of the level of organic maturation in coal, carbonaceous shale and siltstone in the Canadian Rocky Mountains do not reveal any thermal metamorphism attributable to heat dissipation along thrusts (Bustin 1983). These observations are consistent with the hypothesis that stresses in foreland belts may be unusually low.

A rigid–plastic rheology is not the only one that has been used in the modelling of thrust belts. The assumption that failure in the wedge is governed by the Coulomb–Navier criterion modified for pore fluid pressure (Eqn 5.10) leads to equally satisfactory results (Davis *et al.* 1983). A Coulomb–Navier rheology is probably more realistic than a rigid–plastic one, since conditions in thin-skinned belts are usually above the brittle–ductile transition. However, the overall mechanics of the two models is the same, and is basically analogous to that of a wedge of snow in front of a moving bulldozer: once the critical surface slope is attained as a consequence of horizontal compression, failure results in over-thrusting and formation of a thrust belt.

Active shortening by horizontal tectonic compression is kinematically equivalent to underthrusting of the basement under a zone that acts as a rigid buttress with respect to the sedimentary wedge. Conditions of the latter type are likely to apply to submarine accretionary wedges of scraped oceanic sediments on the landward side of trenches. Thus, the above discussion is relevant to processes occurring not only during continental collision, but also during subduction.

PART III

The microphysical approach
to Earth rheology

9 The atomic basis of deformation

In order to complete the study of Earth rheology, we now turn our attention to the microphysics of flow in crystalline materials.

Consideration of interatomic lattice forces leads to an estimate of the theoretical strength of crystals many orders of magnitude larger than the observed strength. This discrepancy is accounted for by the presence of lattice defects (vacancies and dislocations).

Lattice defects are the agents of solid-state creep. The connection between macroscopic rheological behaviour and microscopic properties allows the establishment of microphysical constitutive equations relating concentration and mobility of defects to strain rate.

9.1 Lattice parameters and elasticity

The general approach in the first two parts of this book has followed the classical lines of continuum mechanics, i.e. macrorheology. Now we wish to go one step further, and enquire into the atomic basis of the various observed rheologies. To do this we must turn our attention to the *microphysics* of the deformation. The reasons for taking this step are manifold. Although the continuum approach is perfectly adequate to describe the rheological behaviour of the Earth, it does not tell us much about the actual processes controlling deformation. We are naturally curious to understand what happens when a material deforms. We would also like to have a firmer physical basis on which to extrapolate laboratory results to much longer timescales. Furthermore, whatever conclusions about the rheology of the Earth we may reach following the continuum approach must be compatible with the microphysics of deformation. Consequently, the study of the latter is not only of interest in itself, but it is also necessary to strengthen conclusions arrived at by continuum-mechanical analysis. No treatment of the rheology of the

235

Earth is complete without the insights provided by the microphysical approach.

Even if attention is restricted mainly to ductile processes – as is the case here – the relevant literature is very large. The following books may be useful in exploring the background of, and expanding upon, the fundamental concepts of solid-state flow. Pippard (1957) provides one of the most lucid short treatments of classical thermodynamics; Rosenberg (1978) gives an introduction to the physics of the solid state; Cottrell (1953) is a classic on plastic flow in crystals; Hull (1975) provides an elementary account of dislocations; Nicolas & Poirier (1976) discuss deformation and flow in a geological context; finally, Poirier (1985) gives a review of high-temperature deformation processes in crystals.

Earth materials are usually *crystalline solids*, consisting of the periodic repetition of regular atomic patterns in space (the term 'solid' here has no rheological connotation). Whereas the crystalline nature of matter is 'averaged out' in the continuum approach, it is of central importance from the microphysical viewpoint. The crystal structure can be described in terms of a three-dimensional net of straight lines (not necessarily mutually orthogonal) chosen to represent the atomic pattern. The *lattice* is the set of intersections of the lines, which usually coincides with atomic sites. Space is thus subdivided into identical *unit cells*, whose three-dimensional stacking forms the crystal. On the basis of their symmetry and morphology, crystal forms can be subdivided into seven systems and 32 classes. The reader is referred to the aforementioned books for discussion and examples.

For geometric reasons, and because of the *anisotropy* of crystal properties, it is necessary to designate unambiguously planes and directions within a crystal. The identification of *planes* is illustrated in Figure 9.1a. The crystallographic axes (i.e. the axes of the unit cell) are x, y, and z. For simplicity, the unit cell is taken as cubic, with unit dimensions $\overline{OA} = \overline{OB} = \overline{OC} = a$. The plane $A'B'C'$ has intercepts $\overline{OA}'$, $\overline{OB}'$, and $\overline{OC}'$. The *Miller indices* of the plane are given by the reciprocals of the ratios of the intercepts to the corresponding unit dimensions reduced to the three smallest integers (in the case of Fig. 9.1a they are (322)). A plane with intercepts $\overline{OA}$, $\overline{OB}$, and $\overline{OC}$ has Miller indices (111). Other examples are given in the figure: the plane DFBA, for instance, has Miller indices (110); the plane EFBG, (010). In general, Miller indices of a given plane are denoted by (hkl); those of a set of planes of a given type by $\{hkl\}$. An overbar denotes a negative index.

Directions within a crystal are denoted by the components of the equivalent directed segment going through the origin of the unit cell. For instance, the direction $\overline{MN}$ (Fig. 9.1b) is obtained by considering the segment $\overline{OE}$ and forming the three smallest integers from the ratios of

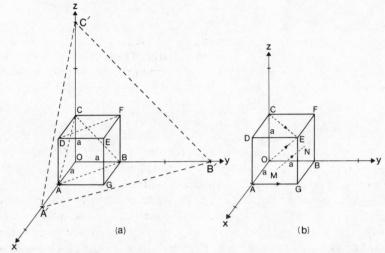

Figure 9.1 Miller indices of (a) planes and (b) directions in the crystal lattice (from Hull 1975).

its components along the crystallographic axes to the corresponding unit dimensions, i.e. in this case [111]. Directions $\overline{CE}$ and $\overline{AG}$ are [110] and [010], respectively. In general, a single direction is denoted by $[uvw]$, a set of equivalent directions by $\langle uvw \rangle$. The convention for negative directions is as for planes.

A crystal is a stable configuration of equilibrium positions of atoms. The interatomic distance (spacing) is of the order of a few ångströms $(1 \text{ Å} = 10^{-10} \text{ m})$. The position of atoms is the result of a balance of attractive and repulsive forces, and the equilibrium spacing is that for which the potential energy of the lattice is minimum. Figure 9.2 shows schematically the variation of the potential energy $U(r)$ with spacing r. The repulsive force gives rise to a positive potential energy R, and the attractive force to a negative potential energy A. Assuming $R \propto r^{-n}$ and $A \propto r^{-m}$, the resultant energy has the shape denoted by U. If $n > m$ (as must be the case in order to have stable equilibrium), U goes through a minimum $-U_0$ for $r = r_0$, which is the equilibrium interatomic spacing. The energy minimum $|U_0|$ is a measure of the work required to remove an atom from the lattice, and is called the *dissociation* or *cohesive energy* of the crystal.

The short-range repulsive force arises from the mutual repulsion of the electron clouds of neighbouring atoms. The attractive force arises from different types of interaction. In *ionic bonding* atoms become ions by the transfer of electrons from the outer shell of one atom to another, and this results in electrostatic attraction between opposite charges. In

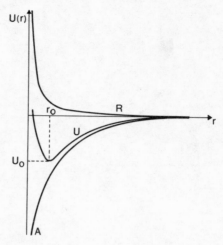

Figure 9.2 Variation of potential energy with interatomic spacing; U_0 is the cohesive energy, r_0 the equilibrium spacing.

covalent bonding the electron clouds are distorted so that neighbouring atoms share them. The bonding is often *mixed* or *ionocovalent* (partly ionic and partly covalent). Other attractive mechanisms, such as van der Waals forces arising from electric dipole moments, are relatively weak.

The *elasticity* of a crystalline solid arises from the action of inter-atomic forces which tend to maintain each atom in its equilibrium lattice position. The compressibility of a material is a measure of the pressure necessary to change its volume (i.e. interatomic spacing), and is therefore related to the lattice potential energy. If k, β, r, n, V, and p are bulk modulus, compressibility, interatomic spacing, number of atoms per molecule, molecular volume, and pressure, respectively, we have

$$k \equiv \frac{1}{\beta} = -V \frac{\mathrm{d}p}{\mathrm{d}V} \qquad V = nr^3$$

The work done by the pressure in compressing the elastic solid is stored as lattice energy, i.e.

$$p \, \mathrm{d}V = -\mathrm{d}U \qquad p = -\mathrm{d}U/\mathrm{d}V$$

and we may therefore write

$$k = V \, \mathrm{d}^2U/\mathrm{d}V^2$$

Using the expression for $V(r)$ and since $\mathrm{d}V = 3nr^2 \, \mathrm{d}r$, the last relation becomes

$$k \equiv \frac{1}{\beta} = \frac{1}{9nr} \frac{d^2U}{dr^2}$$

and consequently the zero-pressure incompressibility k_0 is proportional to the second derivative of $U(r)$ evaluated at r_0, that is

$$k_0 = \frac{1}{9nr_0} \left(\frac{d^2U}{dr^2}\right)_{r=r_0} \tag{9.1}$$

Therefore, given a model of the lattice potential energy, Equation 9.1 allows the estimation of U_0 from measurements of lattice spacing and compressibility. Cohesive energies of minerals are of the order of $10^3 \, \text{kJ mol}^{-1}$.

We saw in Section 3.1 that the general form of Hooke's Law for an anisotropic solid is

$$\sigma_{ij} = C_{ijkl}\varepsilon_{kl}$$

and that the maximum number of independent elastic parameters is 21. This is the number of parameters in crystals with the lowest symmetry, i.e. triclinic. As the degree of symmetry increases, the number of independent elastic parameters decreases, to a minimum of three for cubic crystals (and two for elastically isotropic solids).

The connection between high temperature and ductile flow may be understood by considering the *thermal vibrations* of atoms. As long as the energy of deformation is not sufficient to remove an atom from the potential trough U_0 (the cohesive energy), the material behaviour is elastic. Atoms vibrate about their equilibrium positions with a frequency of about 10^{13} Hz; as temperature increases, so does the amplitude of atomic vibration. (Since the lattice energy is asymmetric about its minimum value, the vibrations are *anharmonic* and consequently a change in temperature causes a change in the mean distance between atoms, resulting in thermal expansion.) The energy associated with the oscillatory motion is the internal thermal energy of the material. Each atom has six *degrees of freedom,* its total energy being the sum of three kinetic terms and three potential terms, associated with vibrations in orthogonal directions. By the *Principle of Equipartition of Energy,* these six terms are on the average the same and equal to $\frac{1}{2}(kT)$, where $k = 1.3805 \times 10^{-23} \, \text{J K}^{-1}$ is the Boltzmann constant and T is the absolute temperature. The mean value of the kinetic energy associated with each atom is therefore $\frac{3}{2}(kT)$. The distribution of kinetic energy ψ of n atoms is given by the *Maxwell–Boltzmann statistics:*

$$dn(\psi) = \frac{2n}{\pi^{1/2}} \frac{\psi^{1/2}}{(kT)^{3/2}} \exp\left(-\frac{\psi}{kT}\right) d\psi \qquad (9.2)$$

where $dn(\psi)$ is the number of atoms with kinetic energy between ψ and $\psi + d\psi$. As temperature increases, so does the proportion of atoms with high kinetic energy. The likelihood of some atoms being displaced from their equilibrium position by an applied force is therefore larger at higher temperatures. At absolute temperatures of the order of one-third to one-half of the melting point, most crystalline solids creep in a ductile manner.

The derivation of the Maxwell–Boltzmann distribution for kinetic energy is based on the assumptions that the energy state of an atom is not influenced by its neighbours, and that the atoms are distinguishable (which is the case in a solid, since each atom can be designed by the spatial co-ordinates of its equilibrium position). We shall derive one form of the Maxwell–Boltzmann statistics in Section 9.3, when dealing with point defects in crystals. Distributions of Maxwell–Boltzmann type, although not correct quantum-mechanically, are entirely satisfactory for high-temperature flow processes, and play an important part in the microphysics of deformation.

9.2 Theoretical and observed shear strength

When a single crystal undergoes ductile deformation (i.e. suffers a permanent change of shape), some atoms within it move to new equilibrium positions. Experimental observation shows that ductile deformation is generally achieved by *translational slip,* whereby one part of the crystal slides on a plane with respect to another part, and crystal orientations in the slipped and unslipped parts are preserved (see Fig. 9.3a). A *slip system* is defined by the *slip plane* across which atomic displacement occurs, and by the *slip direction,* i.e. the direction of displacement. A slip system is denoted by the Miller indices for the slip plane and the slip direction, $(hkl)[uvw]$. Slip planes often coincide with planes of dense packing in the crystal structure; slip directions are commonly along directions of close packing. Slip can occur along any equivalent planes and directions of a given system, so that a crystal with one activated slip system behaves like a 'pack of cards'. More than one slip system can be present in a crystal; their number and properties depend on the crystal structure and chemistry, and on extrinsic rheological parameters.

The theoretical shear stress necessary to cause slip is a measure of the

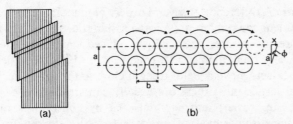

Figure 9.3 (a) Ductile deformation of a single crystal by slip along a set of glide planes; (b) simultaneous slip of an atomic plane (top) over another (bottom).

expected strength of the crystal. It can be calculated on the assumption that plastic deformation occurs by simultaneous slip of atoms across the slip plane in an otherwise perfect lattice. The situation is depicted in Figure 9.3b, where separation between atomic planes and interatomic spacing within a plane are denoted by a and b, respectively. Slip requires the displacement of the atoms in the top plane by an amount b in the positive x-direction to their new equilibrium positions. If $\phi = x/a$ is the shear strain for a small displacement x, then the shear stress is (by Hooke's Law)

$$\tau = \mu x/a$$

where μ is the shear modulus. However, the shear stress must have a periodicity related to the interatomic spacing, i.e.

$$\tau = c \sin (2\pi x/b)$$

Equating the two above expressions for small values of the sine, the constant c can be determined. The shear stress is

$$\tau = \frac{\mu b}{2\pi a} \sin \frac{2\pi x}{b} \qquad (9.3)$$

The critical shear stress is the maximum τ; taking further $a \simeq b$,

$$\tau_c \simeq \mu/2\pi \qquad (9.4)$$

Equation 9.4 predicts that the critical shear stress for slip is within one order of magnitude of the shear modulus of the material. A refinement in the calculation, taking into account more-realistic estimates of the shearing force as a function of displacement, gives a similar result ($\tau_c \simeq \mu/30$). However, the observed strengths of crystalline materials are of the order of $10^{-4}\mu$ or less, and can be vanishingly small at high

241

temperature. A difference of several orders of magnitude between theory and experiment can only mean that the assumptions on which the theory is based are inaccurate. Artificially produced crystals in the shape of fibres with very small diameter ('whiskers'), in which the lattice contains few or no defects, have a yield strength close to the theoretical strength. The crystal lattice must therefore usually contain *defects* of some sort that make slip easier; and slip must be by *consecutive* rather than simultaneous displacement of atoms across the slip plane. We shall examine the principal types of defects in the remainder of this chapter. Here, we consider further some properties of the observed strength in single crystals and polycrystals.

The stress applied to a single crystal can be resolved to find the shear stress along any slip plane of a given orientation. Slip on the system $(hkl)[uvw]$ takes place when this stress (the *critical resolved shear stress*) reaches a value which is characteristic of the material and of the slip system, and which further depends on temperature and strain rate. The strength of a crystal therefore depends not only on its material properties and environmental conditions, but also on its orientation with respect to the applied stress. In general, as temperature increases and strain rate decreases, so the strength tends to decrease. It should be noted that, although time does not appear explicitly in the phenomenological equations for plasticity (cf. Section 5.5), a perfectly plastic material deformed at a fixed strain rate flows under constant stress in a way indistinguishable from that of a viscoplastic or viscous material. The most important difference between 'plastic' and 'viscous' flow is the presence of a finite yield strength in the former, which becomes vanishingly small in the latter.

The ductile flow of a polycrystal, as opposed to that of a single crystal, is subject to additional conditions of compatibility if the continuity of the material is to be preserved. Considering, for simplicity, a monomineralic polycrystal consisting of randomly oriented grains with only one slip system, it is clear that the deformation of individual grains and of the aggregate as a whole are incompatible. For continuity to be preserved, it must be possible to define the strain tensor ε_{ij} over the space domain occupied by the polycrystal. Since there are six strain components and $\varepsilon_{kk} = 0$ (ductile flow is a constant-volume deformation), continuity can in theory be maintained only if the grains have five independent slip systems. This is the *von Mises ductility criterion*. In practice the incompatibility between grains can be accommodated by a variety of processes (bending of the lattice, kinking, twinning, polygonization, recrystallization, diffusive flows – see Nicolas & Poirier 1976), and continuity can be maintained even when crystals – as is often the case in low-symmetry minerals – have less than five slip systems.

9.3 Crystal defects: vacancies

The ductile behaviour of crystalline materials becomes understandable when the effects of lattice defects are taken into account. A defect can be defined as a *break in the periodicity of the lattice*. All natural crystals, and most synthetic ones, contain defects of one sort or another, which can be conveniently classified as follows.

(a) *Point defects* occur when the lattice periodicity is broken at a point. Figure 9.4 shows schematically the possible types of point defects. The most important is a *vacancy,* which arises when a lattice site is empty. The opposite case, when an extra atom is present between lattice sites, is termed a *self-interstitial.* Impurity atoms (of a different atomic species) may also be present, either as substitutional impurities or as interstitials.

(b) *Line defects* or *dislocations* are lattice imperfections occurring along a line within the crystal. They are the most common agents of ductile deformation, and will be considered in Section 9.5.

(c) *Planar defects* are two-dimensional breaks in lattice periodicity. *Subgrain boundaries,* across which crystallographic planes are slightly misoriented, are a typical example.

Diffusive flow of vacancies and the consequent mass transfer play an important role in high-temperature creep. In this section we consider the thermodynamics of vacancy formation and the temperature-dependence of vacancy concentration; in the next section we shall address the problem of vacancy migration.

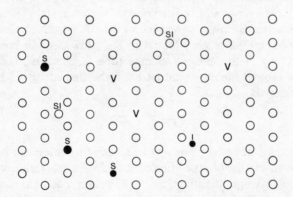

Figure 9.4 Point defects in crystals (open circles denote atoms occupying the nodes of the lattice): V, vacancies; SI, self-interstitials; S, substitutional impurities; I, interstitial impurities.

For a vacancy to be formed, an atom must escape from its lattice site, i.e. its kinetic energy must be larger than the cohesive energy, $\psi > \psi_0 \equiv |U_0|$ (see Section 9.1). The variation of vacancy concentration with temperature can therefore be simply inferred from the Maxwell–Boltzmann distribution of kinetic energy (Eqn 9.2). The number of atoms with kinetic energy larger than ψ_0 is given by

$$n(\psi_0) = \frac{2n}{\pi^{1/2}(kT)^{3/2}} \int_{\psi_0}^{\infty} \psi^{1/2} \exp\left(-\frac{\psi}{kT}\right) \mathrm{d}\psi$$

Evaluating the integral and assuming $\psi_0/kT \gg 1$, i.e. that the cohesive energy is much larger than the average kinetic energy, we obtain

$$n(\psi_0) = 2n\left(\frac{\psi_0}{\pi kT}\right)^{1/2} \exp\left(-\frac{\psi_0}{kT}\right)$$

Identifying $n(\psi_0)$ with the number of vacancies n_V, the atomic fraction of vacancies (ratio of number of vacancies to total number of atomic sites) is

$$N_V \equiv \frac{n_V}{n} = 2\left(\frac{\psi_0}{\pi kT}\right)^{1/2} \exp\left(-\frac{\psi_0}{kT}\right) \tag{9.5}$$

Vacancy concentration therefore depends exponentially on temperature (the temperature-dependence of the pre-exponential term is small in comparison).

The equilibrium concentration of vacancies can also be derived more rigorously from a consideration of the thermodynamics of vacancy formation. The creation of a vacancy is accompanied by a change in the Helmholtz free energy and in the volume of the crystal. The change in free energy arises from the severance of atomic bonds. Denoting by E_f the change in internal energy (equal to the work done in breaking the bonds), and by S_f the entropy change (due to changes in vibrational frequencies around the empty lattice site), the free energy of formation of one vacancy is

$$F_f = E_f - TS_f$$

The volume change consequent upon vacancy formation can be thought of as the sum of one atomic volume Ω and of a relaxation volume V_r (since atoms change slightly their equilibrium positions around the vacant site), i.e.

$$V_f = \Omega + V_r$$

where V_r can be positive or negative according to the balance of inter-atomic forces within the crystal. The enthalpy and the Gibbs free energy of formation are, respectively,

$$H_f = E_f + pV_f$$

$$G_f = H_f - TS_f = E_f + pV_f - TS_f$$

In the above expressions T is absolute temperature and p is pressure.

The equilibrium concentration of vacancies at a given temperature is the one which minimizes the Gibbs free energy of the crystal. If n_V vacancies are formed, then the Gibbs free energy increases by $n_V G_f$. However, the total change G^* is also affected by the entropy of configuration S_c, since there are different ways to distribute the vacancies over all the available atomic sites. Therefore

$$G^* = n_V G_f - TS_c \tag{9.6}$$

where

$$S_c = k \ln q \tag{9.7}$$

q being the number of ways to distribute the vacancies, and k being the Boltzmann constant.

Since there are n atomic sites that can be divided into two classes of sizes n_V and $n - n_V$, the number of possible permutations is (see, for example, Kreyszig 1983)

$$q = \frac{n!}{(n - n_V)!n_V!}$$

Using Stirling's approximation, for large n,

$$n! \simeq (2\pi)^{1/2} n^{(n+1/2)} \exp(-n)$$

we have

$$\ln(n!) \simeq \tfrac{1}{2} \ln(2\pi) + (n + \tfrac{1}{2})\ln n - n \simeq n \ln n - n$$

Therefore the configuration entropy (Eqn 9.7) is

$$S_c = k\{\ln(n!) - \ln[(n - n_V)!] - \ln(n_V!)\}$$

$$\simeq k\left[n_V \ln\left(\frac{n}{n_V} - 1\right) - n \ln\left(1 - \frac{n_V}{n}\right)\right]$$

that is, since $n \gg n_V$,

$$S_c \simeq k n_V [1 + \ln(n/n_V)]$$

Consequently, the total change in Gibbs free energy upon formation of n_V vacancies (Eqn 9.6) is

$$G^* \simeq n_V \left[G_f - kT \left(1 + \ln \frac{n}{n_V} \right) \right] \qquad (9.8)$$

At a given temperature, G^* is minimum when

$$\frac{\partial G^*}{\partial n_V} = 0$$

i.e. for

$$G_f - kT \ln(n/n_V) = 0$$

The equilibrium atomic fraction of vacancies $N_V \equiv n_V/n$ is then

$$N_V = \exp(-G_f/kT) \qquad (9.9)$$

The equilibrium concentration of vacancies therefore has a Maxwell–Boltzmann type of dependence on temperature, as we have already inferred from kinetic energy considerations. This is one of the reasons for the strong temperature-dependence of processes such as creep, which in many cases require some form of vacancy migration. Equation 9.9 is but one particular example of a general type of relation (the *Arrhenius rate equation*) that holds for thermally activated processes. Other instances of its applicability include solid-state diffusion and dislocation migration.

9.4 The laws of diffusion

A vacancy in the crystal lattice can migrate by exchanging its position with that of a neighbouring atom. If a concentration gradient of vacancies is present in the crystal, then a vacancy flux is set up that tends to equalize the vacancy concentration. The direction of the resulting mass diffusion is opposite to that of vacancy diffusion.

All diffusive processes are regulated by the same general laws. In this respect diffusion of matter is not different from heat diffusion (Section

7.4). If C is the concentration of diffusing particles (a function of position and time) and J_i is their flux (number of particles crossing a unit area per unit time), then the latter is proportional to the concentration gradient:

$$J_i = -D_{ij}\, \partial C/\partial x_j \qquad (9.10)$$

Equation 9.10, which arises directly out of experience (as Fourier's Law does) is known as *Fick's First Law of Diffusion*. The parameter D_{ij} is the *diffusion coefficient*, with dimensions $[l^2 t^{-1}]$. We have taken the general case of anisotropic diffusion, where the component of flux in a given direction is affected by all three components of the concentration gradient.

If we now consider an elemental volume – in the continuum-mechanics sense – the total change in flux in the three co-ordinate directions must balance the change in concentration, that is

$$-\frac{\partial J_i}{\partial x_i}\, \mathrm{d}V = \frac{\partial C}{\partial t}\, \mathrm{d}V$$

Dividing by $\mathrm{d}V$ and using Equation 9.10, we obtain

$$\frac{\partial}{\partial x_i}\left(D_{ij}\frac{\partial C}{\partial x_j}\right) = \frac{\partial C}{\partial t}$$

Assuming the diffusion coefficient to be isotropic ($D_{ij} = D$ for $i = j$, $D_{ij} = 0$ for $i \neq j$) and independent of position, the above equation reduces to

$$D\nabla^2 C = \partial C/\partial t \qquad (9.11)$$

which is *Fick's Second Law of Diffusion*. In the case of solid-state vacancy diffusion, J_i is the flux of vacancies and $C(= N_V/\Omega$, where Ω is the atomic volume) is their concentration. The assumption of diffusive isotropy is usually satisfactory.

The vacancy diffusion coefficient in a crystalline material can be characterized in terms of Maxwell–Boltzmann statistics. For simplicity, we consider a cubic monomineralic crystal containing a vacancy fraction N_V. (The results for the ideally simple case are still valid, taking into account some complicating factors, for more-complex crystals.) The diffusion of vacancies takes place by an exchange of position with a neighbouring atom resulting in an overall flux if a vacancy concentration gradient is present. The latter can be set up by a non-hydrostatic stress system. The migration of vacancies can be described as a *random-walk process*. If ν and δl are the jump frequency and jump distance, respectively (the 'jump' is the positional exchange between a vacancy and a nearby atom), then the mean square distance covered by a vacancy during a time δt is given by Einstein's relation

$$\langle d^2 \rangle = \nu \, \delta t \, \delta l^2$$

The vacancy diffusion coefficient may be defined as

$$D_V = \langle d^2 \rangle / 6 \, \delta t = \tfrac{1}{6} \nu \, \delta l^2$$

where the numerical factor takes into account the six possible directions of vibration. The jump frequency is the product of the lattice (Debye) frequency $\nu_D \simeq 10^{13}$ Hz, the number of atoms around a vacancy (the co-ordination number z), and the probability that one of the neighbouring atoms overcomes the potential barrier and jumps into the vacancy. This probability can be described in terms of a Gibbs free energy of migration G_m. The jump frequency is therefore

$$\nu = \nu_D z \, \exp(-G_m/kT)$$

and the diffusion coefficient for a cubic crystal ($z = 6$, $\delta l = a$ where a is the lattice parameter) is given by

$$D_V = \nu_D a^2 \, \exp(-G_m/kT)$$

Using the definition of the Gibbs free energy, this can be written as

$$D_V = \nu_D a^2 \, \exp\left(\frac{S_m}{k}\right) \exp\left(-\frac{H_m}{kT}\right) = D_{0V} \, \exp\left(-\frac{H_m}{kT}\right) \qquad (9.12)$$

where the pre-exponential term D_{0V} is constant. The mobility of vacancies increases rapidly with temperature.

The diffusion of the atoms in the crystal can be described in terms of a *self-diffusion coefficient* D_{SD}. This depends not only on the diffusivity of vacancies, but also on the probability that a vacancy be present near a given atom, which is given by the equilibrium fraction of vacancies, i.e.

$$D_{SD} = N_V D_V$$

Using Equations 9.9 and 9.12, the coefficient of self-diffusion may be written as

$$D_{SD} = \nu_D a^2 \exp\left(-\frac{G}{kT}\right) = \nu_D a^2 \exp\left(\frac{S}{k}\right)\exp\left(-\frac{H}{kT}\right) = D_0 \exp\left(-\frac{H}{kT}\right)$$
(9.13)

where $G = G_f + G_m$, $S = S_f + S_m$, $H = H_f + H_m$ are the Gibbs free energy, the entropy, and the enthalpy of self-diffusion, respectively. Each is the sum of a term connected with vacancy formation and a term connected with vacancy migration. The latter arises from the changes associated with the jump of a vacancy from one lattice site to a neighbouring one.

Self-diffusion is an *activated process,* i.e. energy must be expended to overcome the potential barrier between two configurations, and the diffusion coefficient follows an Arrhenius-type law. To make the effect of pressure explicit, it is customary to write the self-diffusion coefficient (Eqn 9.13) in terms of *activation energy* and *activation volume*

$$D_{SD} = D_0 \exp\left(-\frac{E + pV}{kT}\right)$$
(9.14)

Increasing temperature therefore increases the diffusion coefficient, while increasing pressure decreases it. In practice the activation enthalpy is usually given as energy per mole (kcal mol^{-1} or J mol^{-1}), so that the Boltzmann constant k in the above equations must be replaced by the universal gas constant $R = kN_A = 8.3143$ J mol^{-1}K^{-1} (N_A is the Avogadro Number). The diffusion coefficient is then written as

$$D_{SD} = D_0 \exp\left(-\frac{E + pV}{RT}\right)$$
(9.15)

The activation parameters can be determined experimentally from the diffusion of tracer isotopes in the material of interest. From Equation 9.15 it may be seen that the following relations hold:

$$E = -R \left.\frac{\partial(\ln D_{SD})}{\partial(1/T)}\right|_{p=0} \qquad V = -RT \left.\frac{\partial(\ln D_{SD})}{\partial p}\right|_T$$
(9.16)

Therefore, by measuring the diffusion coefficient as a function of temperature at room pressure, and as a function of pressure at constant temperature, it is possible to estimate activation energy and volume. The data also yield the pre-exponential factor. Diffusivity data are available for several materials. For diffusion of silicon in forsterite, for instance, D_0, E, and V are of the order of 10^{-10} m^{2}s^{-1}, 100 kcal mol^{-1}, and

10^{-5}–10^{-6} m^3 mol^{-1}, respectively (see Section 10.4). For most oxides, the self-diffusion coefficient of oxygen is about 10^{-13}–10^{-15} m^2 s^{-1} at the melting point.

The role of pressure in diffusion can be taken into account also by including its effect on the melting temperature. Instead of Equation 9.15, its empirical equivalent can be used:

$$D_{SD} = D_0 \exp[-cT_m(p)/T] \qquad (9.17)$$

where $T_m(p)$ is the melting temperature at the relevant pressure. The empirical constant c is lower for metals ($c \simeq 18$) than for silicates ($c \simeq 30$). Uncertainties in $T_m(p)$ and c reduce the usefulness of Equation 9.17. The ratio T/T_m is the *homologous temperature;* the diffusion coefficient increases as T/T_m approaches unity.

The above can be strictly applied to simple cubic lattices only. Minerals are much more complex, not only in crystal structure but also because they contain several atomic species, different types of interatomic forces, and impurities (cf. Nicolas & Poirier 1976). In ionic and ionocovalent crystals, concentrations and fluxes of vacancies are subject to conditions of electrical neutrality and zero resultant current. Vacancies are formed in pairs (Schottky defects) to preserve electrical equilibrium. These factors introduce complications, but the conclusions reached in the ideal case are still valid.

Two diffusion regimes can be distinguished according to the atomic fraction of aliovalent impurities (N_I) compared with the atomic fraction of vacancies. *Intrinsic diffusion* applies for $N_I \ll N_V$ (very pure crystals, or high temperature, or both); *extrinsic diffusion* for $N_I \gg N_V$. The predominance of either regime affects the diffusion enthalpies of anions and cations. Another complicating factor is introduced by the imperfect stoichiometry of many crystals, which couples diffusion with the chemical environment (e.g. with the oxygen partial pressure if there is an imbalance of oxygen in one of the sublattices).

When more than one atomic species is able to diffuse, then transport of matter is controlled by the diffusion coefficient of the *slowest-moving* species. For instance, if a crystal has composition A_xB_y, then it can be shown from the condition of electrical equilibrium that the *effective diffusion coefficient* is given by

$$D = \frac{D_A D_B}{n_A D_B + n_B D_A} \qquad (9.18)$$

where D_A and D_B are the diffusion coefficients of A and B, and n_A and n_B are their atomic fractions. If $D_A \gg D_B$, therefore, $D \simeq D_B$. The diffusivity of the slower species (usually, but not always, the larger anion) controls the overall flux of matter.

All of the previous considerations refer to *bulk* (or *lattice*) *self-diffusion,* i.e. to the case when vacancy fluxes occur across the bulk of a crystal. Vacancies can also migrate along linear or planar crystalline defects, such as dislocations (*pipe diffusion*) and grain boundaries (*grain-boundary diffusion*). The diffusion coefficients for these 'short-circuit' processes can be several orders of magnitude larger than for bulk diffusion (cf. Poirier 1985). They become important at relatively low temperatures, when bulk diffusion is slow.

Diffusion is very important for transport processes in crystals, including phase transformation kinetics, recrystallization, and creep. We shall examine its role in ductile flow in Section 10.3, when discussing mechanisms of steady-state creep in polycrystals.

9.5 Crystal defects: dislocations

The most immediate and widespread connection between microphysical properties and ductile flow of crystals is provided by linear defects or dislocations. The idea of dislocations in crystals was introduced independently in 1934 by Orowan, Polanyi, and Taylor, to account for the plastic properties of matter. The two basic observations leading to it are that ductile deformation in crystals is inhomogeneous (i.e. it occurs on selected slip systems), and that observed strength is much less than theoretical strength (Section 9.2). These can be explained by assuming that slip occurs by *consecutive* slip of rows of atoms along the slip plane, rather than by simultaneous slip of all atoms. A now classical analogy is that of a carpet on a floor: the carpet may be moved all at once (simultaneous slip) or a ruck may be made near one edge and then propagated across the carpet (consecutive slip). In both cases the same total displacement is attained, but the required force is much less for consecutive slip. A caterpillar moves by using the same principle.

Dislocations are a necessary consequence of consecutive slip. A *dislocation* may be defined as the line marking the boundary between slipped and unslipped parts on the slip plane (Fig. 9.5). Since by definition a dislocation encloses an area, it must form a closed loop within the crystal, or end at crystal boundaries or triple junctions with two other dislocations. Slip is propagated by outward migration of the dislocation loop.

The most invariant characteristic of a dislocation is its *Burgers vector* which expressess the amount and direction of slip. Both Burgers vector b and dislocation line are contained in the slip plane (we omit indices as

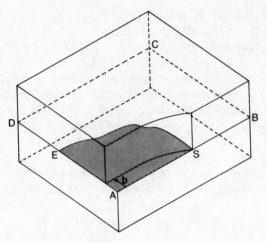

Figure 9.5 A dislocation. The shaded part (AES) of the plane ABCD (slip plane) has slipped by the Burgers vector *b*. ES is the dislocation line.

only the magnitude of *b* enters the relevant equations). In an *edge dislocation* the dislocation line is perpendicular to the Burgers vector; in a *screw dislocation* it is parallel. Mixed dislocations do naturally occur (refer again to Fig. 9.5). Note that an edge dislocation is geometrically less constrained than a screw dislocation, but kinematically more so. An edge dislocation may be of irregular shape as long as it is contained in a plane normal to the Burgers vector; a screw dislocation must be straight and parallel to slip. On the other hand, an edge dislocation can move only in the direction of slip, while a screw dislocation is theoretically free to move on any plane containing the slip direction.

Slip by propagation of edge and screw dislocations is illustrated in Figures 9.6a and b, respectively. When a dislocation of Burgers vector *b* sweeps across an entire slip plane, the two halves of the crystal become displaced by the relative amount *b* in the slip direction. Since the crystal lattice provides a periodic field of force, atoms usually slip from one equilibrium position to another, so that the magnitude of *b* is restricted to a class of discrete values related to lattice spacing, and its direction can be given in terms of crystallographic components.

It can be seen from Figure 9.6 that an edge dislocation is formally equivalent to the insertion of an extra half-plane in the lattice. If the extra half-plane lies above the slip plane, then the edge dislocation is positive; in the opposite case it is negative. In a screw dislocation the atoms form a helix – like a spiral staircase – around the dislocation line. A screw dislocation can be right-handed or left-handed, according to the sense of the helix. Dislocations of the same type but of opposite sign pro-

252

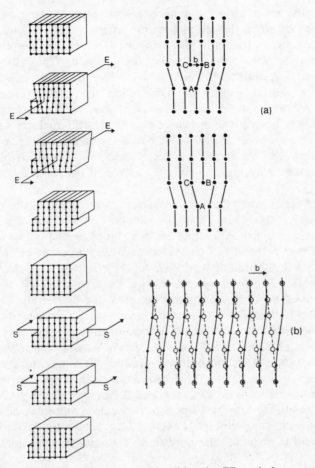

Figure 9.6 (a) Slip by propagation of an edge dislocation EE, equivalent to an extra half-plane in the lattice; (b) slip by propagation of a screw dislocation SS, with atoms forming a helix around the dislocation line. (Representations of slip and of atomic arrangement of edge dislocation from Nicolas & Poirier 1976. Atomic arrangement of screw dislocation reproduced from Hull 1975.)

duce opposite slip if they sweep the crystal in the same direction. If such a pair of dislocations meet on the same slip plane they annihilate each other to form a perfect crystal.

Dislocations can be observed with a number of techniques, such as decoration methods with precipitate particles, TEM (transmission electron microscopy), X-ray diffraction, and field ion microscopy. The most commonly used technique is TEM, which relies on the scattering of electrons in the strained region around a dislocation; its resolution is of the order of 0.4 nm (cf. Hull 1975 for further discussion).

253

Large numbers of dislocations are present in all natural materials. The *dislocation density* is defined as the total length of dislocation lines per unit volume or, equivalently, the number of dislocations per unit area. Dislocation densities are usually very high, ranging from 10^{10} m^{-2} in well-annealed crystals to 10^{15} m^{-2} or more in crystals that have undergone large plastic deformation. Dislocations are present even in the absence of external stress, as they can be built up within the crystal during its growth from a melt or can be nucleated at internal stress concentration centres such as second-phase particles and grain-boundary triple points. However, the increase in dislocation density with deformation requires a mechanism for the multiplication of dislocations in a stressed crystal.

A widespread multiplication mechanism is provided by the *Frank–Read source,* depicted in Figure 9.7. Suppose that an initially straight segment AB of a dislocation line lies in a slip plane and is free to move on it, while its extremities are not (for instance, they could be pinned by nodes in the dislocation network). Under the action of an applied stress, the dislocation line will bow out, shifting the area above the slip plane by an amount equal to the Burgers vector of the dislocation. Once the equilibrium radius of curvature under a given stress is reached (i.e. the segment becomes a semicircle), the dislocation expands outward until the two opposing lobes M and N join. Since they are of opposite sign (they move in opposite directions but must produce the same slip), they annihilate each other, leaving a closed dislocation loop that expands outwards, and a segment joining A and B from which the process can be repeated. In this way, sequences of dislocation loops can be formed at each source.

The critical stress for the operation of Frank–Read sources can be

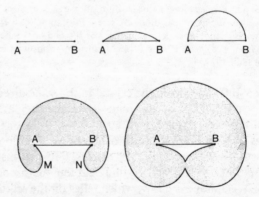

Figure 9.7 A Frank–Read source. The slip plane is the plane of figure; the slipped area is stippled.

254

calculated by applying the theory of elasticity to the crystal (in the continuum-mechanics sense, the term 'dislocation' simply denotes a discontinuity in the displacement). Since atomic planes are distorted near the dislocation, elastic strain is generated. Calculation of strains, stresses, and forces associated with dislocations is therefore possible (cf. Cottrell 1953 and Hull 1975 for details). The critical stress for Frank–Read sources is of the order of $\mu b/l$, where μ is the rigidity and l is the length of the glissile segment. Usually, $b \simeq 10^{-10}$ m (interatomic spacing) and $l \simeq 1$ μm; consequently the critical stress is roughly comparable with the observed strength.

The stress field within a crystal is important for the interaction and mobility of dislocations. The stresses generated by a dislocation are of the type

$$\sigma \propto \mu b/r$$

where r is the distance from the dislocation line. The stored elastic energy per unit volume (Section 3.3) is therefore

$$\Omega_D \propto \mu(b/r)^2$$

that is, the elastic energy per unit length of dislocation is

$$\Omega_L \propto \mu b^2 \qquad (9.19)$$

It is an immediate consequence of Equation 9.19 that dislocations with large Burgers vectors are energetically unfavourable and tend to break into dislocations with smaller vectors. Therefore, b is usually not larger than the lattice spacing. In some instances a dislocation can split or dissociate into parallel *partial dislocations,* which have Burgers vectors that do not correspond to lattice vectors; the sum of the Burgers vectors of the partial dislocations is equal to the Burgers vector of the total dislocation. The area between the leading and the trailing partial is a planar defect across which the lattice does not match (a *stacking fault* – see Fig. 9.8). The dissociation lowers the total elastic energy, and therefore enhances the mobility of dislocations.

Another consequence of Equation 9.19 is that dislocations cannot exist as thermodynamically stable lattice defects, as the Gibbs free energy has no minimum corresponding to any equilibrium concentration of dislocations. A crystal can always reduce its Gibbs free energy by reducing the number of dislocations it contains. This reduction is the basis of processes such as recovery and recrystallization (cf. Sections 10.2, 3 & 6), which enhance the ductility of crystals.

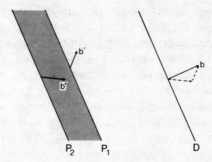

Figure 9.8 Partial dislocations with Burgers vectors b' and b'' separated by a stacking fault (shaded). Total slip by P_1 and P_2 is equivalent to slip by D.

A factor that strongly affects the mobility of dislocations is the presence of *jogs,* which are steps in the dislocation line arising when dislocations cross each other (Fig. 9.9) or by thermal activation. We shall see in the following section that the movement of edge dislocations away from the slip plane (climb) is made easier by the fact that the dislocation is jogged.

Dislocations in crystals can arrange themselves in dislocation walls or *subgrain boundaries,* which are planar defects across which the mismatch in lattice orientation is low ($\leqslant 10°$). A discussion of these features can be found in Section 10.6. Subgrain and grain boundaries play an important role in some high-temperature creep processes.

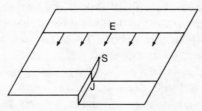

Figure 9.9 Jog J resulting from the intersection of edge dislocation E with screw dislocation S.

9.6 Mobility of dislocations

Migration of dislocations within a crystal is an important agent of ductile deformation. In this section we examine some processes associated with dislocation mobility. Reference may be made to Cottrell (1953), Nicolas & Poirier (1976), and Poirier (1985) for a more detailed treatment.

The existence of a finite strength of crystals (at least at relatively low temperature) demonstrates that dislocation migration and multiplication do not occur before a critical stress is reached. The first problem is therefore to account for this *yield strength* in microphysical terms. However, the yield strength decreases with increasing temperature, and it is not observable in silicates at high homologous temperatures, even at the lowest stress differences available in the laboratory ($\sigma_1 - \sigma_3$ of the order of a few megapascals; see Section 10.4). Such a vanishingly small yield strength reflects the increasing role of recovery processes at higher temperature (cf. Section 10.2). Nevertheless, an examination of the critical stress necessary to move dislocations in their own slip plane clarifies some of the mechanisms affecting dislocation mobility.

We consider first the *conservative motion* of dislocations, not involving diffusive fluxes of matter (non-conservative motion, in which formation and migration of point defects plays an important role, will be studied later). In conservative motion the dislocation migrates on the slip plane by *glide,* whereby the boundary between the slipped and unslipped parts of the crystal changes its position. Resistance to glide occurs by the interaction of the moving dislocation with other defects (impurities, other dislocations, and grain boundaries) and by the frictional resistance of the lattice to motion. The latter is usually termed the *Peierls stress,* and is a consequence of the need to break atomic bonds for the dislocation to glide. The situation is illustrated in Figure 9.10a. Because of the regular nature of the crystal lattice, the force across the slip plane varies periodically during slip. Atoms on one side of the slip plane sufficiently far from the dislocation occupy equilibrium positions relative to atoms on the other side. However, near the dislocation they do not, and when the dislocation attempts to move it is resisted by a restoring force. Consequently, a finite stress (Peierls stress) is necessary for the dislocation to glide; for an ideal lattice it is given by

$$\sigma_P \simeq 2\mu \, \exp(-2\pi w/b) \qquad (9.20)$$

The quantity w is the width of the dislocation, i.e. the distance along the slip plane in the slip direction where the atomic displacement is larger than one-half the maximum displacement. In many materials w is likely to lie between one and ten interatomic spacings. For $w = 3b$, Equation 9.20 gives $\sigma_P \simeq 10^{-8} \mu$, i.e. the Peierls stress is very low and does not control the strength, which must therefore be controlled by the stress necessary to overcome the 'anchoring' of dislocations by other defects. However, in minerals with ionocovalent or covalent atomic bonding the Peierls stress is higher − as shown by the presence of long straight dislocation segments − and may control the strength.

The above reasoning is an oversimplification, since it assumes that a

257

(a)

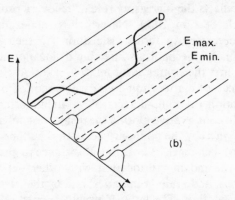

Figure 9.10 (a) Resistance to glide (Peierls stress): PP is the slip plane; the distorted lattice is shown by continuous lines, its equilibrium position by broken lines. (b) Bulge in the dislocation line D, which therefore occupies two energy troughs (E_{min}); glide occurs by lateral spreading of the bulge.

dislocation glides 'rigidly' in the slip plane. The energy field on the slip plane may be envisaged as a sequence of parallel ridges and troughs induced by the periodicity of the lattice; dislocation glide involves the overcoming of these energy barriers. A dislocation line can move from one energy trough to the next all at once, in which case glide involves the simultaneous switching of atomic bonds along a whole row of atoms parallel to the dislocation. However (especially at high temperature) dislocation lines may not lie in their entirety in the same energy trough, but may contain kinks or bulges spanning two or more energy troughs (Fig. 9.10b). The resistance to lateral migration of these irregularities is considerably less than the resistance to motion in the direction normal to the dislocation, and glide occurs preferentially by bulge nucleation and spreading. Consequently, the actual critical stress for dislocation migration is likely to be less than the Peierls stress, the more so at high homologous temperatures. Splitting into partial dislocations (cf. Section 9.5) can also greatly enhance dislocation mobility.

Glide of a dislocation is a *thermally activated rate process* in which both thermal agitation and applied stress help overcome a barrier. A dislocation can be regarded as consisting of a number of 'flow units' (segments of length l) which sweep an area δA on the slip plane when overcoming a barrier. If σ is the applied stress and b the Burgers vector (distance between two consecutive energy troughs), then the work done by the applied stress in the migration of one flow unit is $\sigma b \delta A$. The frequency of slip by a flow unit is therefore

$$\nu' = \nu_0 \exp\left(-\frac{E_0 - \sigma b \delta A}{kT}\right) \tag{9.21}$$

where ν_0 is the attempt frequency and E_0 is the height of the energy barrier.

Two cases may be distinguished, depending on whether the slipped segment is stopped by the next Peierls hill and consequently can jump back into its former position, or whether it travels over a larger distance before being stopped by another barrier. In the first case there is a finite backward jump frequency ν'', and the net rate is

$$\nu = \nu' - \nu'' = \nu_0 \exp\left(-\frac{E_0}{kT}\right)\left[\exp\left(\frac{\sigma b \delta A}{kT}\right) - \exp\left(-\frac{\sigma b \delta A}{kT}\right)\right]$$

$$= 2\nu_0 \exp\left(-\frac{E_0}{kT}\right)\sinh\left(\frac{\sigma b \delta A}{kT}\right)$$

which for $\sigma b \delta A \ll kT$ may be written as

$$\nu \simeq 2\nu_0 \frac{\sigma b \delta A}{kT} \exp\left(-\frac{E_0}{kT}\right) \tag{9.22}$$

In the second case, Equation 9.21 gives the net rate directly. Since each successful jump generates an elementary strain, the above rates (Eqns 9.21 & 22) are related to the overall strain rate of the crystal.

Although the previous considerations are of general applicability, there is an important difference between edge and screw dislocations with respect to glide. In the former glide is limited to a plane containing the dislocation and its Burgers vector; in the latter, since dislocation and Burgers vector are parallel, there can be more than one glide plane containing the dislocation. Consequently, screw dislocations may move from one glide plane to another (see Fig. 9.11). This process, termed *cross-slip*, is common at high temperature or high applied stress, or both, and under certain conditions it may be important in determining the rheology of crystalline materials.

Both glide and cross-slip are conservative types of motion. At higher temperatures ($T \gtrsim \frac{1}{2} T_m$), where diffusion is effective, *non-conservative*

259

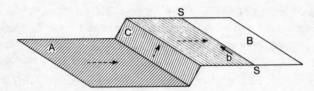

Figure 9.11 Cross-slip of a screw dislocation SS from glide plane A to B; the slipped area is shaded.

motion, involving the formation and migration of vacancies, plays an increasingly important role. Edge dislocations can move out of their slip plane by *climb,* which is a thermally activated process involving the diffusion of vacancies towards or away from the dislocation. Climb requires the removal or addition of a row of atoms from the extra half-plane associated with the dislocation (Fig. 9.12a; screw dislocations have no extra half-plane and therefore cannot climb). This removal or addition is seldom simultaneous, as the dislocation is usually jogged, and it is easier to remove or add a vacancy at a jog, since there are fewer atomic bonds involved (Fig. 9.12b). Jogs therefore act as sources and sinks of vacancies.

A thermodynamic equilibrium concentration of jogs per unit length of dislocation can be defined in a manner analogous to the equilibrium concentration of vacancies (Eqn 9.9). If n_0 is the number of atomic sites along the dislocation of length l, and G_J is the Gibbs free energy of activation for the thermal nucleation of a jog, then the equilibrium number of jogs over l is

$$n_J = n_0 \exp(-G_J/kT)$$

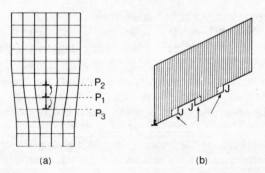

Figure 9.12 (a) Climb of an edge dislocation from glide plane P_1 to plane P_2 (by removal of a row of atoms) or P_3 (by addition of a row of atoms). (b) Jogs in the extra half-plane (shaded) where removal or addition of vacancies preferentially occurs.

If jog nucleation is important, then the Gibbs free energy of activation for climb is

$$G_c = G + G_J \qquad (9.23)$$

where G is the Gibbs free energy for self-diffusion. Equation 9.23 may be written in terms of activation enthalpies or activation energies. If the edge dislocation is already heavily jogged, $E_c = E$; otherwise $E_c > E$ because of the energy of jog formation.

Diffusion-controlled climb is considerably slower than glide. The climb velocity v_c is related to the vacancy flux to or away from the dislocation, which is in turn affected by the applied stress. If the concentration of jogs is high, then the climb velocity is approximately equal to the jog velocity along the dislocation, $v_c \simeq v_J$. The latter can be estimated by considering that, if the flux of vacancies per unit time and unit length of dislocation is ϕ and the length of a jog is b, then a jog receives ϕb vacancies and moves by a distance ϕb^2 per unit time (the reasoning is identical for emission of vacancies). We therefore have

$$v_c \simeq v_J = \phi b^2 \qquad (9.24)$$

The applied stress σ affects the equilibrium concentration of vacancies near the dislocation by increasing the Gibbs free energy of formation of one vacancy by $\sigma \Omega$, where $\Omega \approx b^3$ is the atomic volume ($\sigma \Omega$ is the work involved in removing an atom from the extra half-plane). The vacancy concentration near the dislocation is therefore

$$C_d = \frac{N_d}{\Omega} = \frac{1}{\Omega} \exp\left(-\frac{G_f + \sigma \Omega}{kT}\right) = C \exp\left(-\frac{\sigma \Omega}{kT}\right)$$

where C is the equilibrium concentration of vacancies in the bulk of the crystal, i.e. farther than a characteristic distance λ from the dislocation (compare with Eqn 9.9). The vacancy concentration gradient is then $(C_d - C)/\lambda$, and the flux towards the dislocation is given by Fick's First Law (Eqn 9.10), that is, if $\sigma \Omega \ll kT$,

$$J = -D_V \frac{C_d - C}{\lambda} = -D_V \frac{C}{\lambda} \left[\exp\left(-\frac{\sigma \Omega}{kT}\right) - 1\right] \simeq \frac{D_V C \sigma \Omega}{\lambda kT} \qquad (9.25)$$

The number of vacancies crossing a cylinder of radius b along the unit length of the dislocation is $\phi = 2\pi bJ$. Combining this with Equations 9.24 and 9.25 we have

$$v_c \simeq 2\pi b^3 D_V C \frac{\sigma \Omega}{\lambda kT}$$

But by the definition of diffusion coefficient (Section 9.4) we can write

$$D_V C \simeq D_V N_V / b^3 = D_{SD} / b^3$$

where D_{SD} is the coefficient of self-diffusion. The climb velocity is then

$$v_c \simeq 2\pi D_{SD} \sigma \Omega / \lambda kT \qquad (9.26)$$

The result given by Equation 9.26 – i.e. a linear dependence of climb velocity on stress through the self-diffusion coefficient – is widely used in creep models. However, because of the many assumptions used in its derivation, it is valid only as a first approximation.

The dependence of strain rate on temperature, pressure, and stress (i.e. the rheological constitutive equation) must follow from the thermodynamics of dislocation mobility if creep is related to dislocation movement. What is needed to bridge the gap between microphysics and the macroscopic stress–strain rate relation is a *microphysical constitutive equation* connecting strain rate controlled by dislocation movement with the physical properties of the dislocations. The most widely used microphysical constitutive equation is *Orowan's equation*.

In order to derive Orowan's equation, consider the simple case of a crystal having the shape of a parallelepiped of height h and horizontal cross-sectional area A (volume $V = Ah$). Dislocations are assumed to glide on slip planes parallel to the basal area and with Burgers vector b. If a dislocation sweeps a whole plane, then the resulting shear strain is b/h. If only a part dA of the plane slips, then the shear strain increment is

$$d\varepsilon = \frac{b}{h} \frac{dA}{A} = \frac{bvl \, dt}{V}$$

where $vl \, dt$ is the area swept during dt by a dislocation line of length l moving with velocity v. If n parallel dislocations are present, then the shear strain increment is

$$d\varepsilon = nlbv \, dt / V$$

The total length of mobile dislocations per unit volume, i.e. the density of *mobile* dislocations, is $\rho_m = nl/V$. Therefore we can write

$$d\varepsilon = \rho_m vb \, dt$$

and the shear strain rate is

$$d\varepsilon/dt \equiv \dot{\varepsilon} = \rho_m vb \qquad (9.27)$$

This is the Orowan equation, relating strain rate to density, velocity, and Burgers vector of mobile dislocations. Although it has been derived here for a particularly simple case, it holds for more-complex configurations, and for climb as well as glide. In general, the velocity v is that which applies to the rate-controlling process.

The mobile dislocation density ρ_m may be lower than the total dislocation density ρ_D, since a number of dislocations may be locked or pinned, or form tangles that prevent motion. The stress-dependence of dislocation density (assuming it to be the same for ρ_D and ρ_m) can be easily estimated. If the average distance between dislocations is L, then the density is $\rho \propto L^{-2}$. The internal stress, whose average is taken to be equal to the applied stress, varies as $\mu b/L$ (Section 9.5). The relation between dislocation density and stress is therefore

$$\rho \propto \sigma^2 \tag{9.28}$$

It should be emphasized that Equation 9.28 expresses an average relation only. In practice dislocation density varies widely with stress and temperature. Consequently, use of the density – as determined, for instance, from TEM – to estimate the applied stress during deformation gives weakly constrained results.

If the stress-dependence of dislocation density and velocity are known, then Equation 9.27 allows the prediction of the stress–strain rate relation from microphysical considerations. Several models of dislocation creep (to be discussed in the next chapter) are based on the Orowan equation.

10 Creep of polycrystals at high temperature and pressure

The time-dependent ductile deformation of polycrystals was described in Part I from the continuum-mechanics viewpoint. We now analyse it on the basis of crystal defects and their mobility. Both transient and steady-state creep can be accounted for by suitable models of the underlying microphysical processes.

Steady-state creep, which is of central importance for the rheology of the Earth in the long timescale, may be attained in a number of ways, and its constitutive equation depends on the predominant deformation mechanism. The two most likely stress–strain rate relations in the lower lithosphere and mantle are power-law creep (probably controlled by some form of diffusion-related dislocation climb) and linear diffusion creep with grain boundary sliding (controlled by vacancy diffusion). Both are strongly dependent on temperature and pressure.

Two important processes that may occur during creep are hydrolytic weakening and dynamic recrystallization, which may, under suitable conditions, considerably soften geological materials.

10.1 Attenuation and transient creep

Vacancies and dislocations are the main 'flow units' for ductile deformation in crystals. The materials of the lithosphere and mantle consist of polycrystalline, polymineralic aggregates of randomly or preferentially oriented grains, and consequently their deformation mechanisms are likely to be more complex than those for single crystals. However, the ideas developed for single crystals do apply to polycrystals and go a long way towards accounting for their rheological behaviour.

In this chapter we consider models and results pertaining to the time-dependent deformation of polycrystalline Earth materials. As the emphasis is on the microphysics of flow, we use tensor notation only when necessary to give the complete rheological equations. It is not possible to give a complete treatment of the topic within the scope of this book; rather, basic physical notions are stressed, and the connection between the microscopic and the macroscopic is analysed.

For short times and at small stresses and strains the behaviour of polycrystals is anelastic. This results in recoverable transient creep (Section 4.5) and anelastic attenuation (Section 6.5). (Transient creep may also have an irrecoverable component due to plastic deformation.) *Anelastic relaxation* is the general term for the time-dependent adjustment of a body to a change in an external variable. Transient creep is the manifestation of this relaxation in the time domain; attenuation is its manifestation in the frequency domain. In general, denoting time and frequency by t and ω, respectively, experimental results on silicates may be represented in the form

$$\varepsilon_t(t) = \varepsilon_0(t/\tau)^\alpha \tag{10.1}$$

$$Q(\omega) = Q_0(\omega\tau)^\alpha \tag{10.2}$$

where ε_t and Q are transient strain and seismic quality factor; ε_0 and Q_0 are material parameters (transient strain for $t = \tau$ and quality factor for $\omega\tau = 1$); and τ and α are a characteristic time and a material coefficient, respectively.

Equations 10.1 and 10.2 can be formally accounted for by linear anelasticity theory with a spectrum of relaxation (actually retardation) times (Anderson & Minster 1979; see Section 6.5). Temperature dependence may be introduced by considering anelasticity – whatever its microphysical details – to be a thermally activated rate process with characteristic time

$$\tau = \tau_0 \exp(H'/kT) \tag{10.3}$$

where τ_0 is the retardation time at infinite temperature and H' is the activation enthalpy of the process. Equations 10.1 and 10.2 then become

$$\varepsilon_t(t) = \varepsilon_0(t/\tau_0)^\alpha \exp(-\alpha H'/kT) \tag{10.4}$$

$$Q(\omega) = Q_0(\omega\tau_0)^\alpha \exp(\alpha H'/kT) \tag{10.5}$$

These relations have been verified experimentally for silicate polycrystals (cf., for example, Berckhemer *et al.* 1982 for dunite and peridotite). The material coefficient α is usually in the range $\frac{1}{4} \leqslant \alpha \leqslant \frac{1}{2}$, and the apparent activation enthalpy H' is well below that for self-diffusion.

The spectrum of retardation times $\tau_1 \leqslant \tau \leqslant \tau_2$ defines an *absorption band* at intermediate temperature (or frequency) where $Q \propto \omega^\alpha$. At high temperature, where retardation times are short (or at low frequency), $Q \propto \omega^{-1}$; at low temperature, with long retardation times (or at high frequency), $Q \propto \omega$. The absorption band is shown schematically in Figure 10.1. Its position on the frequency axis varies with depth in the Earth, since the characteristic times depend on temperature and, to a minor extent, pressure.

Although the determination of Q is a poorly constrained inverse problem, observations are compatible with interpretation in terms of a depth-dependent absorption band. The introduction of a spectrum of retardation times allows a unified interpretation of data for frequencies ranging from 10^{-8} Hz (Chandler wobble) to 10^7 Hz (ultrasonic laboratory experiments). A retardation spectrum can be achieved by a distribution of τ_0 for constant H', a distribution of H' for constant τ_0, or a combination of the two. A distribution of τ_0 leads to a shift of the absorption band with temperature and pressure; a distribution of H' leads to a broadening of the band at lower temperature.

From the microphysical viewpoint a retardation time of the type given by Equation 10.3 may arise from a variety of processes involving lattice defects (cf. the review by Minster 1980). Some processes occur at grain boundaries, with or without the presence of a liquid phase (*intergranular relaxation*); others involve energy dissipation within grains (*intragranular relaxation*). Mechanisms of the latter type, related to dislocation damping, are probably responsible for the so-called high-temperature background attenuation (HTB), resulting in internal friction steadily increasing with temperature. However, no model is unique in the sense of excluding all of the others.

Dislocation damping may be modelled by considering a dislocation of length l and Burgers vector b pinned at both ends (see Minster & Ander-

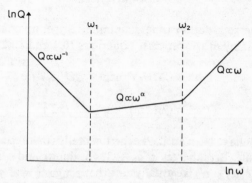

Figure 10.1 Anelastic absorption band. Seismic quality factor as function of frequency.

266

son 1981 for a detailed discussion). If the stress is sufficiently large for dislocation multiplication to occur (i.e. $\sigma \gtrsim 1$–$10\,\mathrm{MPa}$ for $l \gtrsim 1$–10 μm – see Section 9.5), then Frank–Read sources become active and ductile flow occurs. However, at lower stresses, the dislocation bows out by a small amount only. The anelastic strain caused by bow-out may be calculated from the theory of linear bodies (Section 4.6), and is

$$\varepsilon(t) = \varepsilon(\infty)[1 - \exp(-t/\tau)]$$

where $\varepsilon(\infty)$ is its asymptotic value as $t \to \infty$ and τ is the retardation time. The general form of the latter under an applied stress σ is

$$\tau \simeq \sigma l^2/\mu bv$$

where the dislocation velocity is approximately given by (cf. Section 9.6)

$$v \simeq \sigma b^2 D/kT$$

and the diffusion coefficient D is that appropriate to the rate-controlling process. Combining the above two equations, we have for the characteristic time

$$\tau \simeq \frac{kTl^2}{\mu b^3 D} = \frac{kTl^2}{\mu b^3 D_0} \exp\left(\frac{H'}{kT}\right) \tag{10.6}$$

Dislocation bowing may be controlled by a variety of mechanisms – among others, kink diffusion and dragging of point defects – with an activation enthalpy that is poorly constrained but definitely less than that for steady-state creep (cf. Sections 10.3 & 4). Characteristic times estimated from Equation 10.6 are of the right order of magnitude ($1 \leqslant \tau \leqslant 10^6$ s) for the seismic and postseismic band.

Dislocation relaxation by glide permits a simple physical interpretation of the spectrum of retardation times (Eqn 6.33). A comparison of Equations 10.3 and 10.6 shows that $\tau_0 \propto l^2$. Therefore, a distribution of dislocation lengths of the type $F(l) \propto l^{-\beta}$, where $\beta > 0$ depends on the value of α, results in the required $F(\tau)$. The frequency-dependence of the anelasticity of a material can thus be related to the lengths of the dislocations it contains.

The previous considerations hold where time-dependence is the only departure from linear elasticity. This implies a correspondence between stress and equilibrium strain. These conditions are not always fulfilled. Transient creep may have an irrecoverable component arising from plastic flow processes (strain hardening – see Section 10.2). Also, a law

267

such as Equation 10.1 must be restricted to a limited time range, as it predicts an infinite strain rate at $t = 0$ and unbounded creep strain as $t \to \infty$. However, in practice steady-state creep takes over in the long term in most situations of geodynamic interest.

10.2 Steady-state creep: phenomenology

The time-dependent deformation of polycrystals usually consists of a transient stage followed, if conditions are favourable, by a steady-state stage during which strain increases at a constant rate under constant stress (Section 4.5). We now focus on steady-state flow, considering first the role of experiments and the phenomenological approach (cf. Poirier 1985 for a lengthier discussion of the topics treated in this section).

Various types of mechanical tests are possible in the ductile regime, according to the geometry of the sample and the load. Usually a cylindrical sample of size large relative to that of individual grains (to ensure averaging out of slip-system anisotropies) is loaded axially in compression or tension, either with or without confining pressure (cf. Section 5.2). Similar experiments can be performed on single crystals, in which case the orientation of the load with respect to the crystallographic axes is important. Other types of test configurations (bending, indentation, and torsion) are not as common.

According to the experimental set-up, different categories of tests may be distinguished (we denote by σ and ε the stress and strain, respectively, relevant to the experimental situation, e.g. axial compression and longitudinal strain in a uniaxial compression test). In *creep tests,* the sample is loaded at constant stress and the strain–time curve $\varepsilon(t)$ is obtained. In *constant strain rate tests,* as the name implies, the sample is deformed at a constant rate and the stress is measured as a function of strain, $\sigma(\varepsilon)$. In *stress relaxation tests* the deformation is held constant and the consequent decrease in stress $\sigma(t)$ is measured.

Typical creep and stress–strain curves are shown in Figure 10.2 (cf. also Sections 4.5 & 5.1). The transient stage of creep results in a decrease of strain rate with time (Fig. 10.2a). This is reflected in the stress–strain curve as an increase of stress with increasing strain (Fig. 10.2b), i.e. work (or strain) hardening.

The steady-state properties of a material may be described by an *empirical constitutive equation* relating strain rate to stress, temperature, and pressure (time must not explicitly appear for the constitutive equation to be an equation of state). The steady-state strain rate is empirically defined as the constant slope of the linear part of the creep curve. The identification of steady-state is not a simple problem: what may appear

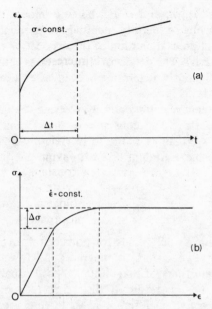

Figure 10.2 (a) Creep test; Δt is the time required to achieve steady state. (b) Constant strain rate test; $\Delta \sigma$ is the increase in stress caused by work hardening.

as a straight line over a narrow interval of time may not be a true straight line when viewed over a larger interval. Usually, however, empirical steady-state is attained in most materials at absolute temperatures above one-third to one-half of the melting point. The experimental results are well described by the so-called *Dorn equation:*

$$\dot{\varepsilon} = A_{\mathrm{D}} \sigma^n \exp(-H/RT) \qquad (10.7)$$

where A_{D}, n, and H are empirically determined 'constants', assumed not to vary with stress and (p, T)-conditions. However, it can be seen that A_{D} must be a function of T and p, if for no other reason than it must contain the elastic parameters of the material for dimensional reasons, and these are (p, T)-dependent. The stress exponent n and the creep activation enthalpy H may also vary with stress and temperature. Their values for some minerals and rocks are given in Section 10.4. Generally, $1 < n < 5$, and H is of the order of 30–130 kcal mol^{-1} (activation quantities are given per mole – in the sequel we shall use the joule rather than the calorie, 1 cal $\simeq 4.18$ J). As $n > 1$ except in few cases of extremely fine-grained aggregates, most materials under laboratory conditions (not only rocks but also metals, ceramics, and ice) exhibit a 'power-law' rheology (compare with Section 4.4).

The parameters A_D, n, and H can be determined from experiments, but it would be wrong to identify them with a particular microphysical mechanism without an examination of the dislocation microstructure of the deformed material. The details of the processes controlling creep are often not unambiguously determined, even after examination of the microstructure.

The stress exponent n is obtained by plotting the variation of strain rate with applied stress (at constant p and T) on logarithmic co-ordinates. If the creep law is of the form given by Equation 10.7, then the data points define a straight line with slope

$$\frac{\partial (\ln \dot{\varepsilon})}{\partial (\ln \sigma)}\bigg|_{p,T} = n \tag{10.8}$$

A linear relationship usually holds for polycrystals over a limited stress interval (typically, two orders of magnitude). However, if the stress ranges over several orders of magnitude, then n usually increases with stress. This may be the result of different creep mechanisms, with different n, being active in different stress intervals; or of mechanisms with a stress-dependent activation enthalpy taking over at higher stresses, which lead to an exponential dependence of strain rate on stress (power-law breakdown – see Fig. 10.3 and Section 10.3).

The creep activation parameters can be determined from creep experiments in a manner analogous to the determination of self-diffusion parameters discussed in Section 9.4. Since $H = E + pV$, where E and V are creep activation energy and volume, respectively, it is an immediate consequence of Equation 10.7 that

$$-R \frac{\partial (\ln \dot{\varepsilon}/A_D)}{\partial (1/T)}\bigg|_{\substack{\sigma \\ p=0}} = E \tag{10.9}$$

$$-RT \frac{\partial (\ln \dot{\varepsilon}/A_D)}{\partial p}\bigg|_{\sigma,T} = V \tag{10.10}$$

The determination of E therefore entails the measurement of the strain rate at constant stress, zero pressure, and different temperatures; then E is obtained from the slope of the resulting Arrhenius plot (Fig. 10.4a). Experiments over wide temperature ranges often give an apparent activation energy which varies with temperature; this is possibly a consequence of the superposition of two or more mechanisms having different activation energies. In order to obtain V, stress and temperature are kept constant and the strain rate is measured at various pressures (Fig. 10.4b).

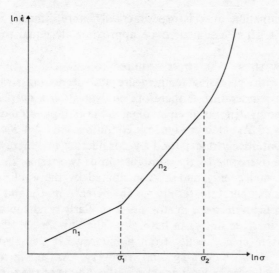

Figure 10.3 Typical experimental stress–strain rate curve for silicate polycrystals. At low stress ($\sigma < \sigma_1$) the rheology is described by a Dorn equation with low stress exponent ($1 \leq n \leq 2$); for $\sigma_1 < \sigma < \sigma_2$, by a Dorn equation with higher stress exponent (usually, $2.5 \leq n \leq 4.5$); at high stress ($\sigma > \sigma_2$) the relation becomes exponential (power-law breakdown).

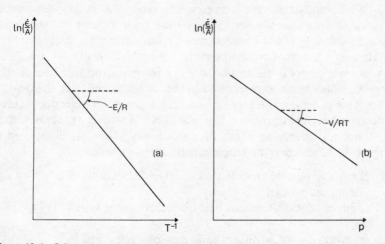

Figure 10.4 Schematic Arrhenius plots for the determination of (a) activation energy and (b) activation volume.

271

The determination of V is experimentally more difficult, and for this reason V is often assumed to be approximately equal to the atomic volume.

The flow stress (the stress required to attain a given strain rate) decreases with increasing temperature and decreasing strain rate. An increase in temperature is therefore equivalent to a decrease in strain rate. This equivalence is often invoked in the extrapolation of laboratory results ($\dot{\varepsilon} \geqslant 10^{-8}\,\mathrm{s}^{-1}$) to geodynamic conditions ($10^{-15} \leqslant \dot{\varepsilon} \leqslant 10^{-10}\,\mathrm{s}^{-1}$). It can be intuitively understood by considering the energy barrier for creep to be overcome by the combination of two terms, E_T contributed by the thermal energy and E_σ contributed by the applied stress. An increase in temperature results in an increase in E_T and therefore a decrease in E_σ; a decrease in strain rate similarly results in a decrease in E_σ and an increase in E_T. In both cases the flow stress decreases.

Steady-state creep results from a balance between two competing processes: strain hardening and recovery. As an introduction to microphysical creep models, we sketch the outline of the formal theories of hardening and recovery.

The dislocation density, and therefore the interference among dislocations, tends to increase as deformation progresses. Dislocations gliding on intersecting slip planes, or meeting intra- and intercrystalline obstacles, form tangles and pile-ups which make further deformation more difficult. With reference to the Orowan equation (Eqn 9.27), both the density of mobile dislocations and their velocity decrease as the number of tangles and pile-ups increase. Consequently the strain rate under constant stress decreases with time, i.e. the stress required to maintain a constant strain rate increases with time. This is the phenomenon known as *strain hardening*. Steady-state creep cannot be attained if strain hardening is not counteracted by some softening process.

However, a strain-hardened state is thermodynamically unstable, since an increase in total dislocation density increases the free energy of a crystal, which tends to revert to a lower energy by reducing the density. This is the process of *recovery* which, as it is thermally activated, is more effective at high temperature. There are several recovery processes, most (but not all) involving climb of dislocations. They are illustrated in Figure 10.5, and may be schematically listed as follows:

(a) climb and mutual annihilation of edge dislocations of opposite sign gliding on parallel slip planes;
(b) climb of edge dislocations past obstacles against which they are piled up;
(c) cross-slip by screw dislocations past obstacles; and
(d) rearrangement of dislocations of the same sign into dislocation walls (polygonization) separating subgrains with lower dislocation density.

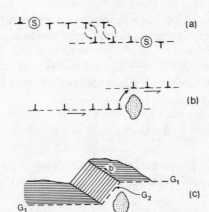

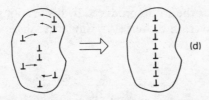

Figure 10.5 Recovery processes: (a) climb and mutual annihilation of edge dislocations; (b) climb of edge dislocations past obstacles; (c) cross-slip of screw dislocations; (d) polygonization. Sources are denoted by S; obstacles are stippled.

Strain hardening increases the average internal stress; recovery decreases it. We may define a *hardening coefficient h* and a *recovery coefficient r*, proportional to the strain hardening and recovery rate, respectively. Then the variation of internal stress $\Delta\sigma^*$ during a time interval dt during which the strain changes by $d\varepsilon$ is

$$\Delta\sigma^* = h\,d\varepsilon - r\,dt \qquad (10.11)$$

both h and r being reckoned positive. If $\Delta\sigma^* > 0$ (low T/T_m or high $\dot{\varepsilon}$, or both), then strain hardening predominates; if $\Delta\sigma^* = 0$ (high T/T_m or low $\dot{\varepsilon}$, or both), then the hardening and recovery rate balance each other and steady-state creep results. (Accelerating creep, where $\Delta\sigma^* < 0$, is also possible, leading to an instability where the stress required to maintain deformation at a constant rate decreases with time.)

For $\Delta\sigma^* = 0$ (steady-state creep), Equation 10.11 yields

$$\dot{\varepsilon} = r/h \qquad (10.12)$$

273

which is the *Bailey–Orowan equation*. It is a statement of the fact that the internal stress is constant during steady-state creep.

The Bailey–Orowan equation can be written in a form similar to the Orowan equation. The hardening and recovery rates are related to the change in internal stress with strain and its decrease with time at constant deformation, respectively, i.e.

$$h = d\sigma/d\varepsilon \qquad r = -d\sigma/dt \qquad (10.13)$$

where the average internal stress is taken as equal to the applied stress. Introducing the mobile dislocation density ρ_m, the hardening coefficient can be written as

$$h = \frac{d\sigma}{d\rho_m} \frac{d\rho_m}{d\varepsilon} = \frac{1}{b\Delta l} \frac{d\sigma}{d\rho_m} \qquad (10.14)$$

since $\varepsilon = \rho_m b \, \Delta l$ is the strain produced by dislocations of Burgers vector b travelling an average distance Δl. Similarly, the recovery rate can be expressed as

$$r = -\frac{d\sigma}{d\rho_m} \frac{d\rho_m}{dt} = \frac{\rho_m}{t^*} \frac{d\sigma}{d\rho_m}$$

where a first-order kinetics has been assumed for dislocation density during recovery, $d\rho_m/dt = -\rho_m/t^*$, with t^* denoting the 'lifetime' of a dislocation. Assuming $t^* = l^*/v$, where l^* is the height of the obstacle and v the speed at which it is overcome, the recovery rate becomes

$$r = \frac{\rho_m v}{l^*} \frac{d\sigma}{d\rho_m} \qquad (10.15)$$

Making use of Equations 10.14 and 10.15, the Bailey–Orowan equation can be written as

$$\dot{\varepsilon} = \rho_m v b \, \Delta l/l^* \qquad (10.16)$$

which is identical to the Orowan equation except for the factor $\Delta l/l^*$ expressing the ratio of glide distance to climb distance. The Bailey–Orowan and Orowan equations, one expressed in terms of hardening and recovery, and the other in terms of dislocation properties, are equivalent, and either may be used to derive stress–strain rate relations based on microphysical properties.

274

10.3 Steady-state creep: models

The macroscopic rheological equation for silicates at low stresses ($1 \leqslant \sigma \leqslant 100$ MPa), as determined from experiment, is usually of the power-law type, although instances of Newtonian behaviour have been observed, especially near the lower limit of the above stress interval and for very small grain sizes. Straightforward extrapolation of experimental results to lower-lithosphere and mantle conditions involves a difference of several orders of magnitude over strain rate and the estimation of pressure effects beyond the experimental range. The impossibility to duplicate in the laboratory the deformation environments of the interior of the Earth makes it necessary to strengthen experimental results by means of *theoretical creep* (or *rate*) *equations,* which are based on microphysical models and therefore are not limited by experiment.

Among the many books treating microphysical creep models we may mention Gittus (1975), Nicolas & Poirier (1976), Frost & Ashby (1982), and Poirier (1985). Reviews by Weertman (1970), Stocker & Ashby (1973), and Ashby & Verrall (1978), among others, are particularly useful for placing the problem in a geophysical context.

We derive the one-dimensional form of the rate equation (i.e. for uniaxial or shear stress); this procedure does not burden the argument with heavy notation. The tensor form of the most important equations is presented at the end of the section. Also, we use the symbol $\sim$ (in this section only) to denote dimensionally correct proportionality; that is, a proportionality that would become an equality with the insertion of a dimensionless coefficient. The value of the latter depends on the details of the assumed geometry; its consideration would needlessly complicate the physical argument, but we shall give it when necessary.

Creep mechanisms (except *cataclasis,* i.e. creep by repeated breaking and rolling of grains, which physically is fracture and of little importance at high temperature) may be classified as follows.

(a) *Glide-controlled creep,* rate-controlled by the overcoming of obstacles to dislocation glide; the stress–strain rate relation is exponential in most cases.
(b) *Creep by diffusion-controlled dislocation climb,* where obstacles are overcome by climb of edge dislocation; it leads to a power-law rheology.
(c) *Harper–Dorn creep,* a particular case of climb-controlled flow in which dislocation density is stress-independent; it leads to a Newtonian rheology.
(d) *Creep by cross-slip of screw dislocations,* where obstacles are overcome by thermally activated cross-slip; the stress dependence of strain rate is poorly known.

(e) *Diffusion creep,* where flow is achieved by migration of vacancies between grain boundaries; grain-boundary sliding, fluid phase transport, and pressure solution are particular cases; the stress–strain rate relation is Newtonian.

(f) *Superplasticity,* either structural or transformational (i.e. associated with a phase change), which leads to a first or second power dependence of strain rate on stress, according to the details of the microphysical mechanism.

In all cases, the rate equation is of the form

$$\dot{\varepsilon} = f(\sigma, p, T, \{M_i\}, \{S_i\}) \tag{10.17}$$

where $\{M_i\}$ are the material parameters (lattice distances, atomic volumes, elastic moduli, and diffusion coefficients), and $\{S_i\}$ are the state variables describing the microstructural state of the material (dislocation density, dislocation properties, and grain and subgrain size). Under the assumption of steady state, the time-rate of change of the state variables is assumed to vanish, so that $\{S_i\}$ no longer appear explicitly in the rate equation, but are given as functions of the external variables. Then Equation 10.17 describes the rheology fully.

We now review the rate equations for the various creep mechanisms listed above.

Glide-controlled creep
The rate-controlling step is the thermally activated overcoming of obstacles to glide. Starting from the Orowan equation (Eqn 9.27), we express dislocation density and velocity in terms of stress. Since $\sigma \sim \mu b L^{-1}$ and $\rho_m \sim L^{-2}$, where L is the average distance between dislocations, μ is the rigidity, and b is the Burgers vector, the mobile dislocation density is

$$\rho_m \sim \left(\frac{\sigma}{\mu b}\right)^2 \tag{10.18}$$

The glide velocity depends on the frequency of slip (Section 9.6), and in the general case has the form

$$v \sim b v_0 \exp\left(-\frac{H(\sigma)}{kT}\right)$$

where v_0 is a characteristic frequency, and $H(\sigma)$ is a stress-dependent

276

creep activation enthalpy. The strain rate is therefore

$$\dot{\varepsilon} \sim \nu_0 \left(\frac{\sigma}{\mu}\right)^2 \exp\left(-\frac{H(\sigma)}{kT}\right) \tag{10.19}$$

The form of $H(\sigma)$ varies with the nature and shape of the obstacles. Glide-controlled creep is effective at high stress and relatively low temperature, hence the name 'low-temperature plasticity'. It has a very long transient strain-hardening state, as should be expected, since recovery does not operate. Low-temperature plasticity is of no great importance in geodynamics, where stresses are usually well below those required for it.

Power-law creep
There are several ways in which 'power-law creep', where strain rate is proportional to the nth power of the stress ($2 \leqslant n \leqslant 5$), can arise, and they all involve recovery. A very general model is based on climb-controlled dislocation glide (*Weertman creep*), in which the rate-controlling process is the climb of edge dislocations (itself diffusion-controlled and consequently active at high temperatures), and the strain-producing process is glide. It can be derived from the Orowan equation (Eqn 9.27), with the usual relation for dislocation density (Eqn 10.18) and taking the climb velocity to be diffusion-controlled (Eqn 9.26), i.e.

$$v \equiv v_c \sim D_{SD} \frac{\sigma\Omega}{\lambda kT}$$

where D_{SD} is the coefficient of the self-diffusion, Ω is the atomic volume, and λ is a characteristic distance. Taking $\lambda \simeq b$, $\Omega \simeq b^3$, we have for the strain rate

$$\dot{\varepsilon} \sim \frac{D_{SD}\mu b}{kT} \left(\frac{\sigma}{\mu}\right)^3 \tag{10.20}$$

In the Weertman model dislocations are assumed to be produced by Frank–Read sources on parallel planes, and the basic recovery process is the annihilation by climb of mutually trapped edge segments. Equation 10.20 is based on the implicit assumption that the density of sources is proportional to the dislocation density, and is therefore stress-dependent; if the density of sources is stress-independent, then the stress exponent becomes $n = 4.5$. Climb-controlled power-law creep therefore

277

generally follows an equation of the form

$$\dot{\varepsilon} \sim \frac{D_{SD}\mu b}{kT} \left(\frac{\sigma}{\mu}\right)^n \qquad (10.21)$$

Many other models for power-law creep are available, most of which are variants of the Weertman model and lead to a stress exponent $n = 3$. Climb-controlled recovery creep is usually associated with a highly organized microstructure, characterized by the presence of *subgrains* and *cells* separated by dislocation walls (see Fig. 10.6). This microstructure can play an important role during creep. For instance, a power-law rheology can arise by diffusional flow coupled with a subgrain or cell microstructure. It will be seen later that diffusional flow of vacancies through the lattice from one grain boundary to another leads to an equation of the type (Eqn 10.28)

$$\dot{\varepsilon} \sim \frac{D_{SD}\mu\Omega}{kTd^2} \left(\frac{\sigma}{\mu}\right)$$

where d is the average grain diameter. If it is assumed that diffusion in the presence of a highly organized microstructure is not between grain boundaries, but between subgrain walls, then the subgrain diameter d^* must be substituted for d in the above equation. Since d^* depends on stress as (cf. Section 10.6)

$$d^* \sim \mu b/\sigma$$

the resulting creep law is identical to Equation 10.20.

A different stress-dependence may arise if dislocations glide freely across cells of diameter d^* and the average cell-wall dislocation segments are shorter than thermal jog separation. Creep is then controlled by

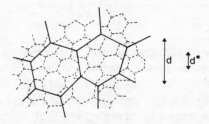

Figure 10.6 Ordered microstructure. Grains of average diameter d and subgrains of average diameter d^*.

double jog nucleation in cell walls, and the rate equation becomes

$$\dot{\varepsilon} \sim \frac{D\mu b}{kT} \left(\frac{\sigma}{\mu}\right)^2 \qquad (10.22)$$

where D is the appropriate diffusion coefficient, which in this case also includes terms related to jog formation and pipe diffusion (cf. Minster & Anderson 1981 for details).

It should be noted that empirical determination of the stress exponent is not sufficient to identify the operative creep mechanism. For instance, a dependence of the type $n = 2$ does not arise solely from cell-wall recovery. Glide-controlled creep may give the same apparent stress dependence if the exponential is only a weak function of the stress (cf. Eqn 10.19). We shall see that cross-slip creep and some forms of superplasticity also result in $n = 2$. Again it must be emphasized that examination of the microstructure is necessary in order to identify the creep mechanisms producing macroscopic flow.

In all of the above power-law creep models (except Eqn 10.22) the strain rate is related to the coefficient of self-diffusion of the slowest-moving species, D_{SD}; therefore the creep activation enthalpy is the same as the activation enthalpy for self-diffusion. The implicit assumption is that conditions of *jog saturation* prevail, i.e. that the climb velocity is controlled by the rate at which vacancies can diffuse to or from dislocations. If the edge dislocations are 'undersaturated', then the creep activation enthalpy will be larger than that for self-diffusion, as some energy is required for jog formation (see Section 9.6). As we shall see in Section 10.4, this is usually the case for olivine, whose properties govern the rheology of the upper mantle. The effective diffusivity may also be affected by diffusion along dislocation cores (pipe or core diffusion).

At high stress ($\sigma \geqslant 10^{-3}\,\mu$) there is a transition from climb-controlled to glide-controlled creep, and the stress-dependence of the strain rate becomes approximately exponential (*power-law breakdown*). Such high stresses are seldom of interest in geodynamics.

Harper–Dorn creep

At low stresses and low dislocation densities ($\rho_m \leqslant 10^8\,\mathrm{m}^{-2}$) Newtonian viscous behaviour is sometimes observed, with characteristics (grain-size independence, lack of grain-boundary effects, and presence of a transient stage) that exclude the possibility of diffusion creep. The mechanism of this Newtonian dislocation creep is not very clear, but probably involves stress-independent dislocation density. If ρ_m is approximately constant and the rate-controlling process is dislocation climb, then the Orowan

279

equation yields

$$\dot{\varepsilon} \sim \left(\frac{\rho_m \Omega}{\lambda}\right) \frac{D_{SD} \mu b}{kT} \left(\frac{\sigma}{\mu}\right) \tag{10.23}$$

Harper–Dorn creep is rarely observed, but could be important at small stresses ($\sigma \leqslant 1-10$ MPa) in minerals where, due to the high Peierls stress, dislocations tend to be long and straight, and are therefore not very responsive to variations in stress. It is the only Newtonian creep mechanism that is independent of grain size.

Cross-slip creep
The predominant recovery mechanism at high temperature is usually assumed to be diffusion-controlled climb of edge dislocations. However, if edge dislocations moved more freely than screw dislocations, then creep would be controlled by cross-slip. Poirier (1976) proposed that this could indeed be the case, as gliding screw segments are usually dissociated and the dissociation constrains the screw dislocation to its glide plane. Cross-slip of a split screw dislocation can occur only if a segment recombines, becoming thus capable of cross-slipping into another glide plane. Contrary to climb, which is thermally activated only indirectly (through vacancy diffusion), cross-slip is directly thermally activated, with an activation enthalpy which is not known, but is possibly a function of stress and stacking fault energy. If Δl_{CS} is the glide distance after unlocking of a screw dislocation, and $\nu_{CS} = t_{CS}^{-1}$ is the frequency of unlocking by cross-slip, then the velocity in the Orowan equation is

$$v \equiv v_{CS} \sim \nu_{CS} \, \Delta l_{CS} \, \exp\left(-\frac{H_{CS}(\sigma)}{kT}\right)$$

Using this, together with the usual expression for the density of dislocations (in this case screw dislocations), we obtain for the strain rate

$$\dot{\varepsilon} \sim \nu_{CS} \, \frac{\Delta l_{CS}}{b} \left(\frac{\sigma}{\mu}\right)^2 \exp\left(-\frac{H_{CS}(\sigma)}{kT}\right) \tag{10.24}$$

If the exponential term is not predominant, Equation 10.24 approximately reduces to a power-law type with $n = 2$. While theoretically possible, cross-slip creep does not appear to be supported by microstructural evidence in silicate polycrystals.

Diffusion creep
Diffusion plays a role in dislocation-climb controlled creep. It can also control creep directly, if grain boundaries are the only sources and sinks

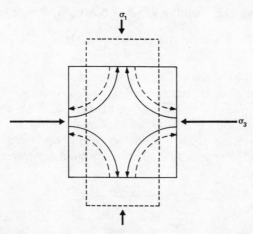

Figure 10.7 Diffusion creep (bulk diffusion) in a single crystal. Vacancy fluxes are denoted by dashed curves, atomic fluxes by continuous curves; the shape changes from square to rectangular.

of vacancies. (This implies efficient diffusion and few dislocations within grains, i.e. high homologous temperature, small grain size, and low stress.)

Diffusion creep is effected by vacancy fluxes between grain boundaries. Consider a single crystal subject to a non-hydrostatic stress field (Fig. 10.7): the vacancy concentration is different near surfaces in relative tension and in relative compression, and the consequent mass transfer generates strain. Recalling the considerations developed in Sections 9.3 and 9.4, the equilibrium concentration of vacancies in an unstressed crystal is

$$C = \frac{N_V}{\Omega} \exp\left(-\frac{G_f}{kT}\right)$$

where Ω is the atomic volume, and N_V and G_f are the atomic fraction and Gibbs free energy of formation of vacancies, respectively. The work done by the applied deviatoric stress σ to extract an atom from the lattice and bring it to a crystal face is $\pm\sigma\Omega$; therefore the vacancy concentrations near surfaces in relative tension and compression are, respectively,

$$C' = C \exp(\sigma\Omega/kT) \qquad C'' = C \exp(-\sigma\Omega/kT) \qquad (10.25)$$

Assuming the vacancy concentration gradient within the crystal to be approximately $\Delta C/d$, where $\Delta C = C' - C''$ and d is the average grain

281

size (diameter), the resulting vacancy flux is, by Fick's First Law,

$$J = -D_V \, \Delta C/d \qquad (10.26)$$

where D_V is the vacancy diffusion coefficient.

The number of atoms transported through a cross-sectional area of the grain per second is $\phi = -Jd^2$ (the minus sign follows from the opposite directions of vacancy and mass flows). Consequently, the volume added to − or substracted from − a surface is $\phi\Omega$, equivalent to a layer of thickness $\phi\Omega/d^2$ per unit time. The corresponding strain increment is

$$d\varepsilon \sim \frac{(\phi\Omega/d^2)}{d} \, dt$$

resulting in a strain rate

$$\dot{\varepsilon} \sim (\Omega/d^3)\phi \qquad (10.27)$$

(the $\sim$ sign takes into account the omission of a dimensionless proportionality factor which depends on the geometry of diffusion).

The quantity ϕ in Equation 10.27 can be expressed in terms of stress using Equations 10.25 and 10.26, i.e.

$$\phi = -Jd^2 = D_V d^2 \frac{\Delta C}{d} = D_V Cd\left[\exp\left(\frac{\sigma\Omega}{kT}\right) - \exp\left(-\frac{\sigma\Omega}{kT}\right)\right]$$

and, since $\sigma\Omega \ll kT$ (low stress and high temperature),

$$\phi \sim \frac{D_V Cd}{kT} \, \sigma\Omega$$

But $D_V C = D_{SD}\Omega^{-1}$, where D_{SD} is the coefficient of self-diffusion (see Section 9.4). Accordingly

$$\phi \sim \frac{D_{SD}d}{kT} \, \sigma$$

and the strain rate (Eqn 10.27) becomes

$$\dot{\varepsilon} \sim \frac{D_{SD}\mu\Omega}{kTd^2} \left(\frac{\sigma}{\mu}\right) \qquad (10.28)$$

Diffusion creep is therefore Newtonian and grain-size dependent. When more than one diffusing species are present, the slowest-moving controls creep.

The above rate equation has been derived assuming that self-diffusion occurs through the lattice (bulk diffusion). However, diffusion can also occur along the grain boundaries, which provide an enhanced diffusivity path (grain-boundary diffusion). The *effective diffusion coefficient* is the sum of a bulk diffusion term and a grain-boundary diffusion term

$$D^* = D_{SD} + (\pi\delta/d)D_{GB} \qquad (10.29)$$

where δ is the grain-boundary width and the factor π comes from geometrical assumptions about grain shape. The creep equation in the general case can therefore be written as

$$\dot{\varepsilon} \sim \frac{D^*\mu\Omega}{kTd^2}\left(\frac{\sigma}{\mu}\right) \qquad (10.30)$$

If bulk diffusion is predominant (*Nabarro–Herring creep*) $D_{SD} \gg (\pi\delta/d)D_{GB}$, then $D^* \simeq D_{SD}$ and Equation 10.30 reduces to 10.28. If grain-boundary diffusion is predominant (*Coble creep*), $(\pi\delta/d)D_{GB} \gg D_{SD}$, then $D^* \simeq (\pi\delta/d)D_{GB}$ and Equation 10.30 becomes

$$\dot{\varepsilon} \sim \frac{D_{GB}\mu\delta\Omega}{kTd^3}\left(\frac{\sigma}{\mu}\right) \qquad (10.31)$$

Grain-boundary diffusion has a lower activation enthalpy than bulk diffusion; it is therefore predominant in the lower part of the temperature range where diffusion creep is effective.

A condition favourable to diffusion creep is a very small grain size, of the order of 10–100 μm at the most. Given the d^{-2} and d^{-3} dependence of Nabarro–Herring and Coble creep, an increase in grain size decreases very significantly the diffusional strain rate, which becomes negligible when compared with other processes.

The 'grain-boundary width' introduced above is not the structural thickness of the grain boundary, but the width of the high-diffusivity path parallel to it. This could be larger than the structural thickness. For metals, the value $\delta = 2b$ is usually adopted. In minerals, however – and in ceramics – δ could be much larger than b, as indicated by scanning TEM of grain boundaries in naturally deformed rocks (White & White 1981). Such large grain-boundary widths would enhance the relative importance of Coble creep.

Diffusion creep is steady-state only if the average grain size (actually, the mean length of the diffusion path) is constant during deformation.

However, returning to Fig. 10.7 it can be seen that the length of the diffusion path increases with increasing strain. In a polycrystal the situation is similar if no grain-boundary sliding is possible (Fig. 10.8): each grain undergoes the same deformation as the aggregate as a whole, and the strain rate decreases with increasing strain. However, this contradicts the evidence for fine-grained materials, in which the transient stage is practically absent and the grains remain approximately equiaxed (equant) during flow. Clearly, grain boundary sliding must occur, to maintain continuity of the material and to allow large deformations of the aggregate while the individual grains suffer only minor deformation.

The basic mechanism in grain-boundary sliding is a neighbour-switching event, as shown in Figure 10.9. Grains change their shape only as necessary to maintain continuity. This is *diffusion-accommodated grain-boundary sliding* or *Ashby–Verrall creep* (see, for example, Ashby & Verrall 1978). The analysis is similar to that of diffusion creep without grain-boundary sliding; the only significant difference is that the average length $\bar{d}$ of the diffusion path is reduced, and therefore the creep rate is enhanced. Writing $\bar{d} = d/\beta, \beta > 1$, the rate equation for Ashby–Verrall creep is

$$\dot{\varepsilon} \sim \beta \, \frac{D^* \mu \Omega}{kTd^2} \left(\frac{\sigma}{\mu} \right) \tag{10.32}$$

where $\beta \simeq 5$ (compare with Eqn 10.30). Diffusion-accommodated grain-boundary sliding is sometimes referred to as 'superplasticity', but without any further qualifications this terminology is rather imprecise.

Steady-state flow by diffusive mass transfer may also occur by *fluid-phase transport*, i.e. by transport of matter in solution in a thin liquid film

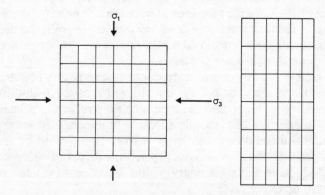

Figure 10.8 Deformation of a polycrystal by diffusion creep without grain-boundary sliding. Each grain undergoes the same deformation as the aggregate (right).

284

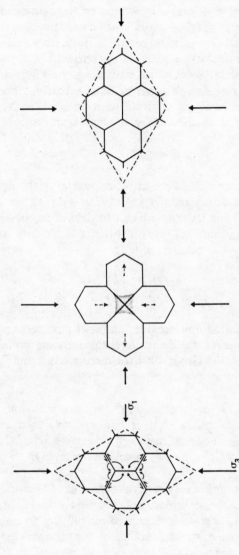

Figure 10.9 Grain-boundary sliding. The initial configuration of the polycrystal is shown on the left; by the diffusive fluxes shown (stipples denote the area where deformation is concentrated) the grains switch neighbours; the final configuration (right) differs from the initial, but the grain shape is unchanged (from Ashby & Verrall 1978, and Gittus 1978).

surrounding the grains. The film could be a partial melt in the upper mantle (cf. Section 7.7; note, however, that it is unlikely that a small fraction of melt would completely wet the grain boundaries), or water in the upper crust at diagenetic and low metamorphic grades. Transport in a water film is referred to as *pressure solution* by geologists and is a ubiquitous phenomenon, judging from the common occurrence of associated microstructures (pits on pebble surfaces in conglomerates, mutual penetration of fossils, and grain overgrowth). Deformation is achieved by migration of soluble material along the liquid film from zones of relative compression to zones of relative tension. The process can be formally regarded as a type of diffusion creep where the effective diffusivity is dominated by diffusion in the liquid, so that

$$D^* \simeq c_0 F D_F \qquad (10.33)$$

where c_0 is the atom fraction of the diffusing species in the liquid, F is the liquid volume fraction, and D_F is the diffusivity of the diffusing species in the liquid. The latter is difficult to determine, but is usually much higher than D_{SD} and D_{GB}, provided that a liquid film completely surrounds the grains. The rate equation is

$$\dot{\varepsilon} \sim \frac{c_0 F D_F \mu \Omega}{k T d^2} \left(\frac{\sigma}{\mu} \right) \qquad (10.34)$$

In other words, the liquid film provides an interconnected net of high-diffusivity paths. However, the details of the process are not altogether clear (cf. Rutter 1983 and Green 1984 for discussion).

Superplasticity
'Superplasticity' is a rather loose term, denoting a behaviour and not related to any particular mechanism. It was originally used to describe very large stable deformation of metals in elongation tests. Conditions favourable to superplasticity are very small grain size, high temperature, and low strain rate.

Purely tensile deformation very rarely occurs in geodynamics. Perhaps the best way to consider superplasticity independently of the state of stress is to regard it as a state of enhanced ductility, where under particular conditions (range of stress or grain size, for instance) the material is softened. From this viewpoint Ashby–Verrall creep (Eqn 10.32) is a type of superplasticity, as the occurrence of grain-boundary sliding softens the material. Indeed, superplastic deformation always results in equant grains, which implies some form of grain-boundary sliding. However – at least in metals – the stress exponent is often $n \simeq 2$. This

286

can be accounted for by assuming that grain-boundary sliding is accommodated not by diffusion, but by climb of dislocations at grain boundaries (see, for example, Mukherjee 1975). The resulting rate equation is

$$\dot{\varepsilon} \sim \frac{D^* \mu \Omega}{kTd^2} \left(\frac{\sigma}{\mu}\right)^2 \tag{10.35}$$

Grain-boundary sliding superplasticity is sometimes called 'structural' superplasticity to distinguish it from *transformation superplasticity,* which denotes the softening of a material while undergoing a phase change. Transformation superplasticity has been observed in several materials, and is manifested as an increase in strain rate (at constant stress) or a decrease in flow stress (at constant strain rate) occurring during the transformation. Consequently, if a polycrystal is repeatedly cycled through the transformation while subject to a deviatoric stress, very large strains can be attained.

The mechanism of transformation superplasticity is not clear, but is probably related to the presence of internal stresses set up by the transformation. Shear stresses large enough for plastic yielding may occur even under hydrostatic pressure in heterogeneous media, as a result of interactions between grains of different orientations and elastic moduli (so-called 'Hertzian' stresses – see Schloessin 1978). A polycrystalline material with anisotropic thermal expansion under thermal cycling softens because of internal stresses between grains (*Cottrell creep*). Something similar probably happens during phase changes. Poirier (1982) proposed a model where the superplastic strain rate is produced by additional dislocations originated by the strain associated with the transformation. Whatever the details, transformation superplasticity could be a relevant flow mechanism in the mantle transition zone (cf. Section 12.3).

As a note to the preceding discussion on creep mechanisms, two points must be made clear. First, we have consistently disregarded the presence of a *threshold stress* below which creep does not occur (for instance, dislocations cannot multiply below a critical stress) because at high temperature this stress is very small, of the order of 1 MPa or less. Secondly, when *grain size* appears in the rate equations it is considered to be an independent variable (cf. Section 10.6 for a discussion of stress-dependent grain size); even so, the assumption of uniform grain size is only an approximation for polycrystals.

We have written the rate equations as dimensionally correct proportionality statements. To give their full tensor form, invariants of strain rate and stress must be introduced, together with geometrical proportionality factors. Stress and strain rate can be interpreted as the square roots of the second invariants of the respective deviatoric tensors (see Eqns 4.20). One need consider only deviators, as hydrostatic stresses do not cause flow. Moreover, because of incompressibility, $\dot{\varepsilon}'_{ij} = \dot{\varepsilon}_{ij}$ (the prime denotes the deviatoric component). The Newtonian diffusion creep law then becomes

$$\dot{\varepsilon}_{ij} = \alpha \, \frac{D^* \mu \Omega}{kTd^2} \left(\frac{\sigma'_{ij}}{\mu} \right) \tag{10.36}$$

where α is a dimensionless factor. The generalized power-law equation may be written as (compare with Eqns 4.24 & 25, and with the arguments presented in Section 4.4)

$$\dot{\varepsilon}_{ij} = \frac{3^{(n+1)/2}}{2} \, A' \, \frac{D_{SD} \mu b}{kT} \left(\frac{\sigma'_E}{\mu} \right)^{n-1} \left(\frac{\sigma'_{ij}}{\mu} \right) \tag{10.37}$$

where A' is a dimensionless factor, and $\sigma'_E = (\frac{1}{2} \sigma'_{ij} \sigma'_{ij})^{1/2}$. Similarly, one could write the equation for creep with stress-dependent activation enthalpy, expressing the latter as $H(\sigma'_E)$.

The numerical parameters α and A' in the above equations have different origins. As diffusion creep is seldom observed in rocks, the value of α is derived on the basis of flow geometry (itself depending on assumptions about grain shape). Usually, $\alpha = 21$ for Nabarro–Herring and Coble creep, and $\alpha = 105$ for Ashby–Verrall creep. These values must be multiplied by two when – as is often the case – $\dot{\gamma}_{ij} = 2\dot{\varepsilon}_{ij}$ is used for the shear strain rate (cf. Eqn 4.9). On the other hand, it is better to consider A' as an unknown to be inferred from the fit of the power-law equation to experimental data; Equation 10.37 has been written for A' determined in uniaxial tests.

The various creep mechanisms are interrelated in different ways. *Concurrent* creep processes operate simultaneously and their strain rates are additive; *alternative* processes do not. For instance, glide-controlled and climb-controlled creep are alternative: either one operates. On the other hand, diffusion creep and dislocation creep (either glide- or climb-controlled) are concurrent: the total strain rate is the sum of the individual strain rates, whose relative importance varies with temperature, stress, and grain size. These considerations will be further developed in Chapter 11, when dealing with deformation maps.

10.4 Creep of rocks and minerals

When one moves from the idealized models reviewed in the previous section to the complexities of natural materials, many additional – and often poorly understood – factors come to the fore. The determination of the steady-state flow properties of rocks and minerals involves not only creep experiments, but also observation of the microstructures associated with flow in both naturally and experimentally deformed specimens. The literature is vast: comprehensive reviews may be found in Nicolas & Poirier (1976) and Kirby (1983). The reader is referred to these two sources for original references.

There are two complementary approaches to the study of creep in the laboratory: the experiments may be performed on a single crystal (natural or artificial), or on polycrystals (monomineralic or polymineralic). In both cases, the temperature is high (typically, $T \geqslant 1300$ K for ultrabasic materials), in order to obtain strain rates $\dot{\varepsilon} \geqslant 10^{-8}$ s^{-1}. Single-crystal experiments are usually performed at room pressure, while experiments on natural aggregates are carried out under confining pressure of the order of 1 GPa. The advantages of using artificial, high-purity single crystals lie mainly in the control over the variables (composition, chemical environment, and crystal orientation), which may allow a closer look at the mechanisms of creep. On the other hand, a polymineralic aggregate is not so precisely characterized; but its laboratory deformation may be more relevant to natural conditions. Empirical creep laws for single crystals and polycrystals are in good agreement, showing that the underlying mechanisms are similar.

The stress difference $(\sigma_1 - \sigma_3)$ of geodynamic interest is in the range 1–100 MPa. At these stresses, and at high temperatures, practically all geological materials under laboratory conditions flow by power-law creep ($2 \leqslant n \leqslant 5$). At higher stresses (the transition depending on the material) the stress-dependence becomes exponential. Stresses lower than 5 MPa are seldom achieved experimentally; there is some evidence that some very fine-grained materials flow linearly ($n \simeq 1$) at the lower end of the laboratory stress range.

Deformation mechanisms and creep laws for olivine and olivine-rich rocks are discussed in this section, as an example of the complexities of the actual rheology when compared with the model rheology. The empirical creep laws of other materials that are probable constituents of the crust are also reviewed. The importance of olivine stems from the fact that it is the main constituent of the upper mantle (cf. Section 6.6), and consequently its properties control the rheology of the lower lithosphere and mantle down to a depth of 400 km. Olivine (Mg, Fe)$_2$SiO$_4$ is a nesosilicate of orthorhombic symmetry. The most widely studied olivine is forsterite (Fo), the magnesium end-member, to which natural olivine is close (Fo$_{90}$). The oxygen ions are arranged in a quasi-hexagonal close-packed structure, given an orthorhombic symmetry by the cation filling (cf. Poirier 1985). The unit cell lattice parameters are [100] 4.76 Å,

[010] 10.21 Å, and [001] 5.99 Å. Since the energy of a dislocation is proportional to b^2, and the only dense plane on which slip does not involve the breaking of Si–O bonds in SiO$_4$ tetrahedra is (010), slip is to be expected in the (010)[001] and (010)[100] systems. Study of experimentally deformed olivines has shown that the slip system (010)[100] is, indeed, fairly common; but so is $\{0kl\}$[100], which is also widespread in naturally deformed olivines (cf. Nicolas & Poirier 1976).

Creep data for stress differences in the range 10–500 MPa can be fitted by a Dorn equation (Eqn 10.7) with stress exponent $n = 3.5 \pm 0.6$, creep activation energy $E = 533 \pm 60$ kJ mol^{-1}, and creep activation volume $V = (17 \pm 4) \times 10^{-6}$ m^3 mol^{-1} (Kirby 1983). Figure 10.10 presents selected data on the creep of olivine, from pure forsterite to natural dunites and peridotites. For $\sigma \leqslant 100$ MPa a stress exponent $n = 3$ fits the

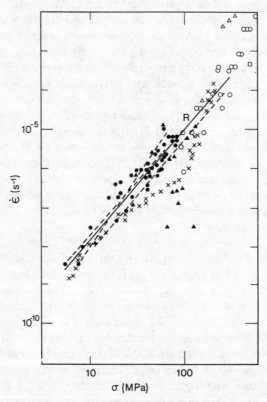

Figure 10.10 Experimental stress–strain rate diagram for olivine (data from various sources). Continuous line (R) gives the fit to the data provided by the theoretical power-law creep equation (from Ranalli 1982).

data well (Ranalli 1982), and even a value $n = 2$ cannot be rejected (Minster & Anderson 1981). In view of the considerations developed in Section 10.3, we write the creep equation as

$$\dot{\varepsilon} = A \, \frac{D_0 \mu b}{kT} \left(\frac{\sigma}{\mu} \right)^3 \exp\left(-\frac{E + pV}{RT} \right) \qquad (10.38)$$

for shear components of stress and strain rate [consequently, $A = (3^{(n+1)/2}/2)A'$; compare with Eqn 10.37]. The coefficient A is dimensionless, contrary to A_D in the Dorn equation. The line R in Figure 10.10 gives the fit of Equation 10.38 to the experimental data, obtained by using the values listed in part (b) of Table 10.1. (The table gives a complete review of the creep and diffusion parameters of olivine, to be used also in Chs 11 & 12.)

A comparison of creep and diffusion parameters shows that silicon is the slowest-moving species at low pressure, and should therefore be rate-controlling if creep under laboratory conditions is diffusion-controlled via dislocation climb. The reason for the slow diffusivity of silicon, despite its small size when compared with oxygen, probably lies in its strong covalent bonding within SiO_4 tetrahedra. However, the activation energy and volume for creep are larger than for self-diffusion. Consequently the relation between diffusion and creep is not as straightforward as was assumed in the derivation of climb-controlled models. The simplest explanation of the discrepancy is afforded by assuming conditions of jog undersaturation, so that the activation enthalpy for creep is the sum of a self-diffusion term and a term related to jog formation, i.e.

$$H = H^* + H_J \qquad (10.39)$$

where H and H^* are the activation enthalpies for creep and self-diffusion, respectively (a notation that we shall use consistently when dealing with creep in the Earth). At zero pressure Equation 10.39 is equivalent to

$$E = E^* + E_J \qquad (10.40)$$

For $E = 540 \, \text{kJ mol}^{-1}$ ($\simeq 130 \, \text{kcal mol}^{-1}$) and $E^* = 376 \, \text{kJ mol}^{-1}$ ($\simeq 90 \, \text{kcal mol}^{-1}$) the jog formation energy is estimated to be $E_J = 164 \, \text{kJ mol}^{-1}$ ($\simeq 40 \, \text{kcal mol}^{-1}$), a not-unrealistic value.

The activation volume affects strongly the pressure-dependence of creep, and is itself pressure-dependent (cf. Section 10.7). Although educated guesses can be made about its variation with depth, it is one of the major sources of uncertainty − together with the effects of phase transitions − in estimates of the creep properties of the mantle.

Table 10.1 Creep and diffusion parameters for olivine.

Parameter		Comments	Reference
(a) *diffusion creep* (O = oxygen, Si = silicon)			
numerical factor	$\alpha = 21$	inferred from creep model	Stocker & Ashby 1973
numerical factor, grain-boundary sliding	$\alpha = 105$	inferred from creep model	Ashby & Verrall 1978
atomic volume, O	$\Omega(O) = 1.23 \times 10^{-29} \text{ m}^3$	$\frac{1}{4}$ of molecular volume	Stocker & Ashby 1973
atomic volume, Si	$\Omega(Si) = 4.92 \times 10^{-29} \text{ m}^3$	molecular volume	Stocker & Ashby 1973
pre-exponential diffusivity, O	$D_0^*(O) = 3.5 \times 10^{-7} \text{ m}^2\text{s}^{-1}$	diffusion measurements	Jaoul et al. 1980
pre-exponential diffusivity, Si	$D_0^*(Si) = 1.5 \times 10^{-10} \text{ m}^2\text{s}^{-1}$	diffusion measurements	Jaoul et al. 1980
activation energy, O (lattice)	$E^*(O) = 3.76 \times 10^5 \text{ J mol}^{-1}$	diffusion measurements	Jaoul et al. 1980
activation energy, Si (lattice)	$E^*(Si) = 3.76 \times 10^5 \text{ J mol}^{-1}$	diffusion measurements	Jaoul et al. 1980
activation energy, grain-boundary diffusion	$\frac{1}{2}E^* < E^*_{GB} < \frac{2}{3}E^*$	estimated	
activation volume, O	$V^*(O) = 11 \times 10^{-6} \text{ m}^3 \text{ mol}^{-1}$	estimated	Minster & Anderson 1981
activation volume, Si	$V^*(Si) = 4 \times 10^{-6} \text{ m}^3 \text{ mol}^{-1}$	estimated	Minster & Anderson 1981
(b) *power-law creep*			
numerical factor	$A' = 4.5$	creep experiments (for elongation under $\sigma_1 - \sigma_3$) average for olivine slip systems	Ashby & Verrall 1978
Burgers vector	$b = 6 \times 10^{-10} \text{ m}$		Stocker & Ashby 1973
pre-exponential diffusivity	$D_0 = 0.1 \text{ m}^2\text{s}^{-1}$	$A'D_0 = 0.45$ fits creep data	Ashby & Verrall 1978
activation energy	$E = 5.40 \times 10^5 \text{ J mol}^{-1}$	fits Arrhenius plot	cf. discussion in Kirby 1983
activation volume	$V = 15 \times 10^{-6} \text{ m}^3 \text{ mol}^{-1}$	fits Arrhenius plot	cf. discussion in Kirby 1983

Another interesting observation is that the strain rate is independent of the chemical environment in pure forsterite, but depends on it in natural olivine (for instance, $\dot{\varepsilon} \propto P_O^{1/6}$, where P_O is the partial pressure of oxygen – cf. Jaoul *et al.* 1980). This dependence shows that flow properties are related to point-defect concentration, which is compatible with diffusion playing some part in creep.

A detailed examination of dislocation microstructures (Gueguen 1979) reveals that pure edge segments are long and straight along the [001] direction, and much more common than screw segments. It is therefore unlikely that creep is cross-slip controlled. Creep could be either glide-controlled through kink formation, or climb-controlled through jog formation (see also the detailed discussion by Hobbs 1983). At the present stage the most likely conclusion is that creep in olivine is controlled by the diffusion-related climb of dislocations (possibly into cell walls) under conditions of jog undersaturation.

The creep properties of other minerals and rocks are not as well known as those of olivine, and therefore their detailed description would be subject to even higher uncertainties. From the empirical viewpoint, creep of practically all rocks that are probable constituents of the crust can be described by a Dorn equation written in the form

$$\dot{\varepsilon} = A_D \sigma^n \exp(-E/RT) \tag{10.41}$$

where σ is the stress difference, and the parameter A_D (of dimensions $[(ml^{-1}t^{-2})^{-n}t^{-1}]$) is assumed to be constant, though in practice it is a weak function of temperature. The pressure-dependence of the exponential term is not included in Equation 10.41, for the very good reason that it is not generally known. However, this limitation is not very severe, as $p \leqslant 1\,\text{GPa}$ in the crust.

Table 10.2 lists values for A_D, n, and E for several rocks of crustal composition, and for rocksalt as an example of an extremely ductile material, taken from the compilation by Kirby (1983). The term 'wet' refers to samples that have not been dried before the experiments, and which therefore contain variable amounts of structural water.

The parameter values listed in Table 10.2 are only indicative, and error limits can be qualitatively assessed to be large. Nevertheless, some order-of-magnitude conclusions are possible. The materials may be subdivided into four groups, according to their 'softness' under crustal conditions, which is in turn reflected in their creep activation energy. Rocksalt is very soft and flows even under upper crustal conditions (as shown by the common occurrence of evaporite-lubricated *décollements* and salt domes). Rocks rich in quartz, such as quartzite and granite, have activation energies of the order of $100–150\,\text{kJ}\,\text{mol}^{-1}$, and are softer than

Table 10.2 Creep parameters for crustal materials (from Ranalli & Murphy 1986; data modified from Kirby 1983).

Material	A $(GPa^{-n}s^{-1})$	n	E $(kJ\,mol^{-1})$
rocksalt	5.0×10^{16}	5.3	102
granite	5.0	3.2	123
granite (wet)	10^2	1.9	137
quartzite	10^2	2.4	156
quartzite (wet)	2.0×10^3	2.3	154
albite rock	1.3×10^6	3.9	234
anorthosite	1.3×10^6	3.2	238
quartz diorite	2.0×10^4	2.4	219
diabase	3.2×10^6	3.4	260

plagioclase-rich rocks (both intermediate and basic), whose activation energies cluster around 200–$250\,kJ\,mol^{-1}$. Ultrabasic rocks such as dunites and peridotites are the hardest under crustal conditions.

The variations of *creep strength* (i.e. flow stress at a given strain rate) with depth for several crustal rocks are shown in Figures 10.11 and

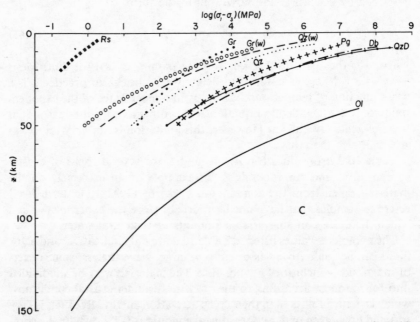

Figure 10.11 Variations of creep strength with depth for several crustal rocks and a 'cold' geotherm. Rs, rocksalt; Gr, granite; Qz, quartzite; Pg, plagioclase-rich rock; QzD, quartz diorite; Ol, peridotite; (w) following the rock symbol denotes wet conditions (from Ranalli & Murphy 1986).

10.12. They have been calculated by solving Equation 10.41 for stress at a fixed strain rate $\dot{\varepsilon} = 10^{-14}\,\mathrm{s}^{-1}$, and either 'cold' (Fig. 10.11) or 'hot' (Fig. 10.12) geotherms. These are defined by assuming the lower boundary of the lithosphere to be the $T = 1500\,\mathrm{K}$ isotherm, and a linear temperature gradient between the upper and lower boundaries. The latter is placed at $z = 150\,\mathrm{km}$ in the 'cold' case, and at $z = 50\,\mathrm{km}$ in the 'hot' case.

Two qualitative results, which do not depend on the degree of accuracy of the creep parameters, are apparent from Figures 10.11 and 10.12. First, the strength as a function of depth (below the brittle–ductile transition) is strongly dependent on temperature; the rheological properties of the lithosphere can therefore be expected to be quite different in different provinces. Secondly, since the lithosphere is compositionally layered, its strength structure is likely to show systematic vertical variations, which could be determinant in localizing layers of tectonic decoupling (see also Section 12.1).

The strength estimated from Equation 10.41 is likely to be an upper limit, as at low stresses and very small grain sizes the flow mechanism may change to diffusion creep with grain boundary sliding, which involves a softening of the material. Fine-grained mylonite zones are

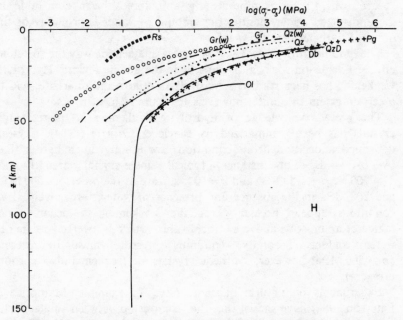

Figure 10.12 Variations of creep strength with depth for crustal rocks and a 'hot' geotherm. Symbols as in Figure 10.11 (from Ranalli & Murphy 1986).

295

therefore softer than coarse-grained rocks. Other factors which can enhance ductility are pressure solution and shear heating. Although first-order models are useful, more-detailed studies must try to incorporate the complexities of the geology.

10.5 Hydrolytic weakening

In the review of creep parameters for crustal materials (Table 10.2) a distinction has been made between 'wet' and 'dry' granite and quartzite. The strength of wet rocks is less than the strength of dry rocks under most crustal conditions; this factor could have important geodynamic consequences.

The presence of water in the upper crust affects the brittle properties of rocks through the action of pore pressure (Section 5.4); this action is widespread, judging from the occurrence of faulting under relatively low stress differences (Section 8.6). In the ductile regime (say, below mid-crustal depths) the effect of water on the rheology can be discussed in terms of changes in the creep strength in a hydrous environment. (The term 'water' is used in this context to denote water-related defects). Hydrous environments are likely to be present in the crust, particularly at levels where hydrous minerals (e.g. amphiboles) become unstable. In the undepleted upper mantle, the amount of water is probably of the order of 0.01 per cent by weight.

The term *hydrolytic weakening* denotes in a generic way the softening effect of water on rocks and minerals in the ductile regime. Hydrolytic weakening has been studied extensively in quartz and quartzite, and to a certain extent in dunite; however, it occurs in most rocks.

The present knowledge of hydrolytic weakening in quartz single crystals has been summarized by Blacic & Christie (1984). Crystals deformed under anhydrous conditions are usually much harder than 'wet' crystals. For instance, typical single-crystal strengths for $T \geqslant 700$ K, $p \simeq 1.5$ GPa, and $\dot\varepsilon \simeq 10^{-5}\,\mathrm{s}^{-1}$ are of the order of 2000 MPa and 200 MPa in the absence and presence of water, respectively. Two conditions appear necessary for the weakening to occur. First, temperature must be above a critical value, which is inversely related to water content. Secondly, confining pressure must be present ($p \geqslant 500$ MPa; however, synthetic crystals are weakened also at room pressure).

Experiments on natural quartzite (see, for example, Mainprice & Paterson 1984) have shown that the relationship between single-crystal and polycrystal weakening is not straightforward. At $p \simeq 300$ MPa, less weakening is observed in the rock than is expected on the basis of synthetic single-crystal behaviour. This is possibly due to the fact that only a small fraction of the total water present is contained in the grains of a polycrystal; the rest may be segregated as films in bubbles or a grain-boundary phase, or both.

A similarly complex picture emerges from studies of the hydrolytic weakening of dunite at $T \simeq 1500$ K and $p \simeq 300$ MPa (Chopra & Paterson 1981, 1984). Dry rocks show no grain-size dependence of their creep properties. Wet samples are softer and show a degree of weakening which depends on grain size. Fine-grained aggregates (average grain size $d \simeq 100$ μm) are weakened more than coarse-grained aggregates ($d \simeq 900$ μm). This seems to imply that the weakening effect increases with grain-boundary area, and is therefore concentrated at grain boundaries.

There is no unambiguous model yet for hydrolytic weakening. The observed (p, T)-dependence of weakening in single crystals is consistent with the variations of water solubility and diffusivity in quartz. A widely quoted mechanism involves the hydrolysis of Si–O bridges, so that the strong Si–O–Si bonds are replaced by the weaker hydrogen bond, Si–OH - - - HO–Si, thereby enhancing the mobility of dislocations. The presence of water may also affect recovery and recrystallization (the latter appears to occur preferentially near water sources such as dehydrating minerals); or, through variations in oxygen partial pressure, it may alter point-defect concentration. Grain-boundary processes may also be active, as shown by the grain-size dependence of hydrolytic weakening observed in natural polycrystals.

In conclusion, it is not possible to extend our imperfect knowledge of hydrolytic weakening in the laboratory to *in situ* conditions, except by saying that the presence of water in particular horizons in the lower crust and upper mantle is likely to decrease the creep strength of the material (see also Tullis *et al.* 1979). Any more-precise statement must await a clarification of the weakening mechanisms.

10.6 Dynamic recrystallization and scaling laws

A polycrystal contains a network of grain and subgrain boundaries whose configuration can evolve during creep and affect the deformation behaviour. Characteristic lengths associated with the configuration are related to flow stress by empirical scaling laws which have been used in geodynamics in the estimation of tectonic stresses in the lithosphere and mantle.

A *grain boundary* between different crystals (or *interface* if the crystals are of different mineral species) is a surface across which the two lattices are misoriented by an angle larger than about $10°$ (*high-angle* boundary). A *subgrain boundary* (*low-angle* boundary) is one across which the misorientation is less. The structural boundary width, i.e. the disordered region between the lattices, is of the order of the interatomic distance. Boundaries and sub-boundaries may be viewed as arrays of dislocations; a discussion of their properties and mobility can be found in Poirier (1985).

The network of grain and subgrain boundaries is a dynamic configuration whose changes may result in the complete recrystallization of the material. Recrystallization can occur both during and after creep, and is an important softening process.

Static recrystallization or *annealing* occurs at high temperature under stress-free conditions in previously strained polycrystals, provided that the strain exceeds a critical value. It consists of a migration of grain boundaries whereby relatively dislocation-free grains grow by consuming grains with higher dislocation densities. The main driving force of the process arises from the difference in the Gibbs free energy per unit volume in contiguous grains with different dislocation densities. Annealing lowers the strain energy and softens the material by decreasing the total density of dislocations.

It is difficult to assess how important annealing is in the Earth. If a material is tectonically deformed at high temperature and then emplaced relatively rapidly in the upper lithosphere, then the microstructure will be 'frozen in' and annealing will be prevented by the low temperature. At deeper levels static recrystallization is certainly a factor to be considered, although a stress-free situation is probably unrealistic.

More relevant to geodynamics is *dynamic* (or *syntectonic*) *recrystallization*, which may take place during the high-temperature deformation of a polycrystal. Dynamic recrystallization is observed in metals and minerals undergoing power-law creep, and is related to the presence and mobility of dislocations within the grains. The softening effect of dynamic recrystallization is evidenced in one or more transient episodes of accelerated flow on the creep curve, or equivalently in fluctuations of flow stress at constant strain rate.

The most common form of microstructural change during dislocation creep is *polygonization*, i.e. the formation of slightly misoriented subgrains separated by dislocation walls (see Fig. 10.6). The arrangement of dislocations in subgrain walls represents a lower-energy configuration accommodating deformation gradients induced by the different orientation of slip systems in neighbouring grains. Complete recrystallization can be achieved in two ways.

(*a*) As a natural consequence of polygonization, by progressive rotation of subgrains until the misorientation is so large ($\geq 10°$) that the subgrains become in effect new grains – this is *subgrain rotation (SGR) recrystallization.*

(*b*) By migration of grain boundaries driven by energy differences related to variations in dislocation density; exactly as in annealing, but occurring during creep – this is *grain boundary migration (GBM) recrystallization.*

SGR and GBM recrystallizations are not mutually exclusive. SGR recrystallization is usually the first to begin. Within the stress and temperature range where dynamic recrystallization occurs (which varies according to material but roughly coincides with that of power-law creep), SGR recrystallization is predominant at low σ and T, while GBM recrystallization takes over at high σ and T (cf. Poirier 1985 for a summary of observations).

Dynamic recrystallization and recovery produce similar softening effects, since both counteract strain hardening; however, their relationship varies with the mechanism of recrystallization. While subgrain rotation is a possible (but not necessary) consequence of polygonization, grain boundary migration is effectively in competition with recovery, in the sense that it occurs if the recovery mechanisms are not efficient, so that sufficient strain energy accumulates to drive crystallization.

Although grain and subgrain boundaries form an evolving network, the characteristic grain or subgrain size is, to a first approximation, a function of stress only. This fact allows the establishment of *empirical scaling laws* relating microstructure to applied stress. The scaling laws for dislocation density ρ, subgrain size d^*, and recrystallized grain size d can be written as

$$\rho b^2 = K_1 (\sigma/\mu)^2 \tag{10.42}$$

$$d^*/b = K_2 (\sigma/\mu)^{-r^*} \tag{10.43}$$

$$d/b = K_3 (\sigma/\mu)^{-r} \tag{10.44}$$

where K_1, K_2, and K_3 are dimensionless proportionality constants depending on the material. The dislocation density in Equation 10.42 refers to the density within subgrains if dislocation walls are present. In Equation 10.43, $r^* \simeq 1$ in most cases. The value of r in Equation 10.44 depends on the recrystallization mechanism. For SGR recrystallization $r \simeq 1$, as it should be; for GBM recrystallization, $r \simeq 1.2$. While the average value of r^* is easy to explain (a characteristic length of the microstructure is inversely proportional to the stress), the value of r is not. The identification of recrystallized grains and the determination of their 'average size'

299

are not simple problems, and these uncertainties may be reflected in the scaling law.

In principle Equations 10.42–10.44 may be used to estimate the flow stress in natural materials from measurements of ρ, d^*, and d. However, the operation is fraught with uncertainties. The unreliability of dislocation density as a stress indicator was mentioned in Section 9.6. The subgrain size is similarly ill-defined, as there appears to be a hierarchy of subgrains, and therefore d^* decreases with increasing resolution (for instance, observation by TEM results in significantly smaller values than optical examination). On the other hand, recrystallized grain size appears to offer some consistency, and is widely used to estimate stress.

A terminology has grown out of the practice. The recrystallized grain or *neoblast* size is a *geological piezometer* or *palaeopiezometer*, i.e. an indicator of the *palaeostress* related to the recrystallization (assumed to be the stress during the last creep episode). To use neoblast size as a palaeopiezometer, it is first necessary to calibrate the scaling law on materials deformed in the laboratory under known applied stress (see, for example, Twiss 1977, Mercier 1980, Ross *et al*. 1980). For instance, neoblast size in dry dunite and peridotite scales with stress approximately as

$$\sigma \simeq 74.5 \; d^{-1} \text{ for SGR recrystallization}$$

$$\sigma \simeq 4.81 \; d^{-0.8} \text{ for GBM recrystallization}$$

where σ is in GPa and d in μm. Figure 10.13 gives the $\sigma(d)$ relation for a few lithospheric materials.

Estimates of tectonic stresses by neoblast palaeopiezometry cluster in the interval 20–200 MPa for crustal shear zones (as determined predominantly from dynamically recrystallized quartzites), and around 10–20 MPa for the lithospheric upper mantle (from peridotites). These values are not unrealistic, in the sense that they are within the bounds estimated by other means (cf., for instance, Section 8.5). However, any further refinement may prove to be very difficult, for various reasons. First, there is the observational problem of correctly identifying neoblasts and of estimating their 'average' size from thin sections. Secondly, neoblast size is sensitive to the recrystallization mechanism, and often this is not known. Thirdly, other factors – such as the presence of a second phase or water – may hinder or enhance recrystallization. Consequently, interpretation of results should not be stretched beyond their intrinsic accuracy, which is most probably about one order of magnitude.

Grain size is not always determined by stress whenever power-law

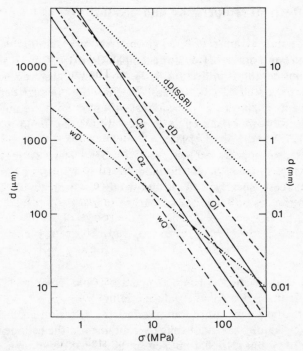

Figure 10.13 Dynamically recrystallized (neoblast) size as a function of stress for various materials (wD, wet dunite; dD, dry dunite; Qz, quartz; wQ, wet quartzite; Ca, calcite; Ol, olivine). SGR denotes subgrain rotation (all the other cases are grain boundary migration). Compiled from various sources.

creep occurs. High temperatures and strains are necessary. Often a sample is only partly recrystallized: it would then be wrong to consider the grain size of the whole sample as a function of stress. When the deformation is effected by diffusion creep, a dislocation-related mechanism such as dynamic recrystallization is unlikely to play a significant role. Therefore grain size can be regarded as an independent variable in diffusion creep.

Theoretically, dynamic recrystallization may change the creep mechanism. If a coarse-grained material is subject to a sufficiently large stress, then it is conceivable that the average grain size will be reduced enough for diffusion creep (whose importance increases as d^{-2} or d^{-3}) to become predominant. However, there is no evidence for such a transition, either in the microstructure or in the activation quantities (Zeuch 1983; cf. also Section 11.3). The effect of dynamic recrystallization consists mainly of softening: recrystallization and recovery are two ways (sometimes alternative, sometimes concurrent) in which strain hardening is counterbalanced by softening and large deformations become possible.

301

10.7 Effects of temperature and pressure

The steady-state strain rate for a given stress and material depends on temperature and pressure (see, for example, Eqns 10.36 & 37, summarizing the considerations in Section 10.3). The temperature dependence of creep has been studied more extensively than the pressure dependence, because of its engineering applications; however, both are important in the Earth. Reviews of the effect of pressure can be found in Ashby & Verrall (1978) and Poirier (1985).

Temperature appears explicitly in the creep equations and, together with pressure, affects the elastic moduli. To a first approximation the moduli increase linearly with p, and decrease linearly with T. For instance, in the case of rigidity one can write

$$\mu = \mu_0 \left[1 - \left(\frac{T - T_0}{\mu_0} \right) \frac{\partial \mu}{\partial T} + \left(\frac{p - p_0}{\mu_0} \right) \frac{\partial \mu}{\partial p} \right]$$

where μ_0 is the value at room temperature and pressure (T_0, p_0). A similar relation holds for the bulk modulus.

Pressure and temperature jointly affect the diffusion coefficient. The dependence is not only that explicitly contained in the exponential term (see Eqn 9.15), but also that implicit in the activation volume. Since the latter is related to the formation of a vacancy, it can be considered as a cavity in a continuous medium. The change in volume of such a cavity with temperature can be neglected when compared with the change with pressure, which can be expressed in terms of the pressure-dependence of the elastic moduli (O'Connell 1977):

$$V = V_0 [1 + (p - p_0)(k_0'/k_0)]^{(-9/4k_0')} \tag{10.45}$$

where k_0 and k_0' are the bulk modulus and its pressure derivative, respectively, at p_0. Since the bulk modulus increases with pressure, V is a decreasing function of p. The variation as expressed by Equation 10.45 turns out to be not very important in the upper mantle, where p is only a small fraction of k_0 for olivine ($k_0 \simeq 1.2 \times 10^{11}$ Pa, $p \leqslant 1.2 \times 10^{10}$ Pa at depths less than 400 km). However, in the lower mantle p reaches values comparable with k_0, and the pressure-dependence of V has important consequences (cf. Section 12.4).

The previous considerations strictly hold only for diffusion-related creep (diffusion creep or diffusion-controlled dislocation creep). However, the creep activation parameters, at least in the case of olivine, are higher than the self-diffusion parameters (see Section 10.4). In the case of activation volume this implies that at least a part of it is not related to vacancy formation and migration. Still – and until evidence to the contrary is found – a pressure-dependence of the type given by Equation 10.45 is justifiable, and leads to self-consistent results when applied to the lower mantle.

The activation volume in a thermodynamic sense represents the pressure-dependence of the Gibbs free energy for creep; that is,

$$V = \frac{\partial G}{\partial p}\bigg|_T$$

The previous considerations do not therefore apply to cases where G is stress-dependent (e.g. glide-controlled creep or cross-slip creep). However, usually $G \propto \sigma\Omega$ or σb^3, and consequently it is possible, at least in principle, to express $V(p)$ from some continuum elastic model of the process (see Section 12.4). In all cases the physical basis of the pressure effect stems from the anharmonicity of the elastic moduli (cf. Poirier 1985).

The pressure-dependence of lattice parameters and volumes (again, their temperature-dependence is negligible in comparison) can be calculated from linear elasticity theory (cf., for example, Ashby & Verrall 1978). For atomic volume and Burgers vector one has

$$\Omega = \Omega_0 \exp\left(-\frac{p - p_0}{k_0}\right)$$

$$b = b_0 \exp\left(-\frac{p - p_0}{3k_0}\right)$$

where subscripted variables, as usual, denote values in the absence of confining pressure.

Pressure and temperature can affect creep in other ways as well (i.e. through their action on dislocation multiplication and density, Peierls stress, and so forth). The discussion above includes the first-order processes likely to be significant in determining (p, T)-related variations in rheology with depth in the mantle. The effects of pressure and temperature are of opposite sign: a material becomes softer with increasing T and harder with increasing p. The result of their combined action, as we shall see in Sections 12.3 and 12.4, is to reduce the variations of rheology with depth and to contain the rheological inhomogeneity of the mantle within limits compatible with mantle-wide convection.

11 Deformation maps and isomechanical groups

Problems of extrapolation are central to the application of rheology to geodynamics. The extrapolation may concern a given material at conditions unattainable in the laboratory (as in the prediction of the rheological behaviour of olivine in the upper mantle) or different materials under scaled conditions (as in the estimation of the properties of lower-mantle phases from the study of analogue materials).

Two techniques that are very useful in this context are the construction of deformation maps and the classification into isomechanical groups. Deformation maps give a unified view of the rheology of a material under varying conditions of stress, strain rate, grain size, temperature, and pressure. Normalization of rheological variables results in the classification of materials into several isomechanical groups, characterized by similar crystal structure, chemical bonding, and dimensionless rheological parameters.

11.1 Construction and use of deformation maps

Several steady-state creep mechanisms (some concurrent, some alternative – cf. Section 10.3) may operate in a given material. In order to predict the rheological behaviour under varying conditions of strain rate, stress, temperature, pressure, and grain size, it is necessary to delimit the (σ, T, p, d)-fields in which a mechanism is predominant, i.e. contributes the largest part of the total observed strain rate. Laboratory experiments do not provide sufficient information, as they sample only a limited range of the variables; therefore *creep fields* can be outlined with a satisfactory degree of confidence only by combining theoretical rate equations and experimental results. *Deformation-mechanism maps* (or, briefly, *deformation maps*) are diagrams which display the fields of operation of the

304

various creep mechanisms and the relations between the variables. They are obviously very important when studying the rheology of the Earth, where extrapolation is often a necessity.

Assuming that there are n possible creep mechanisms, each denoted by the subscript i ($i = 1, 2, \ldots, n$) the strain rate resulting from the ith mechanism has the general form

$$\dot{\varepsilon}_i = f(\sigma, p, T, d) \tag{11.1}$$

where the grain size d is present in some rate equations, but not in others. (The subscript i is not a tensor index but a sequential index; consequently, no summation is implied anywhere). Stress, temperature, and grain size (the latter considered as an independent variable) are often *normalized* as σ/μ, T/T_m, and d/b, respectively. The pressure-dependence in Equation 11.1 can be neglected when the ratio p/k_0 is low, as in the lithosphere (k_0 is the bulk modulus at room temperature and pressure); or both T and p can be given as function of depth z. In either case, the strain rate $\dot{\varepsilon}_i$ becomes a function of three independent variables.

To illustrate the construction of deformation maps, we consider a three-dimensional space (e.g. with co-ordinates σ, T, and d, or their normalized equivalents) in which we must delimit the various creep fields. Deformation maps are simply two-dimensional sections of this three-dimensional space.

Given a set of n creep equations (Eqn 11.1), and once it has been determined which mechanisms are concurrent and which are not (cf. Section 10.3), the procedure for delimiting the domains in (σ, T, d)-space where a given creep mechanism is predominant is, in principle, very simple. First, experimental results are used to obtain values for the material parameters entering the rate equations. Then the various $\dot{\varepsilon}_i$ are calculated and compared in order to determine the largest. A *field boundary* is a surface in (σ, T, d)-space on which the contributions of two creep mechanisms (each predominant on one side of the boundary) are equal. Boundaries are therefore obtained by equating pairs of creep equations, subject to experimental control.

If cataclasis and mechanisms with stress-dependent activation enthalpies are excluded, the most general explicit form of Equation 11.1 is (for shear components)

$$\dot{\varepsilon}_i = \frac{A_i D_i \mu b}{kT} \left(\frac{b}{d}\right)^{m_i} \left(\frac{\sigma}{\mu}\right)^{n_i} \tag{11.2}$$

where A_i is a dimensionless parameter and D_i is a diffusion coefficient; m_i and n_i are the grain size exponent and the stress exponent, respectively. With a proper choice of parameters Equation 11.2 covers all the varieties of power-law creep ($n_i > 1$, $m_i = 0$) and Newtonian creep ($n_i = 1$).

It follows from Equation 11.2 that, if two creep mechanisms are denoted by the subscripts i and j, then the boundary between them is given by

$$\frac{\sigma}{\mu} = \left(\frac{A_j D_j}{A_i D_i}\right)^{1/\Delta n} \left(\frac{d}{b}\right)^{\Delta m/\Delta n}$$

where $\Delta n = n_i - n_j$, $\Delta m = m_i - m_j$. Writing the diffusion coefficient as

$$D_i = D_{0i} \exp(-H_i/RT)$$

the previous equation becomes

$$\frac{\sigma}{\mu} = \left(\frac{A_j D_{0j}}{A_i D_{0i}}\right)^{1/\Delta n} \left(\frac{d}{b}\right)^{\Delta m/\Delta n} \exp\left(\frac{\Delta H}{RT \, \Delta n}\right) \tag{11.3}$$

where $\Delta H = H_i - H_j$ is the difference between the activation enthalpies. As $\Delta H = \Delta E + p \, \Delta V$, the exponential term in Equation 11.3 is pressure-independent if $\Delta V = 0$ or the pressure is low.

Constant strain rate and *constant viscosity* surfaces also follow from Equation 11.2. Except near field boundaries the total strain rate in a given field is approximately given by the strain rate contributed by the predominant creep mechanism $\dot{\varepsilon}_i$. The stress necessary to attain a given strain rate is

$$\frac{\sigma}{\mu} = \left(\frac{\dot{\varepsilon}_i kT}{A_i D_i \mu b}\right)^{1/n_i} \left(\frac{d}{b}\right)^{m_i/n_i} \tag{11.4}$$

Similarly, the viscosity in the ith field is approximately

$$\eta_i = \left(\frac{\sigma}{\mu}\right) \frac{\mu}{2\dot{\varepsilon}_i}$$

and consequently constant viscosity surfaces are given by

$$\frac{\sigma}{\mu} = \left(\frac{kT}{2 A_i D_i b \eta_i}\right)^{1/(n_i - 1)} \left(\frac{d}{b}\right)^{m_i/(n_i - 1)} \tag{11.5}$$

Equations 11.3–11.5 give creep field boundaries, constant strain rate, and constant viscosity surfaces, respectively, in (σ, T, d)-space. However, a three-dimensional deformation space is difficult to visualize, so in practice two-dimensional deformation maps are used. These are sections of deformation space parallel to one of the co-ordinate planes, with the third variable fixed. The most common choice results in (σ, T)-deformation maps, although (σ, d)-maps are also very useful. The former have been applied to geological problems by Stocker & Ashby (1973), Ashby & Verrall (1978), and Ranalli (1977, 1980), among others; the importance of the latter in an Earth science context has been emphasized by Ranalli (1982). A thorough discussion of deformation maps for various materials (mainly of engineering and metallurgical interest) can be found in Frost & Ashby (1982).

Deformation maps are only as reliable as the rate equations and the experimental data used in their construction. Nevertheless, and bearing this limitation in mind, they are a powerful tool in rheological studies, as they provide a unified view of the rheology of a material. Relationships between σ, T, d, $\dot{\varepsilon}$, and η can be read directly off the maps, as well as the various creep fields. More importantly in the context of the study of Earth rheology, the maps allow prediction of creep behaviour outside the experimental range: they are therefore very useful in attempts to estimate the rheology of the Earth from the microphysical viewpoint.

11.2 Deformation maps in stress–temperature space

Both (σ, T)- and (σ, d)-maps can be constructed directly from Equations 11.3, 11.4, and 11.5. Since the most convenient choice of co-ordinates is $(T, \log \sigma/\mu, \log d/b)$, we write the equations for field boundaries, constant strain rate, and constant viscosity surfaces as

$$\Delta n \log(\sigma/\mu) = \log B_1 + \Delta m \log(d/b) + (\Delta H/RT)\log e \quad (11.6)$$

$$n_i \log(\sigma/\mu) = \log B_2 + \log \dot{\varepsilon}_i + m_i \log(d/b) + \log T + (H_i/RT)\log e$$
$$(11.7)$$

$$(n_i - 1)\log(\sigma/\mu) = \log B_3 - \log \eta_i + m_i \log(d/b) + \log T + (H_i/RT)\log e$$
$$(11.8)$$

where

$$B_1 = \frac{A_j D_{0j}}{A_i D_{0i}} \qquad B_2 = \frac{k}{A_i D_{0i}\mu b} \qquad B_3 = \frac{k}{A_i D_{0i}b}$$

Equations 11.6–11.8 (which can be immediately written in terms of homologous temperature T/T_m if desired) give the required curves in (σ, T)-space for $d = $ constant, and in (σ, d)-space for $T = $ constant.

With regard to (σ, T)-deformation maps, and focusing attention to the main creep mechanisms (power-law and diffusion creep, by either bulk or grain-boundary diffusion, including superplasticity), some general properties of the creep fields are evident from inspection of the above equations. The diffusion field predominates at lower stresses, the power-law field at higher stresses. The *transition stress* between the two at fixed grain size varies as $\exp(T^{-1})$, and therefore tends to an asymptotic value for increasing T. This asymptotic value is a function of grain size, $\sigma \propto d^{-1}$. Within the diffusion field, grain-boundary diffusion predominates at lower T, bulk diffusion at higher T; a small grain size increases the importance of grain-boundary diffusion relative to bulk diffusion.

A silicate polycrystal deforms therefore by power-law creep above a grain-size dependent transition stress, and by linear creep below. The power-law field is the one usually sampled by 'low-stress' (i.e. $10 \leqslant \sigma \leqslant 100$ MPa) laboratory experiments. As lower stresses are difficult to attain, and the transition stress increases with decreasing grain size, a Newtonian creep law may be observed in the laboratory only if the polycrystal is very fine-grained ($d \leqslant 10$–100 μm as an order of magnitude). Furthermore, if power-law and diffusion creep are the relevant creep mechanisms in the mantle, then the possibility of Newtonian flow is strongly conditional upon a very small grain size.

Figure 11.1 is an example of a deformation map for olivine with grain size $d = 0.1$ mm, which is within the possible range for the upper mantle. To make the map directly relevant to conditions within the Earth, depth is used instead of temperature as one of the co-ordinates [(σ, z)-diagrams have been termed *applied deformation maps* by Stocker & Ashby 1973]. The use of z as an independent variable is made possible by the existence of $p(z)$ and $T(z)$ relations. In this case, p is the overburden pressure and T varies with exponentially decreasing gradient from 300 K at the surface to 1850 K at $z = 500$ km. (There is no point in extending the map to greater depths, as olivine is not a constituent of the lower mantle.) The creep parameters are as in Table 10.1, except that the diffusivities for power-law creep and bulk diffusion are assumed to be the same, with activation volume equal to the oxygen-ion volume. No superplasticity is assumed in the diffusion field, the grain-boundary width is taken as twice the Burgers vector, and the grain-boundary activation energy as two-thirds of the creep activation energy.

Cataclastic flow and glide-controlled plasticity are the predominant deformation mechanisms at $\sigma \geqslant 8 \times 10^2$ MPa (although not visible at the scale of the map, cataclasis predominates in the uppermost crust at any

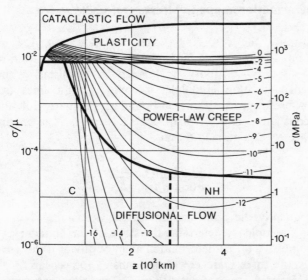

Figure 11.1 (σ, z)-deformation map for polycrystalline olivine with grain size 0.1 mm. Thick lines are creep field boundaries; thin lines, constant strain rate contours (given as powers of 10). C and NH denote Coble and Nabarro–Herring creep, respectively (from Ashby & Verrall 1978).

stress). These stresses are seldom attained in geodynamic processes. Power-law and diffusion creep take over at lower stresses. For the chosen geotherm Coble creep predominates down to a depth of about 80 km. As depth increases, power-law creep occurs above a transition stress whose asymptotic value is of the order of 1–10 MPa. Within the diffusion field Nabarro–Herring creep becomes predominant at $z \geqslant 280$ km. However, both transition stress and depth are functions of creep parameters, especially grain size. The strain-rate contours first decrease rapidly with depth, as the material becomes softer with increasing T; as depth increases, the p-effect counteracts the T-effect, so the contours flatten out and then bend slightly upwards.

An applied deformation map such as that shown in Figure 11.1 carries no claim to absolute accuracy. (For instance, in this particular case the chosen parameters give a rather 'soft' upper mantle; this is a consequence of the low activation volume. Use of a more realistic value of V for power-law creep – cf. Table 10.1 – would change constant strain-rate contours by a factor of 10^{-2}–10^{-3}, bringing them more in line with geophysical estimates.) Still, it is noteworthy that the transition stress between linear and non-linear creep is broadly in the same bracket as the tectonic stress in the upper mantle, as estimated from considerations on convective power (Section 7.9) and neoblast size (Section 10.6).

11.3 Deformation maps in stress–grain size space

In (σ, d)-space the co-ordinate axes are $(\log d/b$, $\log \sigma/\mu)$ and $T =$ constant. In this system field boundaries, constant strain-rate, and constant viscosity contours for creep mechanisms with stress-independent activation enthalpies become straight lines (see Eqns 11.6–8). A useful property of field boundaries is that, while their position depends on p and T, their slope does not. It is an immediate consequence of Equation 11.6 that

$$\frac{\partial (\log \sigma/\mu)}{\partial (\log d/b)} = \frac{\Delta m}{\Delta n} \tag{11.9}$$

and consequently the grain-size and stress exponents of the relevant mechanisms completely determine the slope of the boundary.

An example of a (σ, d)-deformation map for olivine is shown in Figure 11.2. Only the three main creep mechanisms (power-law, Coble, and Nabarro–Herring) are considered; (p, T)-conditions are representative of a depth of approximately 150 km under continents. The diffusivities

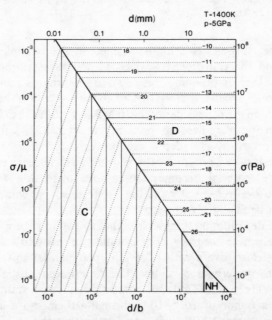

Figure 11.2 (σ, d)-deformation map for polycrystalline olivine. C, NH, and D denote Coble, Nabarro–Herring, and power-law creep, respectively. Thick lines are field boundaries; thin lines, constant-viscosity contours; dotted lines, constant-strain rate contours; the last two as powers of 10 (from Ranalli 1982).

of power-law creep and of bulk diffusion are assumed to be the same, with an activation volume equal to that for oxygen diffusion (cf. Table 10.1); other parameters are as in the (σ, T)-map in the previous section. The three field boundaries meet at a 'triple point' whose position depends only on depth (see Ranalli 1982 for details). The strain-rate contours are more realistic than in the map in Figure 11.1, as the activation volume is larger in this case. Viscosity is independent of grain size in the power-law field, and independent of stress in the diffusion field.

The map shows that at stresses in the 1–10 MPa range, creep is of the power-law type if $d \gtrsim 100$ μm as an order of magnitude, with strain rates and viscosities ($10^{-15} \leqslant \dot{\varepsilon} \leqslant 10^{-13}$ s^{-1} and $10^{20} \leqslant \eta \leqslant 10^{22}$ Pa s, respectively) in the same brackets as inferred from geophysical observation (cf., for example, Section 8.2). Creep becomes linear at smaller grain sizes; the predominant mechanism in the linear field at $z \simeq 150$ km is grain-boundary diffusion, and a decrease in grain size results in a relative softening of the material.

Deformation maps in (σ, d)-space provide a direct way of assessing the possible effects of dynamic recrystallization on creep mechanisms. In Section 10.6 we mentioned the possibility that dynamic recrystallization of an initially coarse aggregate may reduce grain size enough for the creep mechanism to become diffusional. (This *recrystallized diffusion creep* would be non-linear, as grain size would be a function of stress.) We also pointed out that there is usually no evidence that this transition occurs. This can be verified by plotting the neoblast size curve (Eqn 10.44) on a (σ, d)-deformation map (see, for example, Poirier 1985). In the case of olivine the neoblast size curve (both for SGR and GBM recrystallization) lies entirely within the power-law field for realistic values of the variables.

A (σ, d)-map describes the rheology of a material at a given temperature (or depth). Information pertaining to different temperatures can be plotted on a single map: once the triple-point co-ordinates at a given temperature are known so are the creep fields, since their boundaries branch out from the triple point with slopes independent of temperature. The triple-point co-ordinates can be determined immediately from Equation 11.6; they are a linear function of T^{-1} in (log σ/μ, log d/b)-co-ordinates, and they vary on the (σ, d)-plane according to

$$\frac{\partial(\log \sigma/\mu)_{\text{TP}}}{\partial(\log d/b)_{\text{TP}}} = -\frac{2\,\Delta H' - \Delta H''}{2\,\Delta H'} \qquad (11.10)$$

where $\Delta H'$ and $\Delta H''$ are the differences in activation enthalpies between Nabarro–Herring and Coble creep, and power-law and Nabarro–Herring creep, respectively, and the subscript TP refers to the triple point

311

(cf. Ranalli 1982). If the variation of triple-point co-ordinates with pressure is neglected, then Equation 11.10 shows that the rate of change in the position of the triple point in (σ, d)-space is constant and depends only on the activation enthalpies.

A 'telescoped' (σ, d)-map for olivine, obtained by plotting on a single diagram data for different temperatures, is shown in Figure 11.3. The creep parameters are the same as in Figure 11.2; since the activation enthalpies for power-law creep and bulk diffusion are assumed to be equal, the triple point in this particular case moves along a line with slope -1. The chosen temperatures $(900 \leqslant T \leqslant 1800 \text{ K})$ are representative of depths ranging from the middle lithosphere to the mantle transition zone.

The two areas in Figure 11.3 delimited by a square and a rounded rectangle, respectively, show the most probable (σ, d)-ranges for the

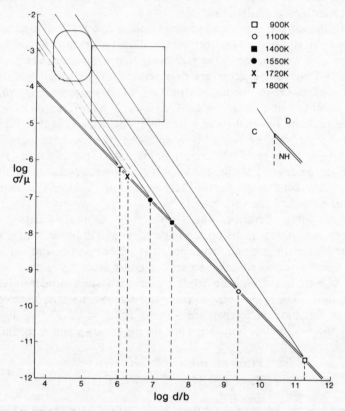

Figure 11.3 'Telescoped' (σ, d)-map for polycrystalline olivine; different symbols denote the position of the 'triple point' at different temperatures. See text for discussion (from Ranalli 1982).

sublithospheric upper mantle ($1 \leqslant \sigma \leqslant 100$ MPa, $0.1 \leqslant d \leqslant 10$ mm) and for shear zones in the lithosphere ($\sigma \geqslant 20$ MPa, $d \leqslant 0.1$ mm). The upper mantle region spans the Coble/power-law creep boundary, and is more likely to be in the power-law field as T increases; lithospheric shear zones, where $T \leqslant 1100$ K, are likely to deform by Coble creep. Occurrence of Ashby–Verrall superplasticity increases the transition stress by a factor between two and three, and would therefore tend to enhance the relative importance of linear creep.

The transition stress between linear and power-law creep can also be examined directly by plotting its value as a function of grain size and temperature (Ranalli 1984). This is particularly convenient in (T^{-1}, log d/b, log σ/μ)-co-ordinates, since then the transition stress surface becomes a plane (see Eqn 11.6). The $\sigma(d, T)$ surface for olivine is shown in Figure 11.4. Creep parameters are as in Table 10.1, with the only simplifying assumption that the pressure-dependence for creep is the same as that for oxygen diffusion. Grain-boundary activation energy is taken as one-half the lattice activation energy, a rather large grain-boundary width is chosen ($\delta = 20b$), and diffusion-accommodated superplasticity is included. The resulting transition stress in the sublithospheric upper mantle is in the range $1 \leqslant \sigma \leqslant 10$ MPa for $T \simeq 1400$–1600 K and $d \simeq 0.1$–1 mm. This result is probably accurate within one or two orders of magnitude. Again, the transition stress is of the same order of magnitude as the estimated tectonic stress. Consequently, the upper mantle may be Newtonian if the grain size is sufficiently small. Subject to this limit on grain size, *upper-mantle rheologies as inferred from*

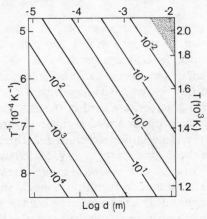

Figure 11.4 Transition stress (MPa) between power-law and linear creep for polycrystalline olivine as a function of temperature and grain size. Stippled area denotes region where linear creep is by bulk diffusion; otherwise, by grain boundary diffusion (from Ranalli 1984).

microphysics and as estimated from postglacial rebound coincide (cf. Section 8.2). However, the closeness of transition stress and tectonic stress may result in the rheology being different in different geodynamic processes, or even changing during a given process if the tectonic stress changes with time (Ranalli 1980).

The above conclusion depends critically on grain size if, as is commonly thought, diffusion and power-law creep are the predominant flow mechanisms in the mantle: under these circumstances a large grain size (say, $d \geq 1$ mm) would place the upper mantle in the power-law field. The only known Newtonian creep mechanism that is independent of grain size is Harper-Dorn creep (see Section 10.3); however, it is not known whether Harper–Dorn creep is a widespread deformation mechanism in the interior of the Earth (cf. Section 12.3 for further discussion).

The preceding examples show that deformation maps are very useful in the study of the rheology of the Earth. The choice of the most effective type of map depends on the particular problem and on which parameter (temperature or grain size) exerts a more critical influence in a given context.

11.4 Isomechanical groups

Deformation maps are available for several materials, mostly of engineering and metallurgical interest (see Frost & Ashby 1982 for a review). They can be used in a systematic comparison of different rheologies, whose ultimate aim is a *rheological classification* of materials.

A comparison of rheological properties, in order to be of general value, must be based on properly scaled, dimensionless ('normalized') variables. For instance, (σ, d)-maps for materials with different melting points should be compared for the same homologous temperature. (A similar situation arises in hydrodynamics, where dimensionless ratios, such as the Rayleigh Number, are used in the analysis of convection – cf. Section 7.8). When properly scaled normalized variables are chosen, polycrystals with the same lattice structure and chemical bonding are seen to behave in a rheologically similar way. Materials with similar properties are said to form an *isomechanical group* (Frost & Ashby 1982). Classification into isomechanical groups, though still incomplete, is potentially very important in geodynamics, since the bulk of the Earth consists of high (p, T)-phases that can be synthesized in the laboratory only in very small quantities or not at all, and therefore inferences on the microrheology of the deep interior have to be based on analogue materials.

314

The procedures for attempting a classification of materials into isomechanical groups are reviewed by Frost & Ashby (1982). Essentially these procedures consist of finding a successful normalization which would emphasize the systematic similarities (and differences) in the rheology. In other words, a successful normalization is one in which the materials in the same isomechanical group and in a similar microstructural state (e.g. grain size) are described by the same rheological diagram or dimensionless rate equation.

Normalization can be achieved in two ways, starting either from experimental data or from rate equations. *Empirical normalization* is the search for normalizing parameters that bring the experimental data for all materials in a given isomechanical group onto the same 'master diagram'. The effectiveness of the normalization is thus measured by the reduction in the scatter of data points. Normalizing parameters that work well for stress and temperature can be modulus-based, energy-based, or temperature-based, e.g. (tildes denote dimensionless variables)

$$\tilde{\sigma} = \frac{\sigma}{\mu} \qquad \tilde{\sigma} = \frac{\sigma}{(H_c/\Omega)} \qquad \tilde{\sigma} = \frac{\sigma}{(kT_m/\Omega)}$$

$$\tilde{T} = \frac{T}{(\mu\Omega/k)} \qquad \tilde{T} = \frac{T}{(H_c/k)} \qquad \tilde{T} = \frac{T}{T_m}$$

where μ is the shear modulus, T_m is the melting temperature, Ω is the atomic or molecular volume, H_c is the cohesive energy, and k is the Boltzmann constant. The two simplest normalizations (in the sense that the scaling parameters are known for most materials) are σ/μ and T/T_m. For strain rate, a useful normalization is

$$\tilde{\dot{\varepsilon}} = \frac{\dot{\varepsilon}}{D_m/\Omega^{2/3}}$$

where D_m is the diffusivity at the melting point. Empirical scaling of pressure has not been investigated to any extent, as pressure effects are usually second-order in engineering applications; however, since pressure has the same dimensions as stress, the two quantities should scale in a similar way.

Normalization of creep laws proceeds from the general rate equation (Eqn 10.17), which, disregarding pressure effects, can be written in dimensionless form as

$$\tilde{\dot{\varepsilon}} = f(\tilde{\sigma},\ \tilde{T},\ \{\tilde{M}_i\},\ \{\tilde{S}_i\}) \tag{11.11}$$

where $\{\tilde{M}_i\}$ and $\{\tilde{S}_i\}$ are dimensionless material parameters and state variables, respectively. It follows from Equation 11.11 that materials in the same microstructural state belong to the same isomechanical group (i.e. obey the same rate equation) if they have identical $\{\tilde{M}_i\}$; that is, *scaled material properties are constant for the members of an isomechanical group.*

As an example, consider Nabarro–Herring creep, whose rate equation in the present context can be written as

$$\dot{\varepsilon} = \alpha\, \frac{\sigma \Omega D_0^*}{kTd^2} \exp\left(-\frac{E^*}{RT}\right)$$

where E^* is the activation energy for self-diffusion. Normalizing stress, temperature, and strain rate by modulus, temperature, and diffusivity, respectively, we obtain

$$\dot{\varepsilon}\, \frac{\Omega^{2/3}}{D_m} = \alpha\, \frac{\Omega^{2/3}}{D_m} \left(\frac{\sigma}{\mu}\right) \frac{\mu\Omega}{kd^2} \left(\frac{T_m}{T}\right) \frac{D_0^*}{T_m} \exp\left(-\frac{E^*}{R(T/T_m)T_m}\right)$$

which becomes, since $D_m = D_0^* \exp(-E^*/RT_m)$,

$$\dot{\varepsilon}\, \frac{\Omega^{2/3}}{D_m} = \alpha \left(\frac{\Omega^{1/3}}{d}\right)^2 \frac{\mu\Omega}{kT_m} \left(\frac{T_m}{T}\right)\left(\frac{\sigma}{\mu}\right) \exp\left[-\frac{E^*}{RT_m}\left(\frac{T_m}{T}-1\right)\right]$$

i.e.

$$\tilde{\dot{\varepsilon}} = \alpha\, \frac{\tilde{M}_1}{\tilde{S}_1^2 \tilde{T}}\, \tilde{\sigma} \exp\left[-\tilde{M}_2\left(\frac{1}{\tilde{T}}-1\right)\right] \tag{11.12}$$

Equation 11.12 is the normalized flow law for Nabarro–Herring creep. The scaled state variables and material parameters are, respectively,

$$\tilde{S}_1 = \frac{d}{\Omega^{1/3}} \qquad \tilde{M}_1 = \frac{\mu\Omega}{kT_m} \qquad \tilde{M}_2 = \frac{E^*}{RT_m}$$

and consequently different materials in the same microstructural state ($\tilde{S}_1$ constant) must have the same $\tilde{M}_1$ and $\tilde{M}_2$ if they are to follow identical rate equations. Normalization proceeds in a similar way for other creep laws, although more dimensionless material parameters may be involved.

The rheological behaviour of materials is therefore strongly characterized by dimensionless combinations of material parameters. Thus, a criterion for assessing the accuracy of classification into iso-mechanical groups is the near-constancy of $\{\tilde{M}_i\}$ for materials belonging to a given group. On this basis Frost & Ashby (1982) distinguished several isomechanical groups. It turns out that not only *crystal structure* but also *chemical bonding* affects the rheology (lattice resistance and activation energies are related to both). For instance, materials with rocksalt structure form four isomechanical groups: alkali halides (e.g. NaCl), oxides (MgO), lead sulphide (PbS), and metal carbides (TiC).

Classification into isomechanical groups has not yet been extended to complex silicates – but, as mentioned previously, could become an important tool in geodynamics. Neither the materials nor the extrinsic conditions pertaining to the deep interior of the Earth can be satisfactorily reproduced in the laboratory. This has led to the use of analogue materials (such as magnesium germanate, Mg_2GeO_4, and fluoride perovskite, $KZnF_3$ – cf. Section 12.3), whose rheological properties are assumed to be indicative of the properties of the actual materials. However, it is not known whether real medium and corresponding analogue belong to the same group. Knowledge of the isomechanical classification would therefore strengthen the basis for *systematics*, i.e. would increase the confidence with which the known properties of some materials can be used to infer the unknown properties of others.

12 Earth rheology from microphysics

An accurate picture of the rheology of the Earth from the continuum-mechanics viewpoint must be consistent with the microphysics of deformation, and both must satisfy observational constraints.

The rheology of the lithosphere is re-examined by introducing explicitly the temperature-dependence of creep parameters and taking into account the rheological stratification induced by compositional layering. This stratification affects not only the flexural behaviour, but also phenomena such as intralithospheric *décollements* and the persistence in time of tectonic lineaments. The occurrence of shear zones can also be accounted for by the effects of viscous dissipation on the microrheology.

The two most likely rheologies for the mantle are power-law creep and Newtonian diffusion creep, although other types of behaviour cannot be excluded: in particular, superplasticity may be associated with polymorphic phase transitions, and linear dislocation creep may be a feasible mechanism in the lower mantle. Both constant-strain rate power-law viscosity and Newtonian viscosity have depth distributions compatible with surface loading data. While the total viscosity increase with depth is insufficient to prevent mantle-wide convection, the layered convection postulated on geochemical grounds leads to a fine viscosity structure that may be beyond the resolving power of the geophysical evidence.

12.1 Rheology of the lithosphere

The rheology of the Earth can be studied from the continuum-mechanics viewpoint, which describes the macroscopic relations between stress, strain, and their time derivatives, and works out their consequences in problems involving deformation, fracture, and flow. Alternatively, it can be studied from the microphysical viewpoint, which analyses processes at the atomic and lattice levels, and derives rate equations from this. The

318

continuum-mechanics approach leads to a picture of the rheology of lithosphere and mantle obtained from geological and geophysical data (for instance, flexure of the lithosphere, Section 8.4; or postglacial rebound, Section 8.2). The microphysical approach is subject to constraints derived from laboratory experiments and microstructural observations of naturally deformed rocks. The two approaches provide complementary views of Earth rheology: in other words, the inferences and results based on continuum mechanics must be compatible with the behaviour expected from the microphysical properties of the relevant materials. In this sense the continuum-mechanics and microphysics models yield a check of the validity of each other; and – ideally – they should result in a unifying view of the rheological properties of the Earth.

In this final chapter we consider some problems in the rheology of the lithosphere and mantle (in the long timescale), to illustrate how this unifying view can be achieved. Although several major questions remain unanswered, the integration of microphysics and continuum mechanics improves our understanding of geodynamic processes, and may lead to some new insights into old problems.

First we examine the rheology of the lithosphere under vertical loads, which was studied in the continuum-mechanics framework in Section 8.4. The main conclusion reached there was that the *thermal structure* is the predominant factor in determining the flexural response: an 'apparent elastic layer' overlies a layer where stress relaxation is significant over geological timescales. The thickness of the apparent elastic layer depends on the thermal age of the lithosphere and the age of the load. The critical depth below which stress relaxation is relatively fast corresponds to an isotherm between 600 and 900 K in the oceanic lithosphere. The flexural rigidity of the continental lithosphere is less well-constrained, as a consequence of its complex thermal history and heterogeneous composition.

Modelling the lithosphere as an elastic or viscoelastic plate gives a satisfactory first-order account of its flexural behaviour. Further progress requires the incorporation of microrheological information (cf. Kirby 1983 for a review). Here we mention some recent studies that explicitly take into account the temperature-dependence of rheological parameters (the pressure in the lithosphere is comparatively low, so its effect is usually negligible).

Temperature in the lithosphere varies from 300 K at the surface to 1500 K at the lithosphere/asthenosphere boundary. The corresponding change in viscosity (and relaxation time) is of several orders of magnitude. Courtney & Beaumont (1983) assumed the lithosphere to have uniform composition and Maxwell rheology, and expressed the

319

variation of viscosity with depth z and thermal age t as

$$\eta(z,\ t) = \eta_a \ \exp\left[\frac{E}{R}\ \left(\frac{1}{T(z,\ t)} - \frac{1}{T_a}\right)\right] \qquad (12.1)$$

where η_a and T_a are, respectively, viscosity and temperature in the half-space below the lithosphere, and E is the activation energy for lithospheric creep. For given values of the parameters and assumed geotherms Equation 12.1 yields the depth-dependence of viscosity at various thermal ages, from which surface deformation and apparent elastic thickness under a specified load can be computed. The *effective elastic thickness* (defined as the thickness of the layer with relaxation time $\tau \geqslant 10^3$ Ma, i.e. $\eta \geqslant 10^6 \eta_a$) decreases with increasing time after loading; but it tends to a limiting value (the *asymptotic elastic thickness*) which increases with the thermal age of the lithosphere at the time of loading. The effects of the thermal age and of time since loading therefore have a common origin in the temperature-dependence of creep.

Flexural data for the oceanic lithosphere (to which the above considerations must be restricted because of the assumption of uniform composition) are satisfactorily matched by the model for $E \leqslant 250$ kJ mol^{-1}. This is a rather low value when compared with the rheology of dry or wet olivine (cf. Section 10.4). The explanation of this discrepancy probably lies in the combination of several factors that have been neglected. Strains in lithospheric flexure are at most of the order of 10^{-2}, and consequently transient creep may play a role; the composition is not actually uniform; and creep may be non-linear. Also, the response of the lithosphere depends on the applied stress: sufficiently large loads (vertical or superimposed lateral) may decrease the effective elastic thickness by causing zones of yielding at the top (brittle) and the bottom (plastic) of the 'elastic' plate.

In order to extend this type of analysis to the continental lithosphere, the rheological effects of compositional layering must be taken into account: lithospheric thickness is controlled by temperature, but crustal thickness is controlled by composition. Kusznir & Karner (1985) incorporated microrheological properties by modelling the lithosphere as a 'quartz-over-olivine' plate with power-law rheology. Flexural stress is computed as a function of depth, time, temperature, and crustal thickness, and is then used to determine effective elastic thickness as a function of age of the lithosphere and time since loading. Elastic thickness relaxes rapidly ($\tau < 1$ Ma) towards its asymptotic value. Since quartz is softer than olivine, a lithosphere with a thick crust has lower flexural rigidity than one with a thin crust, other things being equal. Although observations do not fit any model precisely, it is clear that consideration of composition as well as thermal structure is necessary to unravel the flexural behaviour of continental lithosphere. The situation is complicated further by varying lateral stresses (see, for example, Cloetingh *et al.* 1985).

Most models of lithospheric rheology assume a quartz-rich crust overlying an olivine mantle (see the extensive bibliography in Kirby 1983). In order to obtain a more realistic view of the *rheological stratification* of the lithosphere, Ranalli & Murphy (1986) take into account the creep properties of various crustal rocks as summarized in Table 10.2. The frictional and ductile stress differences are assumed to be given by the modified Anderson theory (Eqn 8.56) and the Dorn equation (Eqn 10.41), respectively, i.e.

$$\sigma = \alpha \rho g z (1 - \lambda) \tag{12.2}$$

$$\sigma = \left(\frac{\dot{\varepsilon}}{A_D}\right)^{1/n} \exp\left(\frac{E}{nRT}\right) \tag{12.3}$$

where σ is the stress difference, α is a numerical parameter depending on the type of faulting, ρ, g, z, and λ are average density, gravity, depth, and pore fluid factor, respectively, and A_D, n, and E are Dorn parameter, stress exponent, and creep activation energy, respectively. The 'strength' at any given depth is the lower of the brittle and ductile stress differences (cf. Section 5.2). In the brittle regime the pore fluid factor is assumed hydrostatic, and the type of faulting is selected on the basis of predominant seismic focal mechanisms; in the ductile regime a strain rate of 10^{-14} s^{-1} is used, and geotherms are constrained to vary linearly between 300 and 1500 K at the upper and lower boundaries of the lithosphere. Different geotherms, resulting in varying lithospheric thickness, are combined with varying crustal thickness and composition to yield the models listed in Table 12.1, broadly representative of different tectonic provinces in continents and oceans.

Lithospheric strength profiles for various tectonic environments are shown in Figures 12.1 and 12.2. Although the uncertainties in rheological parameters are large, the main features of the profiles are established with a reasonable degree of confidence. The depth variation of strength $\sigma(z)$ – defined as the critical stress difference in the brittle regime and the flow stress for a given strain rate in the ductile regime – depends on the temperature gradient (and therefore the lithospheric thickness) and on the crustal thickness and composition. In most cases the rheology has a 'sandwich' structure, with one or more ductile horizons interlayered with brittle horizons, and corresponding oscillations of the $\sigma(z)$ curve.

The models can be checked against geophysical data. Seismic activity,

Table 12.1 Parameters for lithospheric strength profiles.

Model	Lithospheric thickness†(km)	Crustal thickness (km)	Composition‡	Brittle regime§	Type tectonic province
C(s1)	150	40	unlayered	T	Precambrian shields
C(s2)	150	40	layered	T	Precambrian shields
C(c1)	150	60	unlayered	T	continental convergence zones
C(c2)	150	60	layered	T	continental convergence zones
H(1)	50	30	unlayered	N/S	continental extensional zones
H(2)	50	30	layered	N/S	continental extensional zones
O	75	10	basic	T	mature oceanic lithosphere

† Depth at which $T = 1500\,\mathrm{K}$.
‡ Refers to crust (upper mantle is ultrabasic). Unlayered, quartz–granitic; layered, quartz–granitic over intermediate–basic.
§ Predominant type of seismic faulting in the upper crust. T, thrust; N, normal; S, strike-slip.

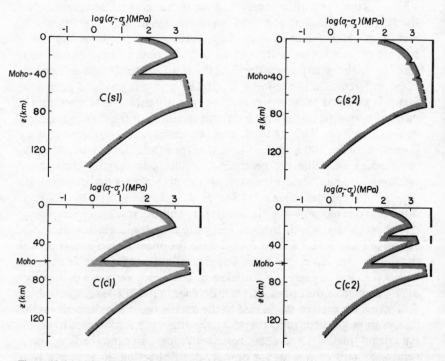

Figure 12.1 Lithospheric strength profiles for 'cold' geotherms and various petrological models of the lithosphere (see Table 12.1). Vertical bars to the right of each profile denote brittle layers where seismic activity may occur (from Ranalli & Murphy 1986).

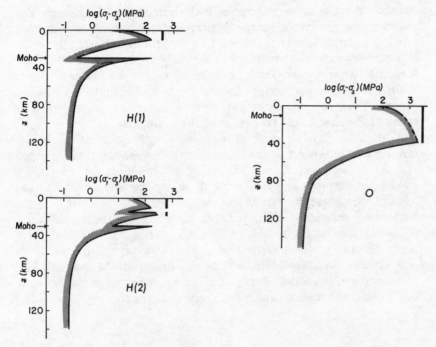

Figure 12.2 Lithospheric strength profiles for 'hot' and 'oceanic' geotherms (refer to Table 12.1)(from Ranalli & Murphy 1986).

when present, is confined to the brittle layers (Sibson 1982, Chen & Molnar 1983). The occurence of ductile layers is marked by minima in seismic wave velocity and electrical resistivity. The existence of intra-lithospheric layers of decoupling as a consequence of the rheology is now commonly accepted (cf., for example, Kirby 1985, White & Bretan 1985).

The rheological stratification of the lithosphere has interesting geo-dynamic consequences. The *total strength* of the lithosphere may be defined as

$$\sigma_L = \int_0^H \sigma(z) \, dz$$

where H is the thickness of the lithosphere [given, for instance, by the depth at which the flow stress is reduced to 1 MPa, and below which no further jumps in $\sigma(z)$ occur]. As expected, σ_L decreases with increasing temperature gradient for a given crustal thickness and composition; however, both thickness and composition affect it, since the ductile

323

strength of rocks usually increases with decreasing SiO_2 content. For instance, if a continental margin is tectonically thickened by obduction of allochthonous flakes (cf. Section 8.7), then the lithosphere after over-thrusting is weaker than before overthrusting, once thermal equilibrium is re-established. Consequently, continental sutures are favourable sites for the resumption of tectonic activity, e.g. rifting and reopening of oceans. Thus, the *permanence of tectonic zones* can be explained in terms of the rheological properties of the lithosphere.

As the microphysical models for the rheology become less schematic, the complex behaviour of the lithosphere under various kinds of loads is seen to be a consequence of its rheological stratification. For instance, yielding in flexure may occur not only at the top and bottom of the 'elastic' lithosphere, but also in some intralithospheric ductile layer if the stresses are sufficiently high; and this will affect the apparent flexural rigidity. Under lateral tectonic loading the behaviour of the lithosphere is not the same as under vertical loading: the low-strength ductile layers may act as decoupling horizons, resulting in large-scale overthrusting, obduction and intralithospheric *décollements*. These problems are only beginning to be studied taking full account of the rheology.

12.2 Shear heating and thermomechanical coupling

Another geodynamic problem that is clarified by consideration of micro-rheological parameters is that of energy dissipation during irrecoverable deformation. This leads to *frictional heating* along faults in the brittle regime, and *viscous shear heating* in the ductile regime. (The term 'frictional heating' is sometimes used for both.)

In this section we consider viscous shear heating, which may be important under certain conditions in mantle flow, and is a major factor in the deformation of ductile shear zones in the lithosphere. Most situations of geodynamic interest can be modelled as one-dimensional shear flows between parallel walls, where the only non-vanishing stress is the shear σ, and the only non-vanishing velocity component v is parallel to the walls (laminar flow); the corresponding strain rate is $\dot{\varepsilon} = \frac{1}{2} \partial v / \partial x$, where x is the co-ordinate axis perpendicular to the walls. Assuming that all dissipated mechanical energy is converted into heat (an upper-bound assumption, as some energy may go into changing the physical or chemical state, or both, of the material), the heat production rate per unit volume is (cf. the considerations developed in Sections 4.3 & 7.8).

$$\dot{W} = \sigma_{ij}\dot{\varepsilon}_{ij} = \sigma \partial v / \partial x$$

The energy-balance equation governing the temperature distribution was derived in Section 7.8 (Eqn 7.31). In the one-dimensional case, including shear heating and neglecting radioactivity, it becomes

$$K \frac{\partial^2 T}{\partial x^2} + \sigma \frac{\partial v}{\partial x} = \rho c \dot{T} \tag{12.4}$$

where K, ρ, and c are thermal conductivity, density, and specific heat (at constant pressure), respectively. Neglecting inertia effects the equation of motion is simply

$$\partial \sigma / \partial x = 0 \tag{12.5}$$

which implies that σ is constant across the shear zone. Together with a rheological equation, Equations 12.4 and 12.5 fully describe the time-dependent one-dimensional shear heating problem.

The solutions $T(x, t)$ and $v(x, t)$ are coupled through the temperature-dependence of the viscosity. However, a first idea of the amount of shear heating can be obtained by considering the steady-state laminar flow of a fluid with constant viscosity η_0 between two walls separated by a distance h and moving at a relative velocity v_0 (*Couette flow* – see, for example, Turcotte & Schubert 1982). In this case Equation 12.4 becomes

$$K \frac{d^2 T}{dx^2} + \eta_0 \left(\frac{v_0}{h}\right)^2 = 0$$

which can be integrated immediately, with boundary conditions $T = T_0$ at $x = 0$ and $T = T_1$ at $x = h$, to give

$$T - T_0 = (T_1 - T_0) \frac{x}{h} \left(1 + \frac{\eta_0 v_0^2}{2K(T_1 - T_0)}\right) - \left(\frac{x}{h}\right)^2 \frac{\eta_0 v_0^2}{2K}$$

Recalling the Prandtl Number (Eqn 7.33), and defining the *Eckert Number* as

$$\mathrm{Ec} = \frac{v_0^2}{c(T_1 - T_0)}$$

the above solution can be written in terms of the dimensionless temperature $\theta \equiv (T - T_0)/(T_1 - T_0)$,

$$\theta = \frac{x}{h} \left(1 + \tfrac{1}{2} \mathrm{Pr}\, \mathrm{Ec}\right) - \left(\frac{x}{h}\right)^2 \tfrac{1}{2} \mathrm{Pr}\, \mathrm{Ec} \tag{12.6}$$

325

In the absence of shear heating ($v_0 = 0$, i.e. $Ec = 0$), the reference temperature profile would be $\theta' = x/h$, so that the excess temperature due to dissipation is

$$\theta'' = \theta - \theta' = \tfrac{1}{2} Pr\ Ec \left(1 - \frac{x}{h}\right) \frac{x}{h}$$

The excess temperature has a maximum $\theta'' = \tfrac{1}{8} Pr\ Ec$ at $x/h = \tfrac{1}{2}$. Taking, for instance, asthenospheric conditions ($\eta_0 \simeq 10^{20}-10^{21}$ Pa s, $v_0 \simeq 10^{-9}$ m s^{-1}, $T_1 - T_0 \simeq 300$ K), the maximum temperature increase due to shear heating is of the order of a few kelvins. However, as can be seen from Equation 12.6, *viscous dissipation increases with increasing viscosity and velocity.* Shear heating is higher in harder materials.

The above considerations neglect the temperature-dependence of the viscosity. Now, it is precisely this dependence that couples temperature and velocity fields (*thermomechanical coupling*): shear heating affects the viscosity, and therefore the strain rate; and the strain rate, in turn, affects shear heating. This feedback process results in *thermal softening* of the material: the decrease in viscosity caused by shear heating concentrates the deformation along the zone of relatively high shear. A condition of equilibrium is established when the heat generated by viscous dissipation is equal to the heat removed by conduction; but deformational instabilities, where the strain rate increases without bound (in practice, until melting occurs), are at least theoretically possible.

To take into account thermomechanical coupling, the rheological rate equation must be used in the energy-balance and motion equations (Eqns 12.4 & 5). The creep law for both power-law and linear processes can be written as

$$\dot{\varepsilon} = \frac{B}{T}\ \sigma^n\ \exp\left(-\frac{H}{RT}\right) \tag{12.7}$$

where B is independent of T, and $n \geqslant 1$ (compare with Eqns 10.36 & 37). Consequently, the viscosity is

$$\eta = \frac{T}{2B}\ \sigma^{1-n}\ \exp\left(\frac{H}{RT}\right) \tag{12.8}$$

which shows that most of the temperature-dependence is contained in the Arrhenius factor. For steady-state Couette flow, Equation 12.4 then becomes

$$K\frac{d^2 T}{dx^2} + \sigma^{n+1}\ \frac{2B}{T}\ \exp\left(-\frac{H}{RT}\right) = 0 \tag{12.9}$$

This equation can be solved analytically for small temperature perturbations, i.e. $T''/T' \ll 1$, where T' and T'' are reference temperature and excess temperature caused by shear heating, respectively (cf. Turcotte & Schubert 1982). Non-dimensionalizing length and temperature as

$$\xi = \frac{x}{h} \qquad \vartheta = \left(\frac{H}{RT'}\right) \frac{T''}{T'}$$

Equation 12.9 reduces to

$$\frac{d^2\vartheta}{d\xi^2} + \frac{H\beta h^2 \exp(-H/RT')}{KRT'^2} \exp(\vartheta) = 0 \qquad (12.10)$$

where

$$\beta = \left(1 - \frac{T''}{T'}\right) \frac{2B}{T'} \sigma^{n+1}$$

is taken as approximately constant (its dependence on temperature is negligible compared with the Arrhenius factor). Defining the dimensionless *Brinkman* (or *Gruntfest*) *Number* as

$$Br = \frac{H\beta h^2 \exp(-H/RT')}{KRT'^2}$$

Equation 12.10 can be written in the simple form

$$\frac{d^2\vartheta}{d\xi^2} + Br \exp \vartheta = 0 \qquad (12.11)$$

The Brinkman Number therefore fully determines the temperature distribution in shear flows subject to viscous dissipation and with temperature-dependent viscosity. In physical terms Br is the ratio between heat generated per unit area by dissipation and conductive heat flux at the reference temperature, i.e. it measures the efficiency of the material to conduct away shear-generated heat.

The solution $\vartheta(Br)$ of Equation 12.11 has been widely discussed in geodynamics, especially in view of assessing the possibility of shear melting (cf. Brun & Cobbold 1980 for a review). The essential point is that ϑ is a multiple-valued function of Br and therefore there are multiple solutions: as Br is related to shear stress and ϑ to the temperature perturbation and so to the strain rate, *different flow regimes are possible for*

327

a given shear stress. At low Br the rate of shear heating is low, and consequently a larger σ is necessary to produce a larger $\dot{\varepsilon}$ (so that ϑ increases with Br). However, a point is reached where shear heating is such that $\dot{\varepsilon}$ increases even for decreasing σ (so that ϑ increases with decreasing Br). Consequently, two different flow regimes are possible for the same shear stress: a low-(T, v) flow (subcritical regime), and a high-(T, v) flow (supercritical regime). A third, still hotter regime is theoretically possible, since the Arrhenius factor tends towards a limiting value for increasing T; however, in practice melting would occur before the necessary conditions were reached.

Supercritical flows may result in a deformational instability if the *applied stress* is constant: then, if the perturbation is sufficiently large, the strain rate could increase 'catastrophically' until melting occurs (a so-called *thermal runaway*). However, an instability cannot occur under *constant-velocity* boundary conditions, since in this case shear heating reduces the stress necessary to maintain the given strain rate, i.e. the viscosity profile adjusts itself to the velocity profile. Thermal runaways are therefore unlikely events in geodynamics.

A geological instance where viscous heating plays an important role is that of localized shear between blocks moving in opposite directions. Such *ductile shear zones* (which can be, for example, downward continuations of brittle faults) are anything from centimetres to kilometres wide, and often show characteristic microstructures denoting exceedingly large deformation (mylonite zones). They obviously represent zones of enhanced ductility within the lithosphere. Their geometry is sketched in Figure 12.3a: two blocks move with relative velocity v_0 along a shear zone centred at $SS(x = 0)$. Shear heating produces a temperature – and therefore a viscosity – anomaly which softens the zone and concentrates

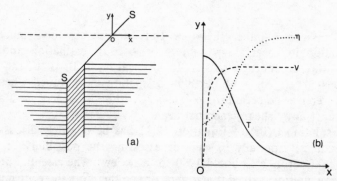

Figure 12.3 (a) Geometry of a vertical shear zone and (b) schematic variations of viscosity, temperature, and velocity (η and T are symmetric with respect to the shear zone, v is antisymmetric).

shear along it. The resulting T, η, and v profiles are shown qualitatively in Figure 12.3b. Away from the shear zone ($|x| \to \infty$), v tends to $\pm v_0/2$, and T and η to their respective ambient values.

The thermomechanical evolution of shear zones has been studied by Yuen et al. (1978) and Fleitout & Froidevaux (1980) for the cases of Newtonian and non-Newtonian rheologies, respectively. All quantities are assumed to be functions of x and t only (in particular, this implies that heat conduction is normal to the shear zone), and inertia effects are neglected, so that the system is governed by Equations 12.4 and 12.5. Introduction of the rheological law (Eqn 12.7) leads to the temperature equation

$$K \frac{\partial^2 T}{\partial x^2} + \sigma^{n+1} \frac{2B}{T} \exp\left(-\frac{H}{RT}\right) = \rho c \dot{T} \qquad (12.12)$$

Integration of Equation 12.7, since $\partial v/\partial x = 2\dot{\varepsilon}$, yields

$$\frac{v_0}{2} = 2B\sigma^n \int_0^\infty T^{-1} \exp\left(-\frac{H}{RT}\right) dx \qquad (12.13)$$

which can be used to eliminate σ from Equation 12.12, i.e.

$$K\frac{\partial^2 T}{\partial x^2} + \left\{ v_0 \left[4B \int_0^\infty T^{-1} \exp\left(-\frac{H}{RT}\right) dx \right]^{-1} \right\}^{(n+1)/n}$$

$$\frac{2B}{T} \exp\left(-\frac{H}{RT}\right) = \rho c \dot{T} \qquad (12.14)$$

Equation 12.14 can be integrated numerically to yield T, and then σ, v, and η can be obtained from Equations 12.13, 12.7 and 12.8, respectively.

Results agree well with geological evidence. For plate-tectonic velocities, viscous heating softens the rock sufficiently to concentrate the velocity gradient in the shear zone. Details depend on rheological parameters, ambient temperature, and velocity, but usually shear zone stresses, strain rates, and viscosities are of the order 10–100 MPa, 10^{-13}–10^{-12} s^{-1}, and 10^{20} Pa s, respectively. The viscosity is thus orders of magnitude less than the ambient viscosity of the lithosphere. The thermal thickness of a shear zone (i.e. the width of the temperature anomaly) is about 20 km, a factor of ten larger than the mechanical thickness. Most of the temperature increase occurs in the first 1 Ma, and thereafter a near-steady state is attained. (The increase in the thermal thickness of the shear zone as $t^{1/2}$ due to thermal diffusivity is probably prevented in nature by upward heat conduction which is not included in the model.)

It is highly unlikely that viscous heating may generate shear melting, as the maximum temperature is well below the solidus temperature. *Ad hoc* conditions (such as perhaps the interlayering of hard and soft rocks; cf. Fleitout & Froidevaux 1980) must apply if melting is to occur by viscous dissipation.

Although thermomechanical coupling is adequate to accommodate high strain rates in a shear zone, an *initial strength defect* must be present: this could be due to a temperature anomaly or to some mechanical weakness. Thermal softening is not the only process which enhances the ductility of rocks. Grain-size reduction, recrystallization, and hydration may also be important. Evidence for mechanical softening must be sought at the microstructural level.

12.3 Rheology of the mantle: microphysical models

The main source of geophysical evidence for the rheology of the mantle lies in relative sea-level, gravitational, and rotational data (cf. Sections 8.2 & 3). To the first order, the evidence is compatible with a mantle with Newtonian viscosity varying from 10^{21} Pa s in the upper mantle to a maximum of 10^{23} Pa s in the lower mantle. The corresponding continuum model over the entire frequency spectrum is that of a Burgers body with Maxwell time $\tau \simeq 10^{3}$ a.

However, problems exist with the *resolving power* of geophysical inversion. In particular, deviations from linearity, transient effects, and the fine details of the variations of viscosity with depth may not alter geophysical observables sufficiently for the Newtonian model to be discarded, yet have significant effects on the pattern of mantle flow and other geodynamic processes. It is therefore important to examine the rheology from the microphysical viewpoint, and to compare the results with the inferences drawn from geophysical evidence.

Of the various creep mechanisms reviewed in Section 10.3, those with stress-dependent activation enthalpy are, for one reason or another, not very likely to predominate in the interior of the Earth. The problem therefore reduces to an evaluation of the relative importance of diffusion creep and dislocation climb-and-glide. The main – and unavoidable – assumption inherent in the microphysical approach is that *the creep processes observable in the laboratory are relevant to mantle conditions*. Some support for this assumption comes from the theoretical rate equations, which are derived without reference to any particular range of extrinsic conditions. However, some important factors are left out – *chemical environment* and its possible effects on the rheology being perhaps the most important.

In this section we discuss the likelihood of various creep mechanisms to be predominant in the mantle, and their possible geodynamic consequences. In Section 12.4, we derive viscosity–depth profiles, and compare them with the results obtained from consideration of convective properties and from surface loading.

For discussion purposes it is useful to consider separately the rheology of upper and lower mantle, and the possible effects of the mantle transition zone. The most common mineral phase in the upper mantle is $(Mg,Fe)_2SiO_4$ olivine. The mineral assemblage goes through a series of phase transitions in the 400–700 km depth range. The lower mantle consists mainly of $(Mg,Fe)SiO_3$ perovskite, with subordinate amounts (20–40 per cent) of $(Mg,Fe)O$ magnesiowüstite if its bulk composition is similar to that of the upper mantle (cf. Section 6.6 for a discussion).

Since the rheological parameters of olivine are known, the upper mantle can be studied making use of deformation maps, as illustrated in Chapter 11. The main conclusion is that the transition stress between Newtonian diffusion creep and power-law creep is in the range $1 \leqslant \sigma \leqslant 10$ MPa for grain sizes that do not scale with stress (Ranalli 1984). The expected tectonic stress in the upper mantle is of the same order of magnitude. It should be noted that realistic upper mantle viscosities (about 10^{20}–10^{21} Pa s) are obtained without recourse to the softening effects of hydration. The linear rheology inferred from geophysical evidence is therefore not at odds with the microphysics of flow, although neither approach yields an unequivocal answer.

The previous considerations are valid for a stress-independent grain size. If this is stress-dependent, then flow may be affected by dynamic recrystallization at sufficiently high stress. The theoretical rate equation for recrystallized diffusion creep can be obtained by introducing the grain size scaling law (Eqn 10.44) into either the Nabarro–Herring or the Coble creep equation. The result can be written in the general form

$$\dot{\varepsilon} = \alpha^* \frac{D^*\mu b}{kT} \left(\frac{\sigma}{\mu}\right)^q \tag{12.15}$$

where α^* is a dimensionless parameter depending on the predominant diffusion process and on the scaling factor for grain size, and D^* is the appropriate diffusion coefficient. The stress exponent is $q = 1 + 2r$ for bulk diffusion, and $q = 1 + 3r$ for grain-boundary diffusion, where r is the stress exponent of the scaling law ($r \simeq 1$ for SGR recrystallization, $r \simeq 1.2$ for GBM recrystallization). In general, therefore, $q > n$, and creep due to dynamic recrystallization will be predominant over power-law creep above a transition stress that can be calculated by comparing Equation 12.15 and the power-law creep equation. Not surprisingly in view of our discussion of the position of the neoblast size curve in (σ, d)-space (cf. Section 11.3), this transition stress for olivine turns out to be orders of magnitude above the range of geodynamic interest. No matter how one looks at it, the stresses required for dynamic recrystallization to change the creep regime appear to be too large to be realized in the mantle.

However, there are other grain size effects that are not dealt with in the previous considerations. Episodes of recrystallization result in transient softening, for which a rigorous theory is lacking. Even in the absence of recrystallization the assumption of a unique grain size is certainly an oversimplification. Polycrystalline aggregates usually show a distribution of grain sizes with finite dispersion. The transition stress between Newtonian and power-law creep is therefore likely to be more diffuse than is predicted on the basis of unique grain size.

Another way to assess the rheology of the upper mantle consists of incorporating various microrheologies in convection models and comparing their consequences with geophysical observables. As one example among many, the thermomechanical model of secondary upper mantle convection beneath the continental lithosphere proposed by Fleitout & Yuen (1984) may be mentioned. Secondary convective rolls are to be expected in the upper mantle, as the Rayleigh Number is amply super-critical (cf. Section 7.9). In the model of Fleitout & Yuen these rolls are assumed to extend from the bottom of the lithosphere, defined as the non-convecting thermal boundary layer (whose thickness is itself determined by the heat transported by convection), to a depth of 600 km, with aspect ratio between 1 and 2. Both Newtonian and power-law (p,T)-dependent rheologies are considered. Realistic values for the viscosity (in the range 10^{20}–10^{21} Pa s) are obtained for a bottom heat flux of about 20 mW m^{-2}, resulting in a lithospheric thickness of the order of 200 km. A comparison of predicted geophysical observables for the two different rheologies shows that free-air gravity and topographic anomalies are approximately a factor of ten smaller in the non-Newtonian case, but their gradients are sharper. Also, the viscosity structure of non-Newtonian convection is more uniform, and deviatoric stresses induced in the lithosphere are lower. Within the framework of the model an upper mantle with power-law rheology seems to agree better with geophysical data.

Uncertainties increase when we consider the transition zone. Doubts persist about the kinetics of the relevant transformations, and the associated rheological and microstructural changes are not known. The flow mechanisms for different phases may not be the same, and one must rely on evidence obtained from analogue materials. Experiments on germanates (Vaughan & Coe 1981) have shown that the transition from

Mg_2GeO_4 olivine to spinel is accompanied by a reduction in grain size of about one order of magnitude, and a concomitant change in creep mechanism. Although both creep laws can be described by a Dorn equation, the olivine phase has $n \simeq 3.5$ and $E \simeq 440 \, kJ \, mol^{-1}$, whereas the spinel phase has $n \simeq 2.0$ and $E \simeq 300 \, kJ \, mol^{-1}$. Microstructural evidence is scarce, but the former appears to flow by power-law creep, while the latter could flow by non-linear superplasticity (i.e. dislocation-accommodated grain-boundary sliding).

It is not clear whether these results are relevant to the mantle transition zone. The experiments were carried out at high stress ($\sigma \geqslant 100 \, MPa$), and the reduction in grain size depends on transformation kinetics. If the ratios of the creep strengths are the same in the germanates and in the silicates, then the olivine $\rightarrow$ spinel transition in the mantle could be associated with a decrease in strength of up to two orders of magnitude. The resulting soft layer just below the transition could act as a decoupling horizon between the thick continental tectosphere required by some seismological and geochemical models (cf. Section 6.6) and the underlying mantle.

As the kinetics of grain growth after the transition is not known, it is impossible to estimate the thickness of the soft layer, which is primarily the result of a balance between the rate of vertical transport of material across the phase boundary and the rate of static grain growth. The softening effect would be most noticeable in subducting lithospheric slabs. Indeed, the lack of seismicity in some slabs below depths of 300–350 km may be related to the strength drop associated with the olivine $\rightarrow$ spinel transformation (Rubie 1984). However, the whole argument rests on the hypothesis of a transformation-induced reduction in grain size, without which the strength actually increases across phase changes (see Section 12.4).

The softening discussed above is consequent upon a phase transition, but is not transformation superplasticity, which refers to the enhanced creep rates sometimes observed to occur during a transition (cf. Section 10.3). Transformation superplasticity still requires vertical transport of material across the phase boundary, but does not depend on a reduction in grain size. Assuming it occurs in mantle materials (for which information is lacking), it may be associated with a flow instability at the depth of the phase boundary. For instance, if flow is mainly by simple shear parallel to the boundary (i.e. horizontal) with superimposed small vertical perturbations, transformation superplasticity will result in larger finite shear strain across the boundary (Parmentier 1981), thus effectively ensuring a degree of decoupling between upper and lower mantle.

Needless to say, the lower mantle presents an even more difficult problem, as it consists of phases whose rheological properties are only vaguely known. One is consequently obliged to make drastic assumptions – i.e. that the creep laws are of the same form as in the upper mantle, and that the deformation mechanisms are comparable with those predominant in the low-pressure phases. Then the problem reduces to assessing the effects of temperature, pressure, and phase changes on the creep parameters (Section 12.4 takes this approach). The models proposed by Karato (1981) and Minster & Anderson (1981) illustrate the difficulties involved. Karato (1981) compared the importance of Newtonian diffusion creep and power-law creep on the assumption that the rheology of the lower mantle is controlled by the properties of (Mg,Fe)O magnesiowüstite. (This implies that the 'framework effect' of the volumetrically predominant phases is negligible.) It is further assumed that the rate-controlling mechanism is oxygen self-diffusion. The diffusion parameters for MgO at room pressure are well-known, and the pressure effect is taken into account by hypothesizing that the activation volume changes with pressure as the crystal volume. None of these assumptions is firmly established. By analogy with NaCl (which has the same crystal structure as MgO), grain size is taken to be stress-dependent for stresses as low as $\sigma/\mu \simeq 10^{-5}$. This effectively excludes Newtonian rheology, since at the relevant tectonic stresses the scaled grain size is too large for diffusion creep to be predominant.

The model of Minster & Anderson (1981) also results in non-Newtonian rheology. The basic assumption is that the dislocation microstructure of mantle materials is highly organized and stress-dependent, with subgrains or cells containing relatively few dislocations separated by dislocation walls. Such a model gives a unified account of the rheology of the whole mantle over the complete frequency spectrum. Attenuation and transient creep are controlled by bowing of intracell dislocations (cf. Section 10.1). Steady-state creep may result from subgrain diffusion creep, dislocation creep by intracell recovery, or dislocation creep by cell-wall recovery. All of these mechanisms are non-Newtonian (diffusion creep controlled by vacancy migration to cell walls is non-Newtonian when the cell diameter scales with stress – see Section 10.3). Minster & Anderson further assume that creep is related to diffusion, the higher activation energy for dislocation creep probably being due to jog formation. According to laboratory experiments silicon is the slowest-diffusing species in olivine at low pressure, and therefore controls the creep rate in the upper mantle; however, oxygen has a larger activation volume, and consequently controls the creep rate at high pressure. The transition depth is estimated to be in the range 300–500 km. By comparing Maxwell times for the rheologies allowed by the model, dislocation creep (with stress exponent $n = 2$ or 3) is found to be more efficient than subgrain diffusion creep. The possibility of Newtonian diffusion creep is precluded by the assumed high degree of stress-dependent organization of the microstructure.

It is interesting to note that most investigations of the microrheology of the lower mantle do not refer to the material of which it is mainly composed, i.e. perovskite. The reason is that very little is known of the rheological properties of perovskite, which has been produced only in very small quantities in the laboratory. Use must again be made of analogue materials, without a firm knowledge of the isomechanical group to which they belong. Poirier *et al.* (1983) investigated the electrical and rheological properties of single-crystal fluoperovskite (KZnF$_3$), which has the same cubic crystal structure as MgSiO$_3$, with the chemical valence of each ion equal to one-half the valence of the corresponding ion in the silicate. Under the high-temperature, low-stress experimental conditions ($T/T_m > 0.8$, $\sigma < 10$ MPa) the strain rate is linearly proportional to the stress. This Newtonian behaviour cannot be the result of diffusion creep, as the single crystals are large; moreover, a transient stage is present, and optical evidence in deformed crystals is characteristic of dislocation creep. Poirier *et al.* conclude that creep occurs by dislocation glide, probably on the {100}⟨110⟩ system, and also possibly on the {001}⟨100⟩ system. The linear stress-dependence is attributed to a dislocation density independent of stress, which by the Orowan equation leads to a Newtonian flow law (Harper–Dorn creep – see Section 10.3). If these experimental results are relevant at all to other perovskites and to lower-mantle conditions, then the possibility is open for the lower mantle to have Newtonian rheology, independently of grain size. However, further experiments and more-specific models for Harper–Dorn creep are required.

The above interpretation is based on the hypothesis that, *at low stress*, dislocations in perovskites do not move and multiply easily, and that dislocation creep is more efficient than diffusion under the relevant conditions (unfortunately, an activation energy for creep could not be determined in the experiments of Poirier *et al.*, since the Arrhenius plot was not a straight line). Another possibility, suggested by Weertman (1978) and Karato (1981), is that the dislocation density is approximately constant *at small strains*. The rheology of the mantle would then be Newtonian in response to surface loading (where strains are of the order of 10^{-4}), and non-Newtonian at the large strains involved in convective flow. Transient relaxation cannot be excluded in the lower mantle (cf. Peltier 1985, Sabadini *et al.* 1985; see also Section 8.3). If the microphysical deformation mechanism for transient small-strain flow is not the same as that for large-strain steady-state flow, then the viscosity inferred from postglacial evidence may be only a lower limit for the long-term viscosity of the mantle.

It is evident from the foregoing discussion that consideration of the microphysics of flow cannot yet resolve the non-uniqueness stemming from the inversion of geophysical data. However, it does provide further insight into the possible role of some flow mechanisms (transformation superplasticity and Harper–Dorn creep, for instance) in large-scale geodynamic processes. The microphysical approach can also be used to estimate directly the variations of viscosity with depth, which can then be compared with geophysical estimates.

12.4 Rheology of the mantle: variation of viscosity with depth

The interest of microphysically-based estimates of mantle viscosity lies not only in a comparison of the results with inferences from surface loading data, but also in an examination of the fine structure of the variations of viscosity with depth. For instance, we may ask what kind of viscosity signature is to be expected if the upper and the lower mantle convect separately: as this implies the occurrence of a mid-mantle thermal boundary layer, the temperature distribution will be different than in the case of mantle-wide convection. Whether this different temperature distribution is reflected in the viscosity distribution depends on the effects of temperature, pressure, and phase changes on the creep parameters.

The two best-known creep mechanisms, and probably the two most likely to occur in the mantle, are non-Newtonian power-law creep and Newtonian diffusion creep. The former is probably a diffusion-related climb-controlled form of dislocation creep; the latter is effected by a combination of bulk and grain-boundary diffusion with a component of grain-boundary sliding. The predominant creep mechanism at any depth is the one resulting in the lower viscosity.

For shear components viscosity is defined as

$$\eta = \left(\frac{\sigma}{\mu}\right)\frac{\mu}{2\dot{\varepsilon}} \qquad (12.16)$$

By substituting the relevant creep equation into Equation 12.16, viscosity is obtained as a function of temperature, pressure, and rheological parameters. In the case of power-law creep (Eqn 10.37) the effective viscosities at constant stress and at constant strain rate are, respectively,

$$\eta_\sigma = \tfrac{1}{2}\frac{kT}{AD_0 b}\left(\frac{\sigma}{\mu}\right)^{1-n}\exp\left(\frac{E+pV}{RT}\right) \qquad (12.17)$$

336

$$\eta_{\dot{\varepsilon}} = \tfrac{1}{2} \left(\frac{kT}{AD_0 b}\right)^{1/n} \left(\frac{\dot{\varepsilon}}{\mu}\right)^{1/n - 1} \exp\left(\frac{E + pV}{nRT}\right) \tag{12.18}$$

The Newtonian viscosity is (from Eqn 10.36)

$$\eta^* = \tfrac{1}{2} \frac{KTd^2}{\alpha\Omega D_0^*} \left[\exp\left(-\frac{E^* + pV^*}{RT}\right) + \frac{\pi\delta}{d} \exp\left(-\frac{E_{GB}^* + pV^*}{RT}\right)\right]^{-1} \tag{12.19}$$

In Equations 12.17–12.19 the symbols have the same meanings as in Section 10.3. An asterisk denotes diffusion creep parameters; their power-law equivalents have no asterisks.

The problem of the estimation of viscosity profiles is twofold. In regions without phase transitions the continuous variation of viscosity with depth $\eta(z)$ must be obtained from the corresponding variations in rheological parameters with temperature and pressure. At phase changes the discontinuities $\Delta\eta$ (if any) must be inferred from the corresponding discontinuities in the parameters. In the former case, the aim is to express the (p,T)-dependence of the creep parameters in terms of known quantities (the elastic moduli and their derivatives, the homologous temperature, and so on). In the latter case, the discontinuities associated with phase changes are derived from systematics, i.e. from a comparative study of materials whose properties are known. The common starting point is a seismological Earth model giving density, pressure, and elastic moduli as functions of depth, together with a thermal model giving the distribution of temperature and related quantities.

We first examine the possible viscosity changes at depths corresponding to polymorphic phase transitions, following the procedure devised by Sammis *et al.* (1977). Flow is assumed to be non-Newtonian ($n = 3$), under constant stress, and with creep activation enthalpy equal to the activation enthalpy for intrinsic oxygen diffusion. However, the argument does not lose its validity for $H > H^*$, provided that the effective diffusion coefficients D and D^* are of the same form and go through similar changes at phase transitions, which should be the case if creep is related to diffusion. The procedure can be extended to constant-strain rate power-law creep and to diffusion creep by using the relevant relations for the viscosities and similar lines of reasoning.

From Equation 12.17, the ratio η_b/η_a between effective viscosities below and above a phase transition (omitting the subscript σ) can be written as

$$\frac{\eta_b}{\eta_a} = \frac{(AD_0 b)_a}{(AD_0 b)_b} \left(\frac{\mu_b}{\mu_a}\right)^2 \exp\left(\frac{\Delta E + p\Delta V}{RT}\right)$$

337

where $\Delta E = E_b - E_a$ and $\Delta V = V_b - V_a$. This ratio can be expressed in terms of the Gibbs free energy by recalling that (see Section 9.4)

$$D_0 = D'_0 \exp(S/R)$$

where S is the activation entropy of the process and D'_0 depends only on lattice spacings and frequencies. Therefore

$$\frac{\eta_b}{\eta_a} = \frac{(AD'_0b)_a}{(AD'_0b)_b} \left(\frac{\mu_b}{\mu_a}\right)^2 \exp\left(-\frac{\Delta S}{R}\right) \exp\left(\frac{\Delta E + p\Delta V}{RT}\right) = \chi \, \exp\left(\frac{\Delta G}{RT}\right)$$

$$(12.20)$$

The viscosity change can consequently be estimated if the quantities

$$\chi = \frac{(AD'_0b)_a}{(AD'_0b)_b} \left(\frac{\mu_b}{\mu_a}\right)^2 \tag{12.21}$$

$$\Delta G = \Delta E + p\Delta V - T\Delta S \tag{12.22}$$

can be expressed as functions of known parameters.

By writing D'_0 in Equation 12.21 in terms of lattice frequency, and using the relation with seismic wave velocities of the latter, Sammis *et al.* (1977) obtained an upper bound for χ. They estimate $\chi \leqslant 3$ across any mantle phase transition. The quantity ΔG (Eqn 12.22) can be evaluated by integrating $dG = V \, dp - S \, dT$, which yields

$$G(p, T) = G_0(T) + \int_0^p V \, dp$$

where the subscript zero denotes the zero-pressure value. Therefore, assuming ΔV to be independent of pressure, the change in the Gibbs free energy is

$$\Delta G = \Delta G_0 + p \, \Delta V$$

But $\Delta G_0 = \Delta E_0 - T \, \Delta S_0 \simeq \Delta E_0$ if the activation entropy is the same for different phases (a reasonable assumption, at least for diffusion – cf. Sammis *et al.* 1977), and consequently Equation 12.20 becomes

$$\frac{\eta_b}{\eta_a} \simeq \chi \, \exp\left(\frac{\Delta E_0 + p \, \Delta V}{RT}\right) \tag{12.23}$$

As activation volume is related to atomic volume, it is likely to be smaller in the high-pressure phase (if the same process controls creep), i.e. $\Delta V \leqslant 0$. Consequently, an upper bound estimate for the viscosity increase across the phase change is

$$\frac{\eta_b}{\eta_a} \lesssim \chi \, \exp\!\left(\frac{\Delta E_0}{RT}\right) \qquad (12.24)$$

The increase in creep activation energy can be inferred from systematics if it is assumed that $\Delta E_0 = \Delta E_0^*$. Using an empirical relation between E_0^* and oxygen-ion packing (volume per oxygen ion) that holds for a variety of materials, ΔE_0^* can be expressed in terms of densities (Sammis *et al.* 1977). Then Equation 12.24 yields an upper bound estimate of the viscosity increase. The maximum decrease in activation volume can be derived from Equation 12.23 on the assumption that $\eta_b/\eta_a \geqslant 1$, i.e. that the viscosity for a given creep process does not decrease across a phase transition.

Sammis *et al.* (1977) applied the above procedure for constant-stress power-law viscosity to pre-PREM seismological models. Ranalli & Fischer (1984) used PREM (see Section 6.6) to estimate increases in both constant-stress and constant-strain rate power-law viscosity, and diffusion-creep viscosity. In no case is the viscosity increase across any transition significantly larger than one order of magnitude, and often it is considerably less. The 220-km discontinuity has minor effects only. Typical changes in activation energy and volume across the 400- and 670-km phase changes are an increase of 15–30 kJ mol^{-1} and a decrease of $1-2 \times 10^{-6}$ m^3 mol^{-1}, respectively. The maximum viscosity increase across the mantle transition zone is unlikely to be by more than a factor of 10^2, and could be by a factor of ten or less. (The discontinuities in viscosity inferred by Ranalli & Fischer 1984 are shown in Figs 12.5–7.)

We now consider the continuous variations $\eta(z)$ in homogeneous regions of the mantle. From Equations 12.17–12.19 the most important depth-dependence of viscosity is seen to be contained in the effective diffusion coefficient. We write D (or, equivalently, D^*) in the form

$$D = D_0' \exp(S/R)\exp(-H/RT) \qquad (12.25)$$

to emphasize that – apart from the explicit temperature-dependence – the two most important factors affecting viscosity variations with depth are the corresponding variations in activation entropy and enthalpy, which we examine separately.

The creep activation enthalpy of a given mineral phase varies with depth in a manner depending on the variation of the activation volume for the process. There are several methods of estimating $V(z)$. It can be assumed to vary with pressure as the volume of a cavity in an elastic matrix (O'Connell 1977; see Section 10.7), or as the volume of the matrix itself (Karato 1981; obviously this procedure yields a smaller variation, since the compressibility of the matrix is less than the effective compressibility of the cavity). Methods based on elastic strain energy models (the so-called Zener–Keyes formulation) and on the empirical correlation of diffusivity with melting temperature (the Weertman formulation) have been reviewed and critically assessed by Sammis *et al.* (1981) and Poirier & Liebermann (1984). They are all implicitly or explicitly based on the assumption that the activation volume is that for vacancy diffusion, but as Poirier & Liebermann (1984) pointed out, this is not a necessary restriction, since the Gibbs free energy of activation can always be related to a strain energy model of the dislocation. In the sequel no distinction is therefore made between creep and diffusion parameters, with the assumption that the relevant ones apply without affecting the argument. Also, specification of isothermal or adiabatic conditions is neglected, as the moduli for one can easily be converted to those for the other.

The elastic strain energy associated with the activation process may be assumed to be purely dilatational (Zener model) or purely shear (Keyes model). The respective activation volumes are denoted by V_k and V_μ. For instance, in the latter case the Gibbs free energy can be expressed as (cf. Poirier & Liebermann 1984)

$$G = \mu \Omega_M$$

where Ω_M is the atomic or molecular volume (per mole). Using the thermodynamic definition of V and the relation between Ω_M and bulk modulus k,

$$V = \frac{\partial G}{\partial p} \qquad \frac{\partial \ln \Omega_M}{\partial p} = \frac{1}{\Omega_M} \frac{\partial \Omega_M}{\partial p} = -\frac{1}{k}$$

the activation volume V_μ is given by

$$V_\mu = G\left(\frac{\partial \ln \mu}{\partial p} - \frac{1}{k}\right) \tag{12.26}$$

Recalling the definition of entropy and the relation between Ω_M and the thermal expansion coefficient α,

$$S = -\frac{\partial G}{\partial T} \qquad \frac{\partial \ln \Omega_M}{\partial T} = \frac{1}{\Omega_M} \frac{\partial \Omega_M}{\partial T} = \alpha$$

the entropy becomes

$$S = -G\left(\frac{\partial \ln \mu}{\partial T} + \alpha\right) \tag{12.27}$$

Combining Equations 12.26 and 12.27 we obtain, after some algebra,

$$V_\mu = H\left(\frac{\partial \ln \mu}{\partial p} - \frac{1}{k}\right)\left(1 - \frac{\partial \ln \mu}{\partial T} - \alpha T\right)^{-1} \tag{12.28}$$

Similarly, for the dilatational model,

$$V_k = H\left(\frac{\partial \ln k}{\partial p} - \frac{1}{k}\right)\left(1 - \frac{\partial \ln k}{\partial T} - \alpha T\right)^{-1} \tag{12.29}$$

Equations 12.28 and 12.29 give the activation volume as a function of elastic moduli and their derivatives, temperature and thermal expansion, and activation enthalpy. Comparing them with data for metals, alkali halides, and olivine, Sammis *et al.* (1981) concluded that V_μ and V_k are a lower and upper estimate, respectively, of the activation volume.

The Keyes and Zener relations reduce to the same if the shear and bulk modulus have identical pressure and temperature derivatives. As Poirier & Liebermann (1984) pointed out, this is equivalent to assuming equality of the Grüneisen parameters for all acoustic modes. The relation between elastic moduli and the thermal Grüneisen parameter γ (defined in Section 7.2) then becomes

$$-\tfrac{1}{6} + \frac{k}{2}\frac{\partial \ln \mu}{\partial p} = -\tfrac{1}{6} + \frac{k}{2}\frac{\partial \ln k}{\partial p} = \gamma$$

and Equation 12.26 (and the corresponding one for V_k) can consequently be written as

$$V_\mu = V_k = V = (2G/k)(\gamma - \tfrac{1}{3}) \tag{12.30}$$

Differentiating Equation 12.30 one obtains

$$\frac{\partial \ln V}{\partial p} = (\gamma - \tfrac{1}{3})^{-1}\frac{\partial \gamma}{\partial p} - \frac{1}{k} \tag{12.31}$$

and, assuming that $\partial \gamma/\partial p = -\gamma/k$ (which is correct if γ varies with pressure as the atomic volume),

$$\frac{\partial \ln V}{\partial p} = -\frac{1}{k} \left[1 + \gamma(\gamma - \tfrac{1}{3})^{-1}\right] \tag{12.32}$$

Equation 12.32 gives the activation volume in terms of bulk modulus and Grüneisen parameter, each of which is approximately known as a function of depth. The assumptions under which it is derived (i.e. that G is entirely elastic strain energy as in the Zener–Keyes formulation, and that the (p,T)-derivatives of shear and bulk modulus are the same) are likely to be approximately correct for the mantle (cf. the discussion in Poirier & Liebermann 1984).

An apparent activation volume V' can be estimated from the empirical correlation between diffusivity and melting temperature (Eqn 9.17). As already noted, the main depth-dependence of viscosity is contained in the effective diffusion coefficient; therefore we can write

$$\eta = \eta_0 \exp(H/RT) = \eta_0 \exp(cT_m/T) \tag{12.33}$$

where $c \simeq 30$ for silicates and oxides. It follows that the apparent activation volume can be expressed in terms of $T_m(p)$ (the so-called Weertman formulation)

$$V' = RT \frac{\partial \ln \eta}{\partial p} = cR \frac{\partial T_m}{\partial p} \tag{12.34}$$

The Zener–Keyes and Weertman formulations are related if the melting temperature follows the Lindemann Law (Eqn 7.25), since, using the equality of the exponents in Equation 12.33 (which implies $H = cRT_m$), Equation 12.34 becomes

$$V' = cRT_m \frac{\partial \ln T_m}{\partial p} = \frac{2H}{k} (\gamma - \tfrac{1}{3}) \tag{12.35}$$

which, when compared with Equation 12.30, yields

$$V'/V = H/G \tag{12.36}$$

The above ratio, when expressed in terms of elastic moduli and their derivatives, is found to be in the range 1.1–1.4 for oxides and silicates, and about 1.2 throughout the lower mantle (Poirier & Liebermann 1984). On the other hand, the Zener–Keyes formulation (in its original form or as a function of the Grüneisen parameter) yields estimates that are practically equal to those obtained from the O'Connell empirical relation (Eqn 10.45). The predicted decrease in activation volume across the lower mantle is about 50 per cent (and this, of course, has to be computed from whatever initial value applies below the transition zone, which will in any case be less than the laboratory value). The decrease in activation volume is therefore a powerful factor limiting the increase in viscosity with depth.

Returning to Equation 12.25 we see that − besides enthalpy − the only other major source of variation in the effective diffusion coefficient could be the activation entropy. For diffusion in metals S^* under standard (p, T)-conditions is approximately equal to the universal gas constant R; and a relation between S^* and V^* based on the Keyes model predicts $S^* \simeq 5R$ in the lower mantle (cf. Sammis et al. 1977). These relations hold for intrinsic diffusion. An empirical estimate of the variation of activation entropy with depth in the lower mantle can be obtained from Equation 12.36. Using the definition of the Gibbs free energy, the entropy can be written as

$$S = \frac{H}{T} \left(1 - \frac{V}{V'} \right) \tag{12.37}$$

In the lower mantle, $V/V' \simeq 0.8$ (constant), and temperature varies approximately from 2000 K at the top to 3000 K at the bottom. The corresponding limits for oxygen diffusion enthalpy and creep activation enthalpy (calculated from the O'Connell formula, with initial values as in Table 10.1 modified by the estimated effects of phase changes) are $6 \times 10^5 \leqslant H^* \leqslant 9 \times 10^5$ J mol^{-1}, and $8 \times 10^5 \leqslant H \leqslant 12 \times 10^5$ J mol^{-1}, respectively. It seems therefore that the activation entropies are approximately constant with depth, with values $S^* \simeq 7R$ and $S \simeq 10R$. On this basis there is no reason to expect any significant first-order change in effective diffusivity with depth, other than that originated by the enthalpy term.

Methods for estimating the maximum viscosity increase at a phase transition (Eqn 12.24) and the viscosity variation with depth in homogeneous regions (by expressing activation volume and enthalpy in terms of elastic and thermal parameters − Eqns 12.28, 29, or 32 − or equivalently, the O'Connell formula) can be used to produce viscosity–depth profiles for the mantle (cf. Sammis et al. 1977, Ranalli & Fischer 1984, Ellsworth et al. 1985). The main conclusions of these studies converge and can be summarized as follows.

(a) The viscosity of the sub-lithospheric upper mantle, both for power-law and Newtonian rheologies, is in the range 10^{20}–10^{21} Pa s.

(b) The total increase in viscosity consequent upon phase transitions is between one and two orders of magnitude.
(c) The viscosity in the lower mantle depends on the rheology and the assumed geotherm, but in any case the variations in constant-strain rate power-law viscosity are lower (about one order of magnitude) than either constant-stress or Newtonian viscosity (two to three orders of magnitude).

The results of Ranalli & Fischer (1984) are shown in Figures 12.4–12.7. Both non-Newtonian and Newtonian viscosities (Eqns 12.17–19) are calculated, using the zero-pressure parameters listed in Table 10.1, and estimating discontinuous viscosity changes in the transition zone by the procedure of Sammis *et al.* (1977); variations of activation volume with depth in homogeneous regions are derived from the O'Connell formula. Elastic moduli, density, and pressure are given by the PREM model. Viscosity profiles are obtained for the two thermal models depicted in Figure 12.4. Both models have identical temperature distribution in the upper mantle, derived by linear interpolation between three fixed points subject to some seismic and experimental control ($T = 1500$ K at $z = 100$ km, 1700 K at 400 km, and 2000 K at 670 km). In the lower mantle, model TBL1 has a uniform geothermal gradient of 0.35 K km^{-1} down to a 150-km thick lower thermal boundary layer

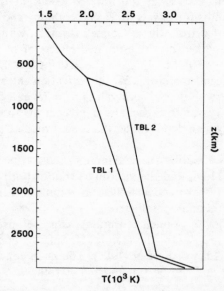

Figure 12.4 Mantle thermal models used in the estimation of viscosity. See text for discussion (from Ranalli & Fischer 1984).

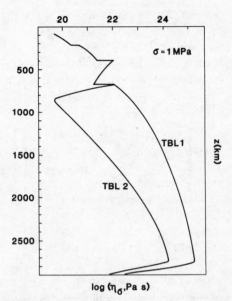

Figure 12.5 Viscosity profiles for a power-law rheology mantle at constant stress (from Ranalli & Fischer 1984).

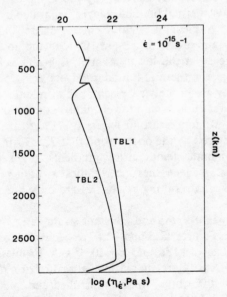

Figure 12.6 Viscosity profiles for a power-law rheology mantle at constant strain rate (from Ranalli & Fischer 1984).

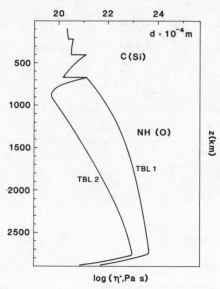

Figure 12.7 Viscosity profiles for a mantle with Newtonian rheology and constant grain size (from Ranalli & Fischer 1984).

across which the temperature increases by 450 K; model TBL2 has a mid-mantle and a lower thermal boundary layer, separated by a region where the geothermal gradient is 0.20 K km^{-1}. These models are very schematic; uncertainties are at least $\pm 100 \text{ K}$ in the upper mantle fixed points, and increase in the lower mantle, as both thickness and total temperature change in thermal boundary layers are poorly constrained. Nevertheless, they are broadly representative of mantle-wide convection (TBL1) and separate upper- and lower-mantle convection (TBL2). Both models result in a temperature in excess of 3100 K at the mantle/core boundary, as required by the observation that $T > T_m$ in the outer core; and the lower-mantle temperature gradients bracket the inferred adiabatic gradients. Given the uncertainties in both temperature and creep parameters, the resulting viscosities are only order-of-magnitude estimates.

Profiles of constant-stress and constant-strain rate power-law viscosities are shown in Figures 12.5 and 12.6, respectively, for representative mantle values of $\sigma = 1 \text{ MPa}$ and $\dot{\varepsilon} = 10^{-15} \text{ s}^{-1}$. Profiles of Newtonian diffusion-creep viscosity are shown in Figure 12.7 for $d = 0.1 \text{ mm}$. (In the upper mantle diffusion is silicon-controlled and mainly along grain boundaries; in the lower mantle oxygen-controlled bulk diffusion is predominant. The transition depth for both mechanisms is situated ap-

346

proximately at the lower boundary of the transition zone for the chosen parameters.)

The following general observations can be made. For power-law rheology, variations in η_σ are much larger than in $\eta_{\dot{\epsilon}}$, as the (p, T)-dependence of the former is much stronger. The total viscosity increase across the whole mantle is about four orders of magnitude if it deforms under constant-stress conditions, and less than two orders of magnitude if it deforms at a constant strain rate. In both cases, about half of the total viscosity increase occurs continuously across the lower mantle. The Newtonian viscosity η^* is intermediate between η_σ and $\eta_{\dot{\epsilon}}$, with an increase of about three orders of magnitude.

The occurrence of a non-adiabatic thermal boundary layer at the base of the mantle results in a sharp decrease in viscosity that could be associated with the generation of convective plumes. A mid-mantle non-adiabatic thermal boundary layer causes a corresponding viscosity decrease over a depth scale of about 200 km, followed by gradual increase governed by the pressure and adiabatic temperature gradients.

The viscosity profiles are subject to the various assumptions and approximations that have been previously discussed. First and foremost, and worth emphasizing again, is the hypothesis that deformation processes in the mantle are the same as in the laboratory, although creep parameters vary with mineral phase and depth. Also conditions of constant stress, strain rate, or grain size are not strictly realized in geodynamics. Nevertheless, it is remarkable that both a *constant-strain rate power-law rheology mantle* and a *Newtonian mantle* satisfy the constraints imposed by geophysical data (see Sections 8.2 & 3). The minimum variation in viscosity is achieved by the former, but the latter is within acceptable bounds. Nowhere, except perhaps in the lowermost mantle, do $\eta_{\dot{\epsilon}}$ and η^* differ by more than one order of magnitude. However, an increase in grain size by a factor of three or more would increase η^* by at least one order of magnitude (cf. Eqn 12.19), and consequently the possibility of a Newtonian mantle (excluding Harper–Dorn creep) is tied with the occurrence of a uniformly fine grain size.

The viscosity increase between the upper and the lower mantle is not sufficient to prevent mantle-wide convection, although the continuous increase in the lower mantle may affect the pattern of flow. As Poirier & Liebermann (1984) pointed out, an isoviscous lower mantle cannot be attained for realistic adiabatic gradients. However, as discussed in Sections 6.6 and 7.9, separate convection in the upper and the lower mantle is a possibility suggested by geochemical evidence. In this case the occurrence of a mid-mantle thermal boundary layer decreases the average viscosity of the lower mantle (because the lower mantle is hotter), but is insufficient to make the lower mantle less viscous than the upper mantle; on the contrary, it brings average upper- and lower-mantle viscosities closer than in the case of mantle-wide convection. So the *large-scale* viscosity structure of the mantle is more uniform in the case of layered convection, while the *detailed* structure is not. The viscosity signature of layered convection may still be beyond the resolving power of geophysical inversion.

Further improvements in the microphysical estimation of viscosity–depth profiles depend on more detailed rheological studies of analogue materials (possibly belonging to the same isomechanical group as perovskite) and on a more accurate knowledge of the kinetics of the relevant phase transformations.

12.5 Overview

The length scale of geodynamic processes varies from 10^{-10} m – a typical value of lattice spacing – to 10^6 m – the linear dimension of a mantle convection cell; similarly, the timescale varies from the 10^{-1} s of seismic body waves to the 10^{17} s of thermal phenomena. Within this huge span, deformation and flow occur at all scales, from minor folds with wavelengths of the order of centimetres, through regional folding and faulting, to plate-tectonic mechanical and thermal processes whose characteristic lengths and times are measured in thousands of kilometres and thousands of millions of years. The common precondition to a satisfactory analysis of these geological processes is an understanding of the rheology of Earth materials and of its dependence on intrinsic and extrinsic parameters – foremost among these being time.

In this book the problem of the rheology of the Earth has been approached from two complementary viewpoints, the continuum-mechanical and the microphysical. Both are necessary in attempting a quantitative description of geodynamic processes; one without the other can only give incomplete answers to incomplete questions. An integration of the two is the only way to an understanding of the dynamic processes affecting the history of the planet.

References

Allègre, C. J., S. R. Hart and J. F. Minster 1983a. Chemical structure and evolution of the mantle and continents determined by inversion of Nd and Sr isotopic data, I. Theoretical methods. *Earth and Planetary Science Letters* **66**, 177–90.

Allègre, C. J., S. R. Hart and J. F. Minster 1983b. Chemical structure and evolution of the mantle and continents determined by inversion of Nd and Sr isotopic data, II. Numerical experiments and discussion. *Earth and Planetary Science Letters* **66**, 191–213.

Anderson, D. L. 1982. Hotspots, polar wander, Mesozoic convection and the geoid. *Nature* **297**, 391–3.

Anderson, D. L. 1984. The Earth as a planet: paradigms and paradoxes. *Science* **223**, 347–55.

Anderson, D. L. and R. S. Hart 1978. *Q* of the Earth. *Journal of Geophysical Research* **83**, 5869–82.

Anderson, D. L. and J. B. Minster 1979. The frequency dependence of *Q* in the Earth and implications for mantle rheology and Chandler wobble. *Geophysical Journal of the Royal Astronomical Society* **58**, 431–40.

Anderson, E. M. 1951. *The dynamics of faulting and dyke formation with applications to Britain*, 2nd edn. Edinburgh: Oliver & Boyd.

Artyushkov, E. V. 1973. Stresses in the lithosphere caused by crustal thickness inhomogeneities. *Journal of Geophysical Research* **78**, 7675–708.

Artyushkov, E. V. 1983. *Geodynamics*. Amsterdam: Elsevier.

Ashby, M. F. and R. A. Verrall 1978. Micromechanisms of flow and fracture, and their relevance to the rheology of the upper mantle. *Philosophical Transactions of the Royal Society London A* **288**, 59–95.

Batchelor, G. K. 1967. *An introduction to fluid dynamics*. Cambridge: Cambridge University Press.

Beaumont, C. 1981. Foreland basins. *Geophysical Journal of the Royal Astronomical Society* **65**, 291–329.

Bell, K., J. Blenkinsop, T. J. S. Cole and D. P. Menagh 1982. Evidence from Sr isotopes for long-lived heterogeneities in the upper mantle. *Nature* **298**, 251–3.

Berckhemer, H., W. Kampfmann, E. Aulbach and H. Schmeling 1982. Shear modulus and *Q* of forsterite and dunite near partial melting from forced-oscillation experiments. *Physics of the Earth and Planetary Interiors* **29**, 30–41.

Birch, F. 1966. Compressibility; elastic constants. In *Handbook of physical constants*, rev. edn, S. P. Clark, Jr. (ed.), Geological Society of America, Memoir **97**, 97–173.

Bird, J. M. (ed.) 1980. *Plate tectonics*. Washington: American Geophysical Union.

Blacic, J. D. and J. M. Christie 1984. Plasticity and hydrolytic weakening of quartz single crystals. *Journal of Geophysical Research* **89**, 4223–39.

Bolt, B. A. 1976. *Nuclear explosions and earthquakes*. New York: W. H. Freeman.

Bolt, B. A. 1982. *Inside the Earth*. New York: W. H. Freeman.

Bott, M. H. P. 1964. Convection in the Earth's mantle and the mechanism of continental drift. *Nature* **202**, 583–4.

Bott, M. H. P. 1982. *The interior of the Earth: its structure, constitution and evolution*, 2nd edn. London: Edward Arnold and New York: Elsevier.

Bott, M. H. P. and N. J. Kusznir 1984. The origin of tectonic stress in the lithosphere. *Tectonophysics* **105**, 1–13.

REFERENCES

Brown, G. C. and A. E. Mussett 1981. *The inaccessible Earth*. London: Allen & Unwin.

Brown, R. L. and P. B. Read 1983. Shuswap terrane of British Columbia: a Mesozoic 'core complex'. *Geology* 11, 164–8.

Brun, J. P. and P. R. Cobbold 1980. Strain heating and thermal softening in continental shear zones: a review. *Journal of Structural Geology* 2, 149–58.

Bullen, K. E. 1963. *An introduction to the theory of seismology*, 3rd edn. Cambridge: Cambridge University Press.

Bustin, R. M. 1983. Heating during thrust faulting in the Rocky Mountains: friction or fiction? *Tectonophysics*, 95, 302–28.

Carslaw, H. S. and J. C. Jaeger 1959. *Conduction of heat in solids*, 2nd edn. Oxford: Clarendon Press.

Cathles, L. M., III 1975. *The viscosity of the Earth's mantle*. Princeton: Princeton University Press.

Chapple, W. M. 1978. Mechanics of thin-skinned fold-and-thrust belts. *Bulletin of the Geological Society of America* 89, 1189–98.

Chen, W. P. and P. Molnar 1983. Focal depths of intracontinental and intraplate earthquakes and their implications for the thermal and mechanical properties of the lithosphere. *Journal of Geophysical Research* 88, 4183–214.

Chopra, P. N. and M. S. Paterson 1981. The experimental deformation of dunite. *Tectonophysics* 78, 453–73.

Chopra, P. N. and M. S. Paterson 1984. The role of water in the deformation of dunite. *Journal of Geophysical Research* 89, 7861–76.

Christensen, U. 1984. Convection with pressure- and temperature-dependent non-Newtonian rheology. *Geophysical Journal of the Royal Astronomical Society* 77, 343–84.

Christensen, U. R. and D. A. Yuen 1984. The interaction of a subducting lithospheric slab with a chemical or phase boundary. *Journal of Geophysical Research* 89, 4389–402.

Cloetingh, S., H. McQueen and K. Lambeck 1985. On a tectonic mechanism for regional sealevel variations. *Earth and Planetary Science Letters* 75, 157–66.

Cook, F. A. and D. L. Turcotte 1981. Parameterized convection and the thermal evolution of the Earth. *Tectonophysics* 75, 1–17.

Cottrell, A. H. 1953. *Dislocations and plastic flow in crystals*. Oxford: Clarendon Press.

Courtney, R. C. and C. Beaumont 1983. Thermally-activated creep and flexure of the oceanic lithosphere. *Nature* 305, 201–4.

Cox, A. (ed.) 1973. *Plate tectonics and geomagnetic reversals*. New York: W. H. Freeman.

Crampin, S. 1984. An introduction to wave propagation in anisotropic media. *Geophysical Journal of the Royal Astronomical Society* 76, 17–28.

Creager, K. C. and T. H. Jordan 1984. Slab penetration into the lower mantle. *Journal of Geophysical Research* 89, 3031–49.

Cserepes, L. 1982. Numerical studies of non-Newtonian mantle convection. *Physics of the Earth and Planetary Interiors* 30, 49–61.

Dalmayrac, B. and P. Molnar 1981. Parallel thrust and normal faulting in Peru and constraints on the state of stress. *Earth and Planetary Science Letters* 55, 473–81.

Davies, G. F. 1984. Geophysical and isotopic constraints on mantle convection: an interim synthesis. *Journal of Geophysical Research* 89, 6017–40.

Davis, D., J. Suppe and F. A. Dahlen 1983. Mechanics of fold-and-thrust belts and accretionary wedges. *Journal of Geophysical Research* 88, 1153–72.

Dewey, J. F. 1982. Plate tectonics and the evolution of the British Isles. *Journal of the Geological Society of London* 139, 371–412.

Dorman, J. 1969. Seismic surface-wave data on the upper mantle. In *The Earth's crust and upper mantle*, P. J. Hart (ed.), American Geophysical Union, Geophysical Monograph **13**, 257–65.

Dziewonski, A. M. 1984. Mapping the lower mantle: determination of lateral heterogeneity in *P* velocity up to degree and order 6. *Journal of Geophysical Research* **89**, 5929–52.

Dziewonski, A. M. and D. L. Anderson 1981. Preliminary reference Earth model. *Physics of the Earth and Planetary Interiors* **25**, 297–356.

Dziewonski, A. M. and E. Boschi (eds) 1980. *Physics of the Earth's interior*. Bologna: Società Italiana di Fisica and Amsterdam: North-Holland.

Dziewonski, A. M., A. L. Hales and E. R. Lapwood 1975. Parametrically simple Earth models consistent with geophysical data. *Physics of the Earth and Planetary Interiors* **10**, 12–48.

Elliott, D. 1976. The motion of thrust sheets. *Journal of Geophysical Research* **81**, 949–63.

Ellsworth, K., G. Schubert and C. G. Sammis 1985. Viscosity profile of the lower mantle. *Geophysical Journal of the Royal Astronomical Society* **83**, 199–213.

England, P. and D. McKenzie 1982. A thin viscous sheet model for continental deformation. *Geophysical Journal of the Royal Astronomical Society* **70**, 295–321.

England, P. and D. McKenzie 1983. Correction to: a thin viscous sheet model for continental deformation. *Geophysical Journal of the Royal Astronomical Society* **73**, 523–32.

Fleitout, L. and C. Froidevaux 1980. Thermal and mechanical evolution of shear zones. *Journal of Structural Geology* **2**, 159–64.

Fleitout, L. and C. Froidevaux 1982. Tectonics and topography for a lithosphere containing density heterogeneities. *Tectonics* **1**, 21–56.

Fleitout, L. and D. A. Yuen 1984. Steady state, secondary convection beneath lithospheric plates with temperature- and pressure-dependent viscosity. *Journal of Geophysical Research* **89**, 9227–44.

Forsyth, D. and S. Uyeda 1975. On the relative importance of the driving forces of plate motion. *Geophysical Journal of the Royal Astronomical Society* **43**, 163–200.

Frost, H. J. and M. F. Ashby 1982. *Deformation-mechanism maps – The plasticity and creep of metals and ceramics*. Oxford: Pergamon Press.

Fung, Y. C. 1977. *A first course in continuum mechanics*, 2nd edn. Englewood Cliffs N.J.: Prentice-Hall.

Garland, G. D. 1979. *Introduction to geophysics*, 2nd edn. Philadelphia: Saunders.

Gibb, R. A. 1983. Model for suturing of Superior and Churchill plates: an example of double indentation tectonics. *Geology* **11**, 413–17.

Gilbert, F. and A. M. Dziewonski 1975. An application of normal mode theory to the retrieval of structural parameters and source mechanisms from seismic spectra. *Philosophical Transactions of the Royal Society London A* **278**, 187–269.

Gittus, J. 1975. *Creep, viscoelasticity and creep fracture in solids*. New York: Wiley.

Gittus, J. 1978. High-temperature deformation of two-phase structures. *Philosophical Transactions of the Royal Society London A* **288**, 121–45.

Gough, D. I. 1983. Electromagnetic geophysics and global tectonics. *Journal of Geophysical Research* **88**, 3367–77.

Gough, D. I. 1984. Mantle upflow under North America and plate dynamics. *Nature* **311**, 428–33.

Green, H. W., II 1984. 'Pressure solution' creep: some causes and mechanisms. *Journal of Geophysical Research* **89**, 4313–18.

Gueguen, Y. 1979. High temperature olivine creep: evidence for control by edge dislocations. *Geophysical Research Letters* **6**, 375–60.

REFERENCES

Hager, B. H. 1984. Subducted slabs and the geoid: constraints on mantle rheology and flow. *Journal of Geophysical Research* **89**, 6003–15.

Hager, B. H. and R. J. O'Connell 1981. A simple global model of plate dynamics and mantle convection. *Journal of Geophysical Research* **86**, 4843–67.

Hager, B. H., R. W. Clayton, M. A. Richards, R. P. Comer and A. M. Dziewonski 1985. Lower mantle heterogeneity, dynamic topography and the geoid. *Nature* **313**, 541–5.

Handin, J. 1966. Strength and ductility. In *Handbook of physical constants*, rev. edn., S. P. Clark, Jr. (ed.), Geological Society of America, Memoir **97**, 223–89.

Hasegawa, H. S., J. Adams and K. Yamazaki 1985. Upper crustal stresses and vertical stress migration in eastern Canada. *Journal of Geophysical Research* **90**, 3637–48.

Haskell, N. A. 1935. Motion of a viscous fluid under a surface load. *Physics* **6**, 265–9.

Hill, R. 1950. *The mathematical theory of plasticity*. Oxford: Clarendon Press.

Hobbs, B. E. 1983. Constraints on the mechanism of deformation of olivine imposed by defect chemistry. *Tectonophysics* **92**, 35–69.

Hobbs, B. E., W. D. Means and P. F. Williams 1976. *An outline of structural geology*. New York: Wiley.

Hoffman, N. R. A. and D. P. McKenzie 1985. The destruction of geochemical heterogeneities by differential fluid motions during mantle convection. *Geophysical Journal of the Royal Astronomical Society* **82**, 163–206.

Hull, D. 1975. *Introduction to dislocations*, 2nd edn. Oxford: Pergamon Press.

Isacks, B., J. Oliver and L. R. Sykes 1968. Seismology and the new global tectonics. *Journal of Geophysical Research* **73**, 5855–99.

Jacobs, J. A. 1974. *A textbook on geonomy*. New York: Halsted Press.

Jacoby, W. R. and H. Schmeling 1982. On the effects of the lithosphere on mantle convection and evolution. *Physics of the Earth and Planetary Interiors* **29**, 305–19.

Jaeger, J. C. and N. G. W. Cook 1979. *Fundamentals of rock mechanics*, 3rd edn. London: Chapman and Hall.

Jaoul, O., C. Froidevaux, W. B. Durham and M. Michaut 1980. Oxygen self-diffusion in forsterite: implications for the high-temperature creep mechanism. *Earth and Planetary Science Letters* **47**, 391–7.

Jeanloz, R. and F. M. Richter 1979. Convection, composition, and the thermal state of the lower mantle. *Journal of Geophysical Research* **84**, 5497–504.

Jeanloz, R. and A. B. Thompson 1983. Phase transitions and mantle discontinuities. *Reviews of Geophysics and Space Physics* **21**, 51–74.

Jeffreys, H. 1963. *Cartesian tensors* (reprinted). Cambridge: Cambridge University Press.

Jeffreys, H. 1976. *The Earth: its origin, history and physical constitution*, 6th edn. Cambridge: Cambridge University Press.

Jeffreys, H. and K. E. Bullen 1940. *Seismological tables*. London: British Association for the Advancement of Science.

Johnson, W., R. Sowerby and J. B. Haddow 1970. *Plane-strain slip-line fields: theory and bibliography*. London: Edward Arnold.

Jordan, T. H. 1978. Composition and development of the continental tectosphere. *Nature* **274**, 544–8.

Jordan, T. H. 1980. Earth structure from seismological observations. In *Physics of the Earth's interior*, A. M. Dziewonski and E. Boschi (eds), 1–40. Bologna: Società Italiana di Fisica and Amsterdam: North-Holland.

Kanamori, H. 1980. The state of stress in the Earth's lithosphere. In *Physics of the Earth's interior*, A. M. Dziewonski and E. Boschi (eds), 531–52. Bologna: Società Italiana di Fisica and Amsterdam: North-Holland.

REFERENCES

Karato, S. 1981. Rheology of the lower mantle. *Physics of the Earth and Planetary Interiors* **24**, 1–14.

Kirby, S. H. 1983. Rheology of the lithosphere. *Reviews of Geophysics and Space Physics* **21**, 1458–87.

Kirby, S. H. 1985. Rock mechanics observations pertinent to the rheology of the continental lithosphere and the localization of strain along shear zones. *Tectonophysics* **119**, 1–27.

Knopoff, L. 1983. The thickness of the lithosphere from the dispersion of surface waves. *Geophysical Journal of the Royal Astronomical Society* **74**, 55–81.

Kreyszig, E. 1983. *Advanced engineering mathematics*, 5th edn. New York: Wiley.

Kusznir, N. J. 1982. Lithosphere response to externally and internally derived stresses: a viscoelastic stress guide with amplification. *Geophysical Journal of the Royal Astronomical Society* **70**, 399–414.

Kusznir, N. and G. Karner 1985. Dependence of the flexural rigidity of the continental lithosphere on rheology and temperature. *Nature* **316**, 138–42.

Lamb, H. 1945. *Hydrodynamics*, 6th edn (reprinted). New York: Dover.

Le Pichon, X., J. Francheteau and J. Bonnin 1973. *Plate tectonics*. Amsterdam: Elsevier.

Loper, D. E. 1985. A simple model of whole-mantle convection. *Journal of Geophysical Research* **90**, 1809–36.

Love, A. E. H. 1944. *A treatise on the mathematical theory of elasticity*, 4th edn (reprinted). New York: Dover.

Mainprice, D. H. and M. S. Paterson 1984. Experimental studies of the role of water in the plasticity of quartzites. *Journal of Geophysical Research* **89**, 4257–69.

McKenzie, D. P. 1972. Plate tectonics. In *The nature of the solid Earth*, E. C. Robertson (ed.), 323–60. New York: McGraw-Hill.

McKenzie, D. and N. Weiss 1975. Speculations on the thermal and tectonic history of the Earth. *Geophysical Journal of the Royal Astronomical Society* **42**, 131–74.

McKenzie, D., A. Watts, B. Parsons and M. Roufosse 1980. Planform of mantle convection beneath the Pacific Ocean. *Nature* **288**, 442–6.

Means, W. D. 1976. *Stress and strain: basic concepts of continuum mechanics for geologists*. New York: Springer.

Melosh, H. J. 1980. Rheology of the Earth: theory and observation. In *Physics of the Earth's interior*, A. M. Dziewonski and E. Boschi (eds), 318–36. Bologna: Società Italiana di Fisica and Amsterdam: North-Holland.

Mercier, J. C. 1980. Magnitude of the continental lithospheric stresses inferred from rheomorphic petrology. *Journal of Geophysical Research* **85**, 6293–303.

Minster, J. B. 1980. Anelasticity and attenuation. In *Physics of the Earth's interior*, A. M. Dziewonski and E. Boschi (eds), 152–212. Bologna: Società Italiana di Fisica and Amsterdam: North-Holland.

Minster, J. B. and D. L. Anderson 1981. A model of dislocation-controlled rheology for the mantle. *Philosophical Transactions of the Royal Society London A* **299**, 319–56.

Morgan, W. J. 1971. Convection plumes in the lower mantle. *Nature* **230**, 42–3.

Mukherjee, A. K. 1975. High-temperature creep. In *Treatise on material science and technology*, vol. 6, R. J. Arsenault (ed.), 163–224. New York: Academic Press.

Murrell, S. A. F. 1981. The rock mechanics of thrust and nappe formation. In *Thrust and nappe tectonics*, K. R. McClay and N. J. Price (eds), Geological Society of London, Special Publication **9**, 99–109.

Nadai, A. 1950, 1963. *Theory of flow and fracture of solids,* 2 vols. New York: McGraw-Hill.

Nicolas, A. and J. P. Poirier 1976. *Crystalline plasticity and solid state flow in metamorphic rocks*. London: Wiley.

O'Connell, R. J. 1977. On the scale of mantle convection. *Tectonophysics* **38**, 119–36.

O'Connell, R. J. and B. H. Hager 1980. On the thermal state of the Earth. In *Physics of the Earth's interior*, A. M. Dziewonski and E. Boschi (eds), 270–317. Bologna: Società Italiana di Fisica and Amsterdam: North-Holland.

Oliver, J., F. Cook and L. Brown 1983. COCORP and the continental crust. *Journal of Geophysical Research* **88**, 3329–47.

Olson, P. 1984. An experimental approach to thermal convection in a two-layered mantle. *Journal of Geophysical Research* **89**, 11293–301.

Olson, P., D. A. Yuen and D. Balsiger 1984. Mixing of passive heterogeneities by mantle convection. *Journal of Geophysical Research* **89**, 425–36.

Orowan, E. 1965. Convection in a non-Newtonian mantle, continental drift, and mountain building. *Philosophical Transactions of the Royal Society London A* **258**, 284–313.

Oxburgh, E. R. 1972. Flake tectonics and continental collision. *Nature* **239**, 202–4.

Oxburgh, E. R. 1980. Mantle mineralogy and dynamics. In *Physics of the Earth's interior*, A. M. Dziewonski and E. Boschi (eds), 247–69. Bologna: Società Italiana di Fisica and Amsterdam: North-Holland.

Parmentier, E. M. 1981. A possible mantle instability due to superplastic deformation associated with phase transitions. *Geophysical Research Letters* **8**, 143–6.

Paterson, M. S. 1978. *Experimental rock deformation – The brittle field*. Berlin: Springer-Verlag.

Paterson, W. S. B. 1981. *The physics of glaciers*, 2nd edn. Oxford: Pergamon Press.

Peltier, W. R. 1980. Mantle convection and viscosity. In *Physics of the Earth's interior*, A. M. Dziewonski and E. Boschi (eds), 362–431. Bologna: Società Italiana di Fisica and Amsterdam: North-Holland.

Peltier, W. R. 1981. Surface plates and thermal plumes: separate scales of the mantle convective circulation. In *Evolution of the Earth*, R. J. O'Connell and W. S. Fyfe (eds), American Geophysical Union and Geological Society of America, Geodynamics Series **5**, 229–48.

Peltier, W. R. 1984. The thickness of the continental lithosphere. *Journal of Geophysical Research* **89**, 11303–16.

Peltier, W. R. 1985. New constraints on transient lower mantle rheology and internal mantle buoyancy from glacial rebound data. *Nature* **318**, 614–17.

Peltier, W. R. and G. T. Jarvis 1982. Whole mantle convection and the thermal evolution of the Earth. *Physics of the Earth and Planetary Interiors* **29**, 281–304.

Peltier, W. R., P. Wu and D. A. Yuen 1981. The viscosities of the Earth's mantle. In *Anelasticity in the Earth*, F. D. Stacey, M. S. Paterson and A. Nicolas (eds), 59–77. American Geophysical Union and Geological Society of America, Geodynamics Series **4**.

Pippard, A. B. 1957. *Elements of classical thermodynamics for advanced students of physics*. Cambridge: Cambridge University Press.

Poirier, J. P. 1976. On the symmetrical role of cross-slip of screw dislocations and climb of edge dislocations as recovery processes controlling high-temperature creep. *Revue de Physique Appliquée* **11**, 731–8.

Poirier, J. P. 1982. On transformation plasticity. *Journal of Geophysical Research* **87**, 6791–7.

Poirier, J. P. 1985. *Creep of crystals – High-temperature deformation processes in metals, ceramics and minerals*. Cambridge: Cambridge University Press.

354

REFERENCES

Poirier, J. P. and R. C. Liebermann 1984. On the activation volume for creep and its variation with depth in the Earth's lower mantle. *Physics of the Earth and Planetary Interiors* **35**, 283–93.

Poirier, J. P., J. Peyronneau, J. Y. Gesland and G. Brebec 1983. Viscosity and conductivity of the lower mantle; an experimental study on a $MgSiO_3$ perovskite analogue, $KZnF_3$. *Physics of the Earth and Planetary Interiors* **32**, 273–87.

Pollack, H. N. and D. S. Chapman 1977. On the regional variation of heat flow, geotherms, and lithospheric thickness. *Tectonophysics* **38**, 279–96.

Price, R. A. 1981. The Cordilleran foreland thrust and fold belt in the southern Canadian Rocky Mountains. In *Thrust and nappe tetonics*, K. R. McClay and N. J. Price (eds), Geological Society of London, Special Publication **9**, 427–48.

Price, R. A. and E. W. Mountjoy 1970. Geologic structure of the Canadian Rocky Mountains between Bow and Athabasca Rivers – a progress report. *Geological Association of Canada, Special Paper* **6**, 7–25.

Quinlan, G. M. and C. Beaumont 1984. Appalachian thrusting, lithospheric flexure and the Paleozoic stratigraphy of the eastern interior of North America. *Canadian Journal of Earth Sciences* **21**, 973–96.

Ramberg, H. 1981. *Gravity, deformation and the Earth's crust*, 2nd edn. London: Academic Press.

Ramsay, J. G. 1967. *Folding and fracturing of rocks*. New York: McGraw-Hill.

Ranalli, G. 1977. Steady-state creep in the mantle. *Annali di Geofisica* **30**, 435–58.

Ranalli G. 1980. Regional models of the steady-state rheology of the upper mantle. In *Earth rheology, isostasy and eustasy*, N. A. Mörner (ed.), 111–23. Chichester: Wiley.

Ranalli, G. 1982. Deformation maps in grain size-stress space as a tool to investigate mantle rheology. *Physics of the Earth and Planetary Interiors* **29**, 42–50.

Ranalli, G. 1984. On the possibility of Newtonian flow in the upper mantle. *Tectonophysics* **108**, 179–92.

Ranalli, G. and B. Fischer 1984. Diffusion creep, dislocation creep, and mantle rheology. *Physics of the Earth and Planetary Interiors* **34**, 77–84.

Ranalli, G. and D. C. Murphy 1986. Rheological stratification of the lithosphere. *Tectonophysics* (in press).

Regan, J. and D. L. Anderson 1984. Anisotropic models of the upper mantle. *Physics of the Earth and Planetary Interiors* **35**, 227–63.

Reiner, M. 1960. *Lectures on theoretical rheology*, 3rd edn. Amsterdam: North-Holland.

Richter, F. M. 1977. On the driving mechanism of plate tectonics. *Tectonophysics* **38**, 61–88.

Richter, F. M. 1979. Focal mechanisms and seismic energy release of deep and intermediate earthquakes in the Tonga–Kermadec region and their bearing on the depth extent of mantle flow. *Journal of Geophysical Research* **84**, 6783–95.

Richter, F. M., and D. P. McKenzie 1981. On some consequences and possible causes of layered mantle convection. *Journal of Geophysical Research* **86**, 6133–42.

Ringwood, A. E. 1975. *Composition and petrology of the Earth's mantle*. New York: McGraw-Hill.

Ringwood, A. E. 1982. Phase transformations and differentiation in subducted lithosphere: implications for mantle dynamics, basalt petrogenesis and crustal evolution. *Journal of Geology* **90**, 611–43.

Rosenberg, H. M. 1978. *The solid state*, 2nd edn. Oxford: Oxford University Press.

Ross, J. V., H. G. Ave'lallemant and N. L. Carter 1980. Stress dependence of recrystallized-grain and subgrain size in olivine. *Tectonophysics* **70**, 39–61.

355

Rubie, D. C. 1984. The olivine → spinel transformation and the rheology of subducting lithosphere. *Nature* **308**, 505–8.

Runcorn, S. K. 1962. Palaeomagnetic evidence for continental drift and its geophysical cause. In *Continental drift*, S. K. Runcorn (ed.), 1–40. New York: Academic Press.

Rutter, E. H. 1983. Pressure solution in nature, theory and experiment. *Journal of the Geological Society of London* **140**, 725–40.

Sabadini, R., D. A. Yuen and E. Boschi 1984. A comparison of the complete and truncated versions of the polar wander equations. *Journal of Geophysical Research* **89**, 7609–20.

Sabadini, R., D. A. Yuen and P. Gasperini 1985. The effects of transient rheology on the interpretation of lower mantle viscosity. *Geophysical Research Letters* **12**, 361–4.

Sammis, C. G., J. C. Smith, G. Schubert and D. A. Yuen 1977. Viscosity–depth profile of the Earth's mantle: effects of polymorphic phase transitions. *Journal of Geophysical Research* **82**, 3747–61.

Sammis, C. G., J. C. Smith and G. Schubert 1981. A critical assessment of estimation methods for activation volume. *Journal of Geophysical Research* **86**, 10707–18.

Scheidegger, A. E. 1982. *Principles of geodynamics*, 3rd edn. Berlin: Springer-Verlag.

Schloessin, H. H. 1978. Stresses and spreading resistance at contacts between spheres at high pressure. *Physics of the Earth and Planetary Interiors* **17**, 22–30.

Schubert, G. 1979. Subsolidus convection in the mantle of terrestrial planets. *Annual Review of Earth and Planetary Sciences* **7**, 289–342.

Sclater, J. G., C. Jaupart and D. Galson 1980. The heat flow through oceanic and continental crust and the heat loss of the Earth. *Reviews of Geophysics and Space Physics* **18**, 269–311.

Shankland, T. J., R. J. O'Connell and H. S. Waff 1981. Geophysical constraints on partial melt in the upper mantle. *Reviews of Geophysics and Space Physics* **19**, 394–406.

Sibson, R. H. 1974. Frictional constraints on thrust, wrench and normal faults. *Nature* **249**, 542–4.

Sibson, R. H. 1982. Fault zone models, heat flow, and the depth distribution of earthquakes in the continental crust of the United States. *Bulletin of the Seismological Society of America* **72**, 151–63.

Sleep, N. H. 1984. Tapping of magmas from ubiquitous mantle heterogeneities: an alternative to mantle plumes? *Journal of Geophysical Research* **89**, 10029–41.

Stacey, F. D. 1977a. *Physics of the Earth*, 2nd edn. New York: Wiley.

Stacey, F. D. 1977b. Applications of thermodynamics to fundamental Earth physics. *Geophysical Surveys* **3**, 175–204.

Stacey, F. D. 1977c. A thermal model of the Earth. *Physics of the Earth and Planetary Interiors* **15**, 341–8.

Stocker, R. L. and M. F. Ashby 1973. On the rheology of the upper mantle. *Reviews of Geophysics and Space Physics* **11**, 391–426.

Sykes, L. R. 1967. Mechanism of earthquakes and nature of faulting on the mid-oceanic ridges. *Journal of Geophysical Research* **72**, 2131–53.

Tapponnier, P. and P. Molnar 1976. Slip-line field theory and large-scale continental tectonics. *Nature* **264**, 319–24.

Thomas, M. D. and R. A. Gibb 1983. Convergent plate tectonics and related faults in the Canadian shield. In *Proceedings of the 4th international conference on basement tectonics*, R. H. Gabrielsen, I. B. Ramberg, D. Roberts and O. Steinlein (eds), 115–34. Salt Lake City: International Basement Tectonics Association.

Toksöz, M. N. and D. L. Anderson 1966. Phase velocities of long-period surface waves and structure of the upper mantle, I. *Journal of Geophysical Research* **71**, 1649–58.

REFERENCES

Tozer, D. C. 1967. Towards a theory of thermal convection in the mantle. In *The Earth's mantle*, T. K. Gaskell (ed.), 325–53. New York: Academic Press.

Tozer, D. C. 1981. The mechanical and electrical properties of the Earth's asthenosphere. *Physics of the Earth and Planetary Interiors* **25**, 280–96.

Tullis, J. A., G. L. Shelton and R. A. Yund 1979. Pressure dependence of rock strength: implications for hydrolytic weakening. *Bulletin de Minéralogie* **102**, 110–14.

Turcotte, D. L. 1983. Mechanisms of crustal deformation. *Journal of the Geological Society of London* **140**, 701–24.

Turcotte, D. L. and G. Schubert 1982. *Geodynamics – Applications of continuum physics to geological problems*. New York: Wiley.

Twiss, R. J. 1977. Theory and applicability of a recrystallized grain size paleopiezometer. *Pure and Applied Geophysics* **115**, 227–44.

Vaughan, P. J. and R. S. Coe 1981. Creep mechanism in Mg_2GeO_4: effects of a phase transition. *Journal of Geophysical Research* **86**, 389–404.

Wagner, C. A., F. J. Lerch, J. E. Brownd and J. A. Richardson 1977. Improvement in the geopotential derived from satellite and surface data (GEM 7 and 8). *Journal of Geophysical Research* **82**, 901–27.

Walcott, R. I. 1970. Flexural rigidity, thickness and viscosity of the lithosphere. *Journal of Geophysical Research* **75**, 3941–54.

Walcott, R. I. 1972. Late Quaternary vertical movements in eastern North America: quantitative evidence for glacio-isostatic rebound. *Reviews of Geophysics and Space Physics* **10**, 849–84.

Walcott, R. I. 1973. Structure of the Earth from glacio-isostatic rebound. *Annual Review of Earth and Planetary Sciences* **1**, 15–37.

Walcott, R. I. 1980. Rheological models and observational data of glacio-isostatic rebound. In *Earth rheology, isostasy and eustasy*, N. A. Mörner (ed.), 3–10. Chichester: Wiley.

Watts, A. B., G. D. Karner and M. S. Steckler 1982. Lithospheric flexure and the evolution of sedimentary basins. *Philosophical Transactions of the Royal Society London A* **305**, 249–81.

Weertman, J. 1970. The creep strength of the Earth's mantle. *Reviews of Geophysics and Space Physics* **8**, 145–68.

Weertman, J. 1978. Creep laws for the mantle of the Earth. *Philosophical Transactions of the Royal Society London A* **288**, 9–26.

White, J. C. and S. H. White 1981. On the structure of grain boundaries in tectonites. *Tectonophysics* **78**, 613–28.

White, S. H. and P. G. Bretan 1985. Rheological controls on the geometry of deep faults and the tectonic delamination of the continental crust. *Tectonics* **4**, 303–9.

Woodhouse, J. H. and A. M. Dziewonski 1984. Mapping the upper mantle: three-dimensional modelling of Earth structure by inversion of seismic waveforms. *Journal of Geophysical Research* **89**, 5953–86.

Wu, P. and W. R. Peltier 1983. Glacial isostatic adjustment and the free-air gravity anomaly as a constraint on deep mantle viscosity. *Geophysical Journal of the Royal Astronomical Society* **74**, 377–449.

Wyllie, P. J. 1971. *The dynamic Earth: textbook in geosciences*. New York: Wiley.

Yuen, D. A., L. Fleitout, G. Schubert and C. Froidevaux 1978. Shear deformation zones along major transform faults and subducting slabs. *Geophysical Journal of the Royal Astronomical Society* **54**, 93–119.

REFERENCES

Yuen, D. A., R. Sabadini and E. Boschi 1982. Mantle rheology from a geodynamical standpoint. *Rivista del Nuovo Cimento* **5**(8), 1–35.

Zeuch, D. H. 1983. On the inter-relationship between grain size sensitive creep and dynamic recrystallization of olivine. *Tectonophysics* **93**, 151–68.

Zoback, M. D. 1983. State of stress in the lithosphere. *Reviews of Geophysics and Space Physics* **21**, 1503–11.

Acknowledgements

Thanks are extended to the following authors and publishers for permission to use previously published illustrations (numbers refer to figures in this book). All figures have been redrafted. Complete references can be found in the bibliography.

Figs 5.2 & 5.4: M. S. Paterson, *Experimental rock deformation – the brittle field*, Springer-Verlag 1978. Figure 5.3: J. C. Jaeger & N. G. W. Cook, *Fundamentals of rock mechanics*, 3rd edn, Chapman & Hall 1979. Figs 6.3 & 6.9: B. A. Bolt, *Nuclear explosions and earthquakes*, W. H. Freeman & Company; copyright © 1976. Figs 6.4 & 6.11: B. A. Bolt, *Inside the Earth*, W. H. Freeman & Company; copyright © 1982. Fig. 6.7: T. H. Jordan, in *Physics of the Earth's interior*, A. M. Dziewonski & E. Boschi (eds), 1–40, Società Italiana di Fisica 1980. Figs 6.8 and 7.8: F. D. Stacey, *Physics of the Earth*, 2nd edn; copyright 1977 by John Wiley & Sons, Inc.; reproduced by permission. Fig. 6.10: (a) M. N. Toksöz & D. L. Anderson (1966), *Journal of Geophysical Research* 71, 1649–58, copyright by the American Geophysical Union; (b) J. Dorman, in *The Earth's crust and upper mantle*, P. J. Hart (ed.), 257–65, 1969, copyright by the American Geophysical Union. Fig. 6.12: D. L. Anderson & R. S. Hart (1978), *Journal of Geophysical Research* 83, 5869–82; copyright by the American Geophysical Union. Fig. 6.13: A. M. Dziewonski & D. L. Anderson (1981), *Physics of the Earth and Planetary Interiors* 25, 297–356; copyright by Elsevier Science Publishers. Figs 7.2 & 7.5: H. N. Pollack & D. S. Chapman (1977), *Tectonophysics* 38, 279–96; copyright by Elsevier Science Publishers. Figs 7.3 & 7.4: D. L. Turcotte & G. Schubert, *Geodynamics – applications of continuum physics to geological problems*; copyright © 1982 by John Wiley and Sons, Inc; reproduced by permission. Fig. 7.9: G. F. Davies (1984), *Journal of Geophysical Research* 89, 6017–40; copyright by the American Geophysical Union. Fig. 8.1: C.A. Wagner *et al.* (1977), *Journal of Geophysical Research* 82, 901–27; copyright by the American Geophysical Union. Fig. 8.3: (a) W. R. Peltier, in *Physics of the Earth's interior*, A. M. Dziewonski & E. Boschi (eds), 362–431, Società Italiana di Fisica 1980; (b) R. I. Walcott, *Annual Review of Earth and Planetary Sciences* 1, 15–37, copyright © 1973 by Annual Reviews Inc., reproduced with permission; (c) R. I. Walcott (1972), *Reviews of Geophysics and Space Physics* 10, 849–84, copyright by the American Geophysical Union. Fig. 8.4: D. A. Yuen *et al.*, *Rivista del Nuovo Cimento* 5 (8), 1–35; Società Italiana di Fisica 1982. Figs 8.6 & 8.7: D. Forsyth & S. Uyeda (1975), *Geophysical Journal of the Royal Astronomical Society* 43, 163–200. Fig. 8.9: L. R. Sykes (1967), *Journal of Geophysical Research* 72, 2131–53; copyright by the American Geophysical Union. Fig. 8.10: D. I. Gough, *Nature* 311, 428–33; copyright © 1984 Macmillan Journals Ltd; reproduced by permission. Fig. 8.12: R. H. Sibson, *Nature* 249, 542–4; copyright © 1974 Macmillan Journals Ltd; reproduced by permission. Fig. 8.13: R. A. Gibb (1983), *Geology* 11, 413–17; published by the Geological Society of America. Fig. 8.14: R. A. Price & E. W. Mountjoy (1970), *Geological Association of Canada*, *Special Paper* 6, 7–25; reproduced with permission. Fig. 8.15: W. M. Chapple (1978), *Bulletin of the Geological Society of America* 89, 1189–98; published by the Geological Society of America. Figs 9.1 and 9.6 (in part): D. Hull, *Introduction to dislocations*; copyright © 1975 Pergamon Press; reproduced with permission. Fig. 9.6 (in part): A. Nicolas & J. P. Poirier, *Crystalline plasticity and solid state flow in metamorphic rocks*; copyright © 1976 by John Wiley & Sons Ltd; reproduced with permission. Figs 10.9 (in part) and 11.1: M. F. Ashby & R. A. Verrall (1978), *Philosophical Transactions of the Royal Society London* A 288, 59–95. Fig. 10.9 (in part): J. Gittus (1978), *Philosophical Transactions of the Royal Society London* A 288, 121–45. Figs 10.10, 11.2 & 11.3: G.

ACKNOWLEDGMENTS

Ranalli (1982), *Physics of the Earth and Planetary Interiors* **29**, 42–50; copyright by Elsevier Science Publishers. Figs 10.11, 10.12, 12.1 & 12.2: G. Ranalli & D. C. Murphy (1986), *Tectonophysics*; copyright by Elsevier Science Publishers. Fig 11.4: G. Ranalli (1984), *Tectonophysics* **108**, 179–92; copyright by Elsevier Science Publishers. Figs 12.4, 5, 6 & 7: G. Ranalli & B. Fischer (1984), *Physics of the Earth and Planetary Interiors* **34**, 77–84; copyright by Elsevier Science Publishers.

Index

361